Complimentary Copy from the
Authors Daniel M. Berman
John T. O'Connor

(by Jan B. Berman, Dan'l not the)

January 10, 1997

Who Owns the Sun?

Who Owns the Sun?

*People, Politics, and the Struggle
for a Solar Economy*

Daniel M. Berman and
John T. O'Connor

CHELSEA GREEN PUBLISHING COMPANY
White River Junction, Vermont

Designed by Jill Shaffer.

Dr. Berman would like to thank the International Institute for Comparative Social Research in Berlin, which helped to defray the expenses of
research for chapter 6, "Labor, Solar, and the Energy Economy." An earlier
version of the chapter was published in German as "Arbeit heute ist bess-
er als Umweltschutz morgen: Gewerkschaften, Umweltschützer und die
Energie-industrie," in *Präventionspolitik: Gesellschaftliche Strategien der
Gesundheitssicherung,* edited by Rolf Rosenbrock, Hagen Kühn, and Bar-
bara Maria Köhler (Berlin: Edition Sigma/Rainer Bohn Verlag, 1994).

Printed in the United States of America
99 98 97 96   1 2 3 4 5

Library of Congress Cataloguing-in-Publication Data
Berman, Daniel M.
        Who owns the sun? : people, politics, and the struggle for a solar
    economy / Daniel M. Berman and John T. O'Connor.
        p.   cm.
    Includes bibliographical references and index.
    ISBN 0-930031-86-5
        1. Solar energy—Economic aspects—United States. 2. Fossil fuels—
    United States. 3. Energy industries—United States. I. O-Connor, John T.
    HD9681.U62B47   1996
    333.792'3'0973 — dc20   96-43593

We dedicate this book to
the five most important people in our lives:
Lorna Enero Berman, Jacob Leo Enero Berman, David Enero Berman,
Carolyn Mugar, and Marge Tabankin

CONTENTS

Acknowledgments

P EOPLE TOO NUMEROUS to mention helped us in myriad ways to write *Who Owns the Sun?* We wish to give special thanks to Eugene Coyle, Marge Tabankin, Frances Graham, Helmut Hildebrandt, Don Loweburg, Chris Keavney, Bob Ross, Ed Smeloff, Bethany Weidner, Joan Cordova, Wilson González, Patrick Gorman, Tom Jensen, Richard Miller, Michael Potts, Joel Ventresca, Dave Penn, Dick Leonard, Peter Bahouth, Paul Gipe, the International Institute for Comparative Social Research in Berlin, and the Turner Foundation in Atlanta for their support and encouragement. Any errors of fact or interpretation are ours alone.

Nobody ever writes a completely new book. The writings and public expressions of Kenneth Boulding, Clóvis Brigagão, Ernest Callenbach, Barry Commoner, Gore Vidal, Ralph Cavanagh, Alan Dalton, Frederick Douglass, Mark Dowie, Carlos Drummond de Andrade, Robert Engler, Douglas Foy, Lois Gibbs, Robert Gottlieb, Tom Hayden, Hazel Henderson, Denis Hayes, Jesse Jackson, David Katz, Martin Luther King Jr., Hagen Kühn, Thomas Jefferson, Milton Friedman, Winona LaDuke, Stuart Leiderman, Tom Lent, Amory Lovins, Hanswerner Mackwitz, Dieter Maier, Anthony Mazzocchi, Bob Moses, Ralph Nader,

Tom Paine, Henri Pezerat, Ray Reece, Scott Ridley, David Roe, Richard Rudolph, John Schaeffer, Hermann Scheer, Annie Thébaud-Mony, Carl Weinberg, Henning Wriedt, and Daniel Yergin helped us in different ways to think about a solar future. There are also inspiring stories about energy and democracy from Australia; Massachusetts; New Mexico; Austin, Texas; Chicago; and many other places, which we didn't tell, so someone else will have to tell those stories.

Ian Baldwin at Chelsea Green had the vision to see where we were going and the courage to take on a controversial manuscript. For this we will always be grateful. We hope that all writers have the good fortune to work with editors like Ben Watson and Jim Schley. They have been "cruel only to be kind" in pushing our assumptions and evidence to their logical conclusions and in chaining us to our writing machines until our stories were told. Lorna Enero Berman supplied invaluable editorial insights that kept the ideas flowing together smoothly. Neither she nor Carolyn Mugar ever faltered in their patience and support of this arduous project over the years. Last but not least, we have to apologize to Jacob and David Berman for their father's countless phone interruptions, bad temper days, and hours at the computer, and are grateful for their boundless love and understanding. Thank you all again.

October, 1996   Daniel Berman and John O'Connor

Foreword

THIS WELL-DOCUMENTED, passionate book by Berman and O'Connor is a jolting reminder of how the resolves and outcries of the 1970s to make the U.S.A., in Richard Nixon's words, "energy independent" or energy self-sufficient by 1980, collapsed—and already practical solar energy was marginalized by the giant fossil-fuel and nuclear power corporations. Yet all the reasons for self-sufficiency rooted in local democratic structures and decision-making remain compelling and even more dynamic than two decades ago.

From the Gulf War to the greenhouse effect and the political power of the energy companies over Washington and the state capitals; from nuclear power's radioactive waste pileups to routine oil spills; from expensive taxpayer subsidies for nuclear and fossil fuels to the lulling effect of temporary fossil-fuel surpluses; from the harmony ideology of certain major environmental groups toward fuel and electric companies, and the abandonment of civic movement for renewable-energy self-sufficiency (in deference to the supremacy of the oil, gas, coal, and nuclear moguls) to the abdication of the media's reporting on energy matters and the similar indifference of industry-indentured elected officials to what is civically right (in contrast to their obeisance to the corporate might), and the induced complacency of the public when the gasoline lines disappeared, *Who Owns the Sun?* lets the sunshine in. With the accelerating appetite for conventional fuel spurred by the rising Asian economies, trends will become more pronounced in the direction of environmental and community tragedies.

What sets the authors apart from other contemporary energy writers is their scope in bridging the technical with the political, and the way they consider the corporate power grabs—new and old—along with the need for democratic models of energy governance. They give their informed opinions forthrightly, and make their recommendations for action and change explicitly.

Berman and O'Connor make their most motivating contributions by chronicling how the fossil-atomic-electric industries bend governments to their will and turn public policies against solar energy, how the oil companies gained control of the bulk of photovoltaic companies, how the electric monopolies are twisting—with few exceptions, such as the public utility in Sacramento, California (SMUD)—the varieties of applicable solar power and conservation so that corporations make the decisions as to when and how these energy sources are to be used. I say "motivating" because the authors show readers the valiant innovators and entrepreneurs who are driving solar power further, in spite of so many non-technical obstacles, promoted by vested interests, which have kept the efforts of these pioneers largely in the pilot stage instead of the rapidly diffusing stage.

Renewable solar energy is a commonwealth. It is not owned by Exxon or Peabody Coal. But its availability for heating, lighting, and cooling America can be controlled by these and other non-renewable fuel corporations. Capital and credit can be withheld from the millions of Americans who in polls show an overwhelming preference to go solar. Solar technology, like new cars, requires a capital-credit infrastructure. Berman and O'Connor see the future as not just a solar-energy future but a consumer-sovereign future where people control and produce their own power from that solar commonwealth. This is not a utopian quest; it is a quest for the necessary amount of democracy to make the available technology deployable.

In their final chapters, "Solar Homesteaders or Solar Sharecroppers?" and "Fighting for a New Solar Society," the authors make clear that the best things in life often only come with better democracy and civic engagement.

As one who has fought against nuclear power—and for the energy self-sufficiency of solar power, in all its manifestations—I find this book to be an agitating road map of what went wrong with our national energy policies, and also a map for how to make them serve the planet's people and the future generations who will inherit the Earth. This book can reactivate the solar movement in America if people read and use its contents.

Who Owns the Sun? stirs the sense of important lost opportunities and makes us indignant enough to recover and propel many initiatives for a solar world.

Washington, D.C.   Ralph Nader
August, 1996

Introduction

*"Keep in mind always the present you
are constructing. It should be the future
you want." —Ola*

—ALICE WALKER, *Temple of My Familiar*

BEFORE THE 1973 ARAB OIL EMBARGO, energy was taken for granted; afterwards, millions of Americans cursed in gas lines and watched their winter heating bills quadruple. Suddenly every segment of the energy realm was up for grabs: nuclear vs. coal, natural gas vs. oil, and conservation and solar vs. all of the above. For the first time since the turn of the century—when coal miners struck for union recognition and when hundreds of cities took control of private electric utilities—the questions of energy, fuels, transportation, and the role of labor in the energy economy all became subjects of white-hot national debate and mass technological creativity.[1] With a somber face, President Richard M. Nixon promised that "Project Independence" would free America from energy imports by 1980.[2]

As oil and utility companies doubled and redoubled their prices, electricity consumption and utility income peaked. Skeptical citizens and respectable foundations began to question whether ever-increasing energy use would really be America's yellow brick road to paradise. Five years later, President Jimmy Carter put a solar water heater on the White House roof, turned down the heat, and donned a sweater to demonstrate to the nation his assertion that addressing the energy crisis was the "moral equivalent of war."

This uproar over energy supplies and use disturbed the giant energy corporations, because public debate of this magnitude threatened corporate control over the definition of the problem. The public began to remember that utilities and oil companies were human arrangements subject to human governance rather than immutable natural phenomena like the stars and the changing of the seasons. Throughout the country, energy became the flashpoint for a grassroots citizens' movement that questioned the legitimacy of corporate control. Formerly placid commissions and courts became theaters of political protest where outraged citizens faced down their corporate antagonists. The Clamshell Alliance on the East Coast and the Abalone Alliance in California organized mass occupations of nuclear construction sites, and filled local jails with thousands of demonstrators chanting "No Nukes."[3] Congress banned lead in gasoline and mandated more stringent mileage standards, and California and other states passed energy-saving building codes. Grassroots activists in the South tried to make the federal Tennessee Valley Authority (TVA) accountable to its ratepayers in the region, in order to prevent the TVA from mortgaging its future to build nuclear plants.[4] In New England, clean-energy and anti-nuclear activists enlisted local governments to create publicly owned and locally controlled energy authorities that would free them from dependence on privately owned energy monopolies. In the Pacific Northwest and on New York's Long Island, citizens stopped construction of billions of dollars worth of nuclear plants, and sometimes they left Wall Street bankers holding the bag. Energy self-sufficiency and local control seemed just around the corner.

But the big transformation never came. In the 1990s, the uproar over energy has subsided to a murmur. The solar water heater has been removed from the White House roof, and fossil-fuel proponents have won the presidency as well as most of the recent energy wars. Oil and natural gas appear to be plentiful and cheap, contradicting the predictions of the oil-shock years. Saddam Hussein's bid to take Kuwait was put down in flames, and Saudi Arabia—protected by American military might—coordinates its oil prices with the United States. Public power has survived, but in a defensive posture, and utility holding companies, like oil companies, have freed themselves from legal restrictions on investment and are going multinational. The greatest victory of the oil and coal companies, electric utilities, automakers, and road-builders has been to turn their former environmental adversaries into collaborators. For mainstream American environmental groups, confrontation is out and green capitalism is in. At present, the dominant idea is to give financial incentives and government subsidies to energy corporations in order to persuade them to invest in efficiency and renewable-energy technologies and "unsell" their major products, fuel and electricity.

The response of the fossil-fuel giants has been to take this incentive money and reinvest it all over the world, in more oil fields, more automobile factories, and more coal-fired power plants. Meanwhile, utility companies have begun to invade the formerly independent field of off-the-grid photovoltaics, without a peep of protest from their newfound environmental allies. The use of fossil fuels continues to increase inexorably and drive the adoption of solar and other renewables ever farther into the future. An alternative political movement dedicated to solar energy has been reduced to begging utilities for a share of the business. More than ever, the world is hooked on cars and oil, natural gas, and coal.

Yet the arguments for decentralized solar technology are more valid today than ever. The love of freedom and democracy, the fear of a disastrous greenhouse effect, the threat of continuing energy wars and environmental catastrophes, and a massive oil and auto trade deficit—all necessitate a conversion to solar self-sufficiency.*

Who Owns the Sun? examines the business strategies behind the public rhetoric of the energy giants, and looks at the social reformers and businesspeople who are trying to democratize and decentralize the energy business. Our primary thesis is that local ownership and democratic control of energy are the necessary, if not sufficient, conditions for a solar economy. We pay special attention to the century-old and too often unappreciated public-power movement in the United States. Given the new wave of deregulation in electric power and the falling price of

---

* Our use of the term "solar" includes wind power and other renewable-energy sources, since all of these are derived from the sun.

photovoltaics and wind turbines, we argue that municipally owned electric companies and citizen electric cooperatives constitute an appropriate model for the governance of energy.

The publicly owned Sacramento Municipal Utility District—governed by an elected board—successfully shut down a nuclear power plant and began to turn that city toward the sun. Today's solar technologies, as Sacramento is beginning to demonstrate, make it possible for ordinary citizens and communities to capture the solar energy they need with photovoltaics, solar water heaters, passive solar construction, wind turbines, and even solar roof shingles—without paying obeisance to the commercial energy cartels and oil-producing, authoritarian regimes.

For the founders of the nation, freedom or liberty had two meanings: the private freedom for citizens to do as they liked with their own property, and the public freedom for citizens to take their ideas to the commons for discussion, debate, and decision. It is this public freedom that is the essence of democracy. If properly conceived and implemented, solar energy could unite these two freedoms in a new commonwealth and a new pursuit of happiness.

Who Owns the Sun?

CHAPTER 1

Solar America:
A Dream Deferred

I N A MILESTONE ARTICLE reprinted in 1970 for the first nationwide Earth Day teach-in, the economist Kenneth Boulding contrasted the "cowboy" economy with the "spaceman" economy. "In the cowboy economy," he wrote, "consumption is regarded as a good thing and production likewise, and the success of the economy is meassured by the amount of the throughput from the 'factors of production.' ..." In the spaceman economy, by contrast, "the essential measure of the success of the economy is not production and consumption at all, but the nature, extent, quality, and complexity of the total capital stock...." Thus in the spaceman economy, which posits limited stocks of material goods and a vast but constant input of solar energy, the primary concern is maintaining stocks of the necessities for life, and "any technological change which results in the maintenance of a given total stock with ... less production and consumption is clearly a gain."[1]

In the 1970s, solar energy—as natural as the sun in your face—seemed to be the answer to Boulding's call for adoption of the spaceman's sensibility. Working to free the country from the grip of a nefarious foreign energy cartel became an act of the highest patriotism.

Suddenly "solar" and "build your own" became national rallying cries, and hundreds of thousands of people began to invest in, construct, or purchase solar heating systems and windmills, in an outpouring of mass technological enthusiasm unparalleled since millions had gone crazy over the Model T Ford a half century earlier.[2] Do-it-yourself magazines such as *Popular Mechanics* were full of energy projects.[3] By May 1975, over five thousand people had made grant inquiries to the newly formed Energy Research and Development Administration. Within a year, at least three thousand new solar firms had been incorporated by an eclectic mix of engineers, labor and antinuclear organizers, energy techies, feminists, skilled tradespeople, New Agers, and hippie pot-farmers. Ordinary Americans trying to keep a few more dollars in their pockets became fascinated by the potential of "soft" energy, and political activists believed that solar would be the key to genuine economic democracy. For these people, the *Whole Earth Catalog*, with its seductive vision of local self-reliance, had replaced the Sears Roebuck catalogue as the ultimate wishbook. Even the federal Department of Energy (DOE) got into the act. DOE's new Solar Energy Research Institute published *Reaching Up, Reaching Out*, a high-spirited and amusing "guide to organizing local solar events," written by seven women, which exuberantly laid out dozens of future solar energy scenarios.[4] The projects in the book's "events sampler" included "Solar Energy Fair and Celebration," "The Children's Solar Hour," "Solar Water Heater Workshop," and "Montclair Future Power: A Program to Enable a Community to Create Its Own Energy Future." A section entitled "Dress Up—Dial Down" related how the Santa Fe League of Women Voters had organized "a luncheon fashion show and bazaar [to] emphasize warmth in wardrobe and diet for the fashion-conscious and conservation-minded." Unpretentious, idiosyncratic, and passionate, *Reaching Up, Reaching Out* reflected the excitement of the growing mass movement for solar energy at the end of the 1970s.

The solar high spirits of the 1970s had captured the imagination of the nation. Solar water and space heating, solar electricity, wind power, and energy conservation were perceived as options that could be adopted by anyone with a mechanical bent, a technical imagination, a decent set of tools, and perhaps good karma. Ninety-four percent of Americans, according to a Harris poll, favored rapid development of solar.[5] What the energy activists lacked was access to the capital to undertake the necessary R&D. But this new solar movement, coupled with a renewed demand for public ownership and control of energy, provoked fear among the utility monopolies and fossil-fuel corporations, and the investment banks that backed them. "Bypass" and "death spiral" soon became terms in utility jargon: if too many customers began to bypass the grid by generating their own electricity, the utilities would be forced to raise the rates for the remaining customers to cover their fixed costs. Even more customers would be driven to energy independence, touching off a death spiral into bankruptcy for the utility. If communities or businesses—or even individual citizens—could make their own electricity and heat their own homes and water with a properly designed house and a few thousand dollars worth of solar panels and hardware, neither giant power plants nor long-distance transmission lines would be necessary, and the utilities' "natural" monopoly and government-guaranteed profits would shrivel like morning glories in the noonday sun. People might even begin to run their cars and motorcycles on solar electricity collected from their own rooftops, and stored in a publicly controlled electrical system—like rainwater in a community reservoir!

The Ford Foundation jumped into the fray in 1974 with a $4 million Energy Policy Report entitled *A Time to Choose*, which mocked the "more is better" philosophy of the power companies, and vigorously recom-

Energy Scare-Tactics

Merely to discuss "zero energy growth" is to unleash a torrent of indignant advertising paid for by major industrial interests which benefit from growth in energy consumption. A typical utility company ad shows a bell-bottomed, well-heeled protester carrying a sign: Generate Less Energy.

"Sure," the ad replies. "And generate galloping unemployment."

Another utility company ad shows an embattled housewife with her arms around her washing machine. The headline: "Try telling the lady she'll have to start washing by hand."

Is there any truth to this scare advertising intended to perpetuate the seemingly inexorable growth in U.S. energy consumption? The answer is no ... insulating homes and buildings and making cars that get better mileage are no threat to anyone—except perhaps to the energy company salesmen.

A Time to Choose, 1974[6]

mended a Zero Energy Growth strategy, one that emphasized energy efficiency and conservation rather than crash programs to expand fuel production.[7] The report attacked in no uncertain terms the massive federal support of nuclear research and development to the exclusion of energy efficiency. Infuriated conservatives quickly commissioned a think tank to produce a book-length rejoinder called *No Time to Confuse*.[8]

The wide appeal of solar energy and the legitimacy provided by books like *A Time to Choose* created a tricky PR problem for the energy barons: they had to kill the enthusiasm for solar, discredit its advocates, but also domesticate it for their own purposes—meanwhile diverting attention from their own shortcomings. Yet many solar advocates were capitalists—the kind of rugged individualists whom Americans (especially Republicans) professed to admire the most. Even worse for the utilities was the fact that some of the most widely applicable solar energy technologies, including passive solar design, solar water heating, and photovoltaics, were inherently decentralized. Proponents of these technologies presented them as the basis for small-is-beautiful economic democracy.[9] If large numbers of customers were to disconnect from the system in favor of on-site generation and conservation, the utilities would be unable to maintain the costly electric power grid. The whole system might go into a "death spiral," reminiscent of the demise of electric trolleys and passenger trains two generations earlier, which had occurred partly as a result of deliberate efforts by General Motors.[10]

The goal of the solar energy patriots of the mid-1970s—who made up an improbable concatenation, including anti-nuke organizers promoting sunpower as a clean alternative; technological revolutionaries from the counterculture; egalitarian, anti-capitalist worker collectives; and open-minded capitalist entrepreneurs—was to start small enterprises producing superior solar collectors and photovoltaic (PV) systems.[11] Their theory was that the network of small businesses and co-ops would grow so extensive and vital that the energy monopolists would be driven from the temple. So the little guys scrambled forward, often a half-step ahead of bankruptcy, in their challenge to the idea that energy could only be reliably provided by huge power plants and protected corporate monopolies. The song "Blackout Blues" by Stuart Leiderman, a polymathic biologist and songwriter working in the

Ozarks, expressed the sensibility of those trying to use the struggle over energy to build a new way of life in the backwoods and even in the city, premised on the old American ideals of independence and self-reliance:

They tell me there's an awful blackout coming
But things can't get much darker than they are.
If I can't get some juice that I can pay for,
I'm gonna make my own and cut their wire.

Don't let those power people soak or jive you;
A friend don't dig that deep into your purse.
Just generate your own and live without them,
And unplug yourself from that old electric curse.[12]

In 1976, Leiderman ran for Congress from Missouri on regional and environmental issues, primarily against the Meremec Dam project of the U.S. Army Corps of Engineers, in the Ozarks. Waged from a 1946 Chevy with guitar in hand, Leiderman's "Free Rivers, Free People" campaign helped consolidate opposition that led to deauthorization of the dam.

The idea was to just get out there and agitate for change. The Farm, a communal settlement in Summertown, Tennessee, was founded by ex-hippies from San Francisco who believed in "clean air, healthy babies, honest work, nonviolence, safe energy, cheap transportation, and rock and roll." The Farm's members built solar-heated dwellings from school-buses and castaway lumber, and experimented extensively with solar water heaters, photovoltaics, windmills, and micro-hydroelectric turbines. One 1,400-square-foot, superinsulated house, built for $55,000 in 1979 for a neighbor, incorporated all the best features of passive solar design: "It had a double-paned greenhouse, clerestory windows to the south, earth berm to the north, a Trombe wall, a gravel-bed heat-storage system that cycled hot air from the greenhouse, an active hot-air water-heating system, a 'Persian air conditioner' which carried cool air into the house from underground, drawn by turbines on the roof. . . ." Massive interior brick masonry exposed to the winter sun retained the heat well into the evening. According to the *Sunday Tennessean*, the house was widely copied.[13]

Many 1970s inventors got excited about the possibilities of alcohol as a vehicle fuel, since alcohol is a renewable "solar" resource distilled from plants instead of petroleum. Typical of the 1970s-era grassroots technological patriots was Mike Brown of Berea, Kentucky, who wrote *Brown's Alcohol Motor Fuel Cookbook*, "to enable the average 'Joe' to make his own motor fuel." "It is Mike's sincere hope," wrote his publisher, "that this book will serve to help keep America independent of foreign energy sources."[14] If readers preferred to sell their product as white lightning instead of tractor fuel, there were chapters entitled "How to Build and Operate a Moonshine Still" and "Testing for Proof," and they were forewarned of potential legal hassles in the chapter entitled "Dealing with Big Brother."

For a few years, the naive faith of these activists in their own know-how and in the buying power of other self-reliant Americans knew no bounds. As Mike Brown said at the time, "I have never heard an American farmer say, 'We can't do it.'" Yet, despite a grassroots consensus that the primary U.S. energy problem was inefficient use of fossil energy rather than an overall shortage, the federal government failed to respond.[15] The oil and electric power monopolies had potent allies, including Vice President Nelson Rockefeller, who pushed for gargantuan new subsidies to the energy corporations for drilling and mining, and these calls for subsidies were always couched in the phraseology of the free market.[16]

Faith without capital was handicapped. Some of the most astute political thinkers among the solar activists, including Earth Day organizer Denis Hayes, realized that solar would die on the vine without high-level political, technological, and financial support, so they formed the Solar Lobby to encourage a broad environmental coalition to work together to promote solar energy. The federal Solar Energy Research Institute (SERI) was a specific result of that lobbying effort, though SERI was allowed to languish under the Republican administration of President Gerald Ford. Denis Hayes sharpened and deepened his arguments for his book *Rays of Hope: The Transition to a Post-Petroleum World*, and when Jimmy Carter was elected president, Hayes appeared to have found a president who comprehended the promise of solar energy.[17] SERI was moved to a permanent home in Golden, Colorado, and President Carter appointed Hayes to the institute's head.

To President Carter's credit, his administration issued Corporate Average Fuel Economy (CAFE) standards for the CAFE law passed under President Ford, despite staunch opposition from Detroit and Houston, which reversed for a short time the relentless climb in gasoline consumption. But most of Carter's energy policies gave top-dollar billing for gigantic new development programs in coal, nuclear power, and shale oil. Carter's first solar budget was "identical in amount" to the budget of the outgoing Ford administration, and solar energy advocates had to stage a nationwide "Sun Day" in May of 1978 to embarrass the Carter administration into approving a supplemental $100 million appropriation for renewable energy.[18]

Also in 1978, the Public Utility Regulatory Policies Act (PURPA) opened a crack in the utilities' monopoly over electric power. PURPA mandated that state public utilities commissions could require utilities to buy power from independent producers of different forms of renewable energy: wind, solar, waste-to-energy, and cogeneration (electricity generated as a byproduct of another process, such as manufacturing). When the legal dust had settled, promoters of clean renewables often found themselves in tactical alliance with huge electricity users such as General Motors, who were eager to discipline a utility industry whose rates seemed to be spinning out of control.[19]

By 1979, after the decade's second oil shock (this time precipitated by the Iranian revolution), President Carter proposed a $100-billion crash program for synthetic fuel development, alongside the $3 billion he had "reluctantly agreed to spend for solar development."[20] Although ecologist Barry Commoner assailed Carter's Energy Plan as "falsely reasoned, economically destructive, dangerously dependent on nuclear energy, and repressing the true potential of solar energy," Jimmy Carter was an environmental exemplar by comparison with his two successors in the White House, Ronald Reagan and George Bush.[21]

Hanging a Meter On the Sun

THE ENERGY CORPORATIONS claimed to have no objection per se to the use of solar energy or farm-produced alcohol. Yet the electric utilities and oil companies, which in the aftermath of the oil embargos had quickly bought a major percentage of the world's coal and uranium supplies, were unlikely to let backyard tinkerers and scraggly ecotopians capture the sun's rays and the nation's imagination, and thereby threaten the cartels' hegemony over energy services.[22] While they could not halt federal support for research on renewable energy sources, the electrical equipment and oil giants secured the majority of federal research handouts, then moved swiftly to buy out the best of the new companies to bring them under control.

Ray Reece's important (and nearly forgotten) book *The Sun Betrayed*—written before Ronald Reagan's election to the presidency—argues that neither Democrats nor Republicans dared defy the private energy corporations.[23] Reece documents in convincing detail the near-total "corporate seizure of U.S. solar energy development" through control of the newly formed Energy Research and Development Administration (ERDA) inside the Department of Energy. Under President Reagan, the Solar Division of ERDA was entrusted to Robert Hirsch, former Director of Controlled Thermonuclear Research at the government's Atomic Energy Commission, and he immediately infuriated solar advocates by trying to scale down the status of a projected $50-million-per-year Solar Energy Research Laboratory to an insignificant "solar management group."[24] Hirsch later moved on to a job with Exxon.

In 1976, the Electric Power Research Institute (EPRI)—an R&D organization founded in 1972 by electric utilities to forestall a complete federal takeover of electric power research[25]—signed a Memorandum of Understanding with ERDA, which the *EPRI Journal* (July/August 1976) treated as a "milestone" in its history.[26] Both ERDA and EPRI were to establish joint "operations groups . . . for each discrete area, or group of areas, of common technical interest to achieve a parity of representation throughout the breadth of the relationship." Under the "information exchange" provision of this memorandum, EPRI and the American Gas Association gained practical oversight and a tacit veto over all R&D contracts made by ERDA, through a joint "senior management overview" process to be undertaken in annual meetings between the president of EPRI and the administrator of ERDA. The terms of the agreement gave EPRI the right of "technical evaluation at reasonable intervals" of all projects funded by ERDA. Through EPRI, therefore, any member utility company could find out any information it desired about any energy projects funded by ERDA, even if the products under development would be in direct competition with what the utilities were planning. In effect, EPRI had turned ERDA into a taxpayer-funded research arm of the utilities, a pattern which would later repeat itself in other contexts.[27]

Outlandish Battlestar-Galactica research projects were funded, which proposed to beam power back to earth from miles-wide photovoltaic satellites. (Daniel Goldin, chief of the National Aeronautics and Space Administration, has yet to dismiss such proposals completely, assuming the dangers of microwaves to human flesh can be resolved.[28]) Meanwhile, R&D support for solar water heating was almost impossible to secure. ERDA ignored thrifty projects that might have helped create a broad solar market among middle-class and low-income consumers. In 1977, Steve Kenin of Taos, New Mexico, designed a solar space-heating kit of translucent polyurethane stretched over an aluminum frame, which could be installed for $600, paying for itself in fuel savings in less than three years. Kenin's $25,000 proposal to build and monitor a half a dozen of his cheap and simple systems, which worked well in the cold but sunny winters of the New Mexico mountains, was passed over for funding.[29] By contrast, ERDA funded a $216,500 demonstration project that installed solar heating and hot water systems in luxury houses in Denver at $9,840 per unit—over $25,000 each in 1996 dollars.

Through July 1976, EPRI itself had received $84 million for energy research projects.[30] EPRI and its member utilities were granted "irrevocable, non-exclusive, royalty-free license . . . to make, use, and sell . . . any invention or discovery made or conceived in the course of jointly funded efforts and covered by a U. S. patent."[31] Although innovations which they originated couldn't be patented by the utilities, EPRI and the utilities were granted privileged access to the R&D process through this "license" provision. Small businesses, on the other hand, were being advised to sell their ideas to large enterprises or universities or "to forfeit all patent rights" in exchange for federal development grants. When one of his modest grant proposals to ERDA under the aegis of his own small company was rejected, Jerry Plunkett, a small businessman with a Ph.D. in metallurgy from MIT, turned around and submitted the identical grant through the University of Utah and watched it get funded. No one at

ERDA seemed to realize that the agency had already rejected the exact same grant. In a Senate hearing, Plunkett charged that federal participation "has been a direct and unfair subsidization of large firms and universities—at the expense of small businesses . . . [which] has tended to discourage competition, foster monopoly, and reduce the rate of adoption of solar energy."[32] To date, no study has ever effectively assessed the extent to which private interests have acquired the patents to government-sponsored solar technologies with the intent of suppressing them.

Historically, solar's main opponents have been the large energy corporations.[33] California utilities, led by Pacific Gas & Electric and Southern California Edison, have vigorously opposed solar hot water heating, which cuts residential gas usage by 20 to 40 percent in every house where it is installed. Peter Barnes, director of San Francisco's Solar Center, recalls that he spent at least half of his time in Sacramento lobbying against Pacific Gas & Electric and other utilities for the right of people and businesses to own their own solar water heaters. Ultimately the fledgling solar hot water industry—both residential and commercial— was able to persuade California's Public Utilities Commission to allow home and building owners to buy their own solar hot water heaters rather than lease them (like old-time telephones) from the utilities.[34] For a few years, solar space and water heating flourished under the friendly tax laws: by 1984 there were 19,000 solar workers in California, in addition to those working in wind power, and annual industry revenues totalled almost $500 million. Like California, Colorado became a vortex of solar activity, with hundreds of people working full-time to install solar hot water heating systems.

The organizing effort leading up to Sun Day in 1978 represented the high point for the pro-solar citizens' movement. Yet even though solar hot water and wind-energy programs had constituted one of the most successful environmental conversion programs in U.S. history, by the mid-1980s these programs were moribund.[35] The solar movement's initial successes in building alliances with the public, the Carter administration, and local governments—especially in California and Colorado—had only invigorated the efforts by solar opponents to preempt direct access to government procurement markets and R&D monies. A coalition of environmentalists, small businessmen, and skilled installers—

fueled by righteous American enthusiasm, common sense, and self-reliance—failed to survive in the cut-throat arena of transnational investment banks, fossil-fuel corporations, and electric utility companies.

Perhaps the vulnerability of these patriotic solar businesspeople should come as no surprise. In late twentieth-century America, independent small business seemed to flounder nationwide as one activity after another was reduced to a standardized industrial formula. Hadn't Colonel Sanders, founder of Kentucky Fried Chicken, and Ray Kroc, the author of McDonald's worldwide expansion, proved that the simplest and apparently most decentralizable of human activities—frying chicken and cheeseburgers—could be industrialized? As we will show in the following chapter by looking at specific instances of solar advocacy in three bellwether states, the nascent broad-based solar movement failed to overcome rigorous institutional and policital opposition. Where capital is monopolized, control is readily maintained and profits are multiplied. Should the logic of the solar industry be any different than that of the fast-food industry?

By 1996, almost twenty years after the first Sun Day, standardized industrial formulas had sucked most of the democratic pizzazz out of solar energy. Today, the resurgence of solar technologies is strictly managed by the same energy corporations who undermined solar hot water heating in the 1970s and 1980s. If the utilities continue to control the process, most solar energy will be generated at big, expensive, intimidating installations surrounded by barbed-wire fences, like any other utility power plants, even though decentralized rooftop and building-integrated solar designs are often a more practical and economic application of solar technologies.[36] For many people, the sabotage of solar projects (as well as research programs into alternatives to automobiles and freeways) offers proof that the energy crisis has actually been caused by the policies and practices of the energy giants.[37] Solar energy in the absence of economic democracy will be marginal, of interest only to technocrats. Solar power without a grassroots solar movement will be merely another form of business as usual.

CHAPTER 2

Keeping Solar Culture Alive

I N THE HEYDAY OF NATIONWIDE preparations for the 1978 celebration of Sun Day, it may have appeared that the United States was on the verge of wholehearted, broad-based adoption of solar and other renewable technologies as crucial components in the energy mix. Since this never occurred, it is critical to consider the rise and fall of solar advocacy during a watershed period.

In the 1970s and 1980s, major battles took place over solar hot water heating in Florida, Hawaii, and California, as well as in other states. While it is true that solar hot water heaters never were more than a small piece of the total energy picture, in Florida, Hawaii, and California—in addition to the environmental benefits of clean energy—solar heaters could cut a family's utility bill by 20 percent or more, depending on the climate and on whether water had been heated previously with electricity, propane, or natural gas.

Solar water heaters were first available more than a century ago. Until then, hot baths were usually taken in a communal bathhouse. In rural areas, hot water was often the by-product of cooking over a wood or coal fire; water had to be pumped, fuel fetched, a fire kindled and stoked, smoke endured, and ashes hauled out. The coal-gas hot water heaters used in the cities were hardly any better: they were leaky manual devices that often sprayed scalding water or even exploded. Besides, coal-gas cost ten times as much as today's natural gas per unit of heat supplied.

Clarence E. Kemp of Baltimore invented the Climax solar hot water heater in 1891 in an attempt to solve these problems. The Climax consisted of four heavy galvanized iron cylinders painted a dull black and set into an insulated pine box with a glass cover. This arrangement was connected to the home's plumbing system. The black cylinders absorbed heat from the sun's rays and from the solar hot box. (Anyone who has ever parked a car in the sun with the windows closed knows the effect.) Climax's 32-gallon bestseller sold for $25 plus installation—a month's wages for the average worker. But the Climax had its drawbacks, too: it wouldn't heat up fully until the late afternoon, and it had to be drained in autumn for half of the year to keep from freezing.

Climax sales really took off in sunny Southern California, with its mild winters and high-priced fuel. Manufactured by a local licensee, the heaters were sold mostly to the comfortable middle class. "For an investment of $25," wrote Ken Butti and John Perlin, authors of the delightful and exhaustive book *A Golden Thread*, "the average home-owner saved about $9 a year on coal—and more if artificial gas was used for water heating."

William J. Bailey's Day and Night solar heater, first sold in 1909 in Southern California for $180 plus installation, added a large insulated storage tank (often disguised as a chimney) above the collector, which could store enough hot water for bathing in the morning. Such "thermosiphon heaters" work on the principle of the circulation of fluid by natural convection: cold, heavy water sinks to the bottom of the tank and solar collector, forcing out warmer and lighter water, then the cold water is heated in turn. Despite its high initial cost, the convenience was worth it, and the Day and Night heater eventually drove the Climax from the market in California. Sales peaked at over a thousand units in 1920.

But massive discoveries of natural gas in the Los Angeles basin, beginning in 1920, wiped out the solar cost advantage, and gas company marketing finished the job: "The gas company would finance gas water heaters on a monthly basis or let you carry them for a year or two with free installation," reports one retired plumber. In the 1920s, when the word "ecology" didn't even make the *Oxford English Dictionary*, there was no magic associated with solar.[1] Even Bailey himself started manufacturing natural-gas water heaters. These twenty lively years had been a good test-run for solar, yet the last consignment of Day and Night solar water heaters, which were intended for Canton Island, a Pan American airstrip in the South Pacific, never left the San Francisco docks after Japan attacked Pearl Harbor in December 1941.[2]

For twenty years, the solar initiative moved abroad: to Australia, Israel, and Japan, where by 1961 sixty thousand simple solar water heaters were selling per year.[3] In Japan, the government still promotes solar water heating to save on fossil fuels, which are almost totally imported, and in 1991, Tokyo alone had over 1.5 million solar water heaters.[4]

Again, by contrast: in Cyprus, 90 percent of homes have solar water heating systems, and in Spain and Greece, according to Greenpeace, "consumer confidence is built up by an electronic monitoring service which runs through the telephone system. The owner is compensated if the solar system does not deliver a guaranteed amount of heat every month."[5] In Israel, the adoption of solar water heating was triggered by Rina Yissar, a woman of "excellent common sense" who painted an old water tank black so that she could give her baby a hot bath. Her engineer husband, fascinated, scanned the technical literature and built a prototype similar to those used in Florida and California. Tens of thousands of solar hot water heaters were sold until the capture of the Sinai oil fields from Egypt in 1967 made oil and gas cheap. Then, in Israel as elsewhere, for a while "the siren call of fossil fuels led people away from the sun."[6] According to Arthur Shavit, a professor of mechanical engineering at the Technion, Israel's foremost technical university, four simple conditions must be met for solar water heating to become reestablished:

1. Solar water heating must be cheaper than the alternatives;
2. People must know about this option;
3. Government building codes must require use of solar; and
4. The equipment must be readily available.

In Israel, all four conditions have been met.[7] Israeli building codes now require solar water heating in all new buildings. As a result, Israelis install 50,000 new heaters a year at a cost of under $1,000 apiece. Two thirds of all households have solar water heaters, the highest proportion in the world.[8] Without strong building codes, the transition to solar water heating would not have occurred.

Florida: The Sunshine State in Hot Water

THE FLORIDA REAL-ESTATE boom of the early 1920s opened up a true mass market for solar hot water heating: electricity was scarce and costly; natural gas was unavailable; and the much-ballyhooed Florida sun was hot. A builder named Bud Carruthers bought the Florida patent rights to the Day and Night heater and set up the Solar Water Heater Company in Miami. Soon it became impossible to sell a house or hotel without one of his devices. In two years, Carruthers had built a factory that occupied an entire city block in Miami, and he had become one of the biggest businessmen in town. Yet the glory days were short-lived; in 1926, the Solar Water Heater Company was driven out of business by a raging hurricane and the end of the building boom.

The second Florida solar boom took shape in 1935, financed by Federal Housing Administration mortgage programs: 80 percent of homes built in Miami between 1937 and 1941 had solar hot water heaters. In the next five years, between 25,000 and 60,000 new residential heaters were installed (the estimates vary wildly), including 5,000 very large arrays on hotels, factories, and housing projects. By the time World War II began, over half of the residences in Miami heated their water with the sun. A few of these thermosiphon heaters from the early 1940s were still providing hot water in Miami and Key West fifty years later. When World War II started, copper for the heating coils became extremely scarce and new construction almost ceased. The postwar building boom proved no better for solar, as the price of electricity and electric hot water heaters fell precipitously compared with the price of buying and installing solar heaters. The solar industry itself was also to blame for serious design problems: after about ten years of use, the iron tanks tended to rust out.[9]

During the 1950s, only a handful of journalists in the United States questioned the "deepening trouble" that could come from total dependence on fossil fuels, and by the end of the decade in Florida, only a few maintenance specialists remained from what had formerly been the largest solar industry in the world. The Association for Applied Solar Energy Research, founded in 1955, shut down eight years later for lack of funds.[10]

In sunny Florida, solar water heaters today could supply 70 to 90 percent of hot water requirements, yet most households heat their water electrically, consuming an average of 300 kilowatt-hours per month in the process. A quarter of household electricity in Florida goes to hot water heating. The Florida House Foundation has designed and built a house with off-the-shelf components that uses half of the electricity and 40 percent of the water of conventional designs.[11] With electricity at 8 cents per kilowatt-hour, every household with solar water heating would cost the local electric company at least $200 per year in lost revenues.[12] If all of Florida's four million residences that heat with electricity were required to install solar hot water heaters, Florida's annual electricity consumption could be reduced by over 7 percent, even without considering the hot water demand in restaurants, laundries, and other commercial establishments.[13]

Florida constitutes a prime case study of how private utilities sabotage solar where they dominate energy policy. Despite a seventy-year history of leadership in solar hot water technology and the presence of the world-renowned Florida Solar Energy Center, few builders even offer solar water heating as an option, and neither the state of Florida nor its local governments have promoted solar energy.[14] As a result, according to the Florida Solar Industries Association, only 0.5 percent of Florida's energy presently comes from solar and renewable sources, and solar water heating represents well under 1 percent of total energy use.[15]

The Florida Solar Energy Center asserts that the nine Sunshine State ordinances that deal specifically with solar are as likely to hinder as to promote it. The Energy Conservation in Buildings law, for example, "removes solar from consideration in most cases due to rigid economic analyses. [The] law could be amended to remove requirements for rigid economic feasibility, or should require that societal costs be considered in any economic analysis." Past Florida laws that exempted solar appliances from property and sales taxes have never been renewed because of "negative" fiscal impacts.[16] The legislature's *Staff Report on Energy Efficiency Issues,* reluctant to challenge utility viewpoints, touted natural gas rather than solar as "the most effective way . . . to reduce demand and consumption of electricity by about 300 kilowatt-hours per month."[17] And Florida Power and Light—Florida's largest corporation—never mentions solar water heating in its publication *Demand-Side Management Plan for the Nineties.*[18]

Moreover, Florida utilities have consistently failed to offer financial assistance or rebates to customers installing energy-saving solar water heaters. Florida Power and Light used to cover only $160 to $700 of the upfront costs of new solar systems—dollar amounts that have remained unchanged since the program began in the late 1970s. Other private utilities in the state offer no rebates at all.[19] Don Kazimir, a manufacturer and winner of a recent design competition for inexpensive solar hot water heaters, sponsored by Governor Lawton Chiles, could remember only three manufacturers and a few dozen dealers and installers who are still active. Kazimir's prize-winning Pacemaker heater, designed for a household of two, uses existing water tanks, sells for about $700, and can be installed by a non-plumber who is handy with tools.[20] Though it is on sale at several Home Depot outlets, the product has sold better in the Caribbean than in Florida.

By 1990, fewer than one Florida residence in twenty could boast of a solar water heater, and most of those had been installed between 1975 and 1985—before the federal tax credit for solar hot water heating expired. Solar water heater sales continue today, though much reduced. David Block, director of the Florida Solar Energy Center, estimated that 11,850 solar water heaters and 14,500 pool heaters were sold in 1991, a number that could be increased ten or fifteen times with appropriate building codes and financing packages.[21] For residential solar water heaters, the news from Florida mostly got worse during the 1990s. According to the best available estimate, the number of residential solar water heating systems sold continued its downward slide, from 4,100 in 1992 to 2,000 in 1994.[22]

Many solar water heaters in recent years have been installed with financing arranged by the Independent Savings Plan Company (ISPC) of Tampa, whose owners, Bob Schabes and "Ben" Bentley, sell almost exclusively to people interested in retrofits. They have established revolving credit accounts in a hundred Florida retail outlets, which enable homeowners to pay off a pricey $3,000 solar water heater at only $45 a month. Still, says Schabes, 99 percent of the new houses being built in Florida lack solar water heaters, and ultimately he believes that utility opposition is to blame.[23] In its ten-year plan for 1990 through the year 2000, Florida Power and Light tried to eliminate all solar water heater rebates, then backed off when ISPC (and then the Florida Solar Industries Association) threatened to challenge the utility's plan before the state public service commission.[24] In 1993, the giant utility succeeded in cutting its solar water heating rebate to $300, and in May 1996, managed to eliminate the rebate entirely, culminating a long history of anti-solar activity. Florida has no income-tax credit for investments in solar technology, because the state has no income tax. Not surprisingly, says David Block, executive director of the Florida Solar Energy Center, "The penetration of solar water heaters in the new home market is dismal. One percent would probably be an accurate figure."[25]

Yet user surveys show that Floridians who own solar water heaters are satisfied with the device. When asked if they would buy one again, 67 percent said "yes," and 17 percent said "probably." However, because of the lack of easily available credit, 79 percent of Florida solar water system buyers pay cash for their systems, a luxury most homeowners cannot afford, at an average cost of $2,500 to $3,000 per unit.[26]

"The main barrier," argues Professor Robert Blackburn of the Florida Solar Research Center, "has been the high cost of solar units as well as the lack of credit." Encouraged by the Israeli experience, Blackburn believes that mass installation and low dealer markups could reduce the installed price per unit from $3,500 to the $700-to-$1,000 range, and that then "sales would soar."[27] The installation of 200,000 new residential solar water heaters per year could be expected to generate at least 5,000 permanent, full-time jobs in installation and maintenance alone.[28] The demonstration effect would be awe-inspiring, because it would once again make the capture and use of solar energy a visible process to the Florida public and to the world at large.

The Project for an Energy-Efficient Florida is one of the organizations trying to encourage energy efficiency and solar. Funded by a

$70,000 grant from the Energy Foundation (see chapter 5 for a more detailed explanation of the role of this key funder), the Project is affiliated with the Florida chapter of the American Planning Association, and publishes the extremely informative *Florida Energy Reporter* every three months.[29] It is also the main organizer of the Coalition for an Energy-Efficient Florida, which advocates more utility emphasis on efficiency and demand-side management and has valiantly tried to convince utilities that it is in their own self-interest to encourage installation of solar water heaters and other renewable-energy technologies. "We're all in this together, only the utilities among us don't know it yet," said one Project staffer.[30]

Despite the efforts of the coalition, private utilities in Florida have proved more interested in increasing total electric consumption than in conservation. According to a Florida legislative report, of the $806 million in ratepayers' funds spent by the utilities on conservation during the 1980s, almost $700 million went to "load management," in other words, shifting electricity consumption around so that a larger proportion of it is consumed during off-peak hours. The state report also noted that Florida's electricity use had risen 5 percent per year during the 1980s, while population growth averaged just under 3 percent. At this rate of increase, total electrical consumption can be expected to double every fourteen years in Florida.[31]

Florida utilities are now attempting to build two or three new generating plants, and as late as 1993, the chamber of commerce in Crystal River continued to promote the town as a site for a nuclear power plant.[32] These utility expansion proposals are being hotly contested by the Florida Consumer Action Network (FCAN) and other grassroots groups, which argue that utility-orchestrated demand-side management should take the front seat in Florida's energy strategy. Among much of the populace, solar consciousness has sunk so low that solar advocates in the legislature found it necessary to pass a law forbidding municipalities and homeowners' associations from banning solar hot water heaters for aesthetic reasons![33] The municipally owned electric system in Florida's capital city of Tallahassee has been the only Sunshine State utility to show an interest in promoting solar water heating. Since the early 1980s, this utility has made available low-interest loans, averaging $3,100 and repayable over seven years, to homeowners who wish to install solar water heating systems. In September 1991, the five-member commission that governs the Tallahassee Electric Department rejected a proposal to build a new fluidized-bed, coal-fired power plant, despite the promise of demonstration funds from the federal Department of Energy. The commission, which is directly elected, was under heavy pressure from citizens who feared the project's high cost and who abhorred the prospect of two hundred railroad gondola cars full of coal rumbling through town each week.[34]

Hawaii: Sunrise for Solar?

TROPICAL HAWAII could be a solar paradise, a prime candidate for total energy self-sufficiency. Solar water heating, photovoltaics, and solar thermal electric power plants could harvest the sunlight, which shines powerfully and consistently, and wind turbines could generate electricity to power electric cars from the steady trade winds. Honolulu's average temperature is 72°F in January and 78°F in July; highs of over 90°F are rare; and it never comes close to freezing, except on top of a few volcanoes. Over most of the islands, the Pacific trade winds keep the air moving, so that buildings properly designed and built require neither air conditioning nor heating, with the possible exception of offices that use computers which are sensitive to moisture, heat, and dust.[35] The most logical place to start is with solar water heating, since most people now heat their water electrically and since water heating eats up 40 percent of the average electric bill in households that lack a solar water heater.[36] Furthermore, the tourist industry, Hawaii's number-one money-earner, consumes torrents of hot water keeping tourists and their bed linen clean.

The state government has endorsed a solar transformation in the past, and it remains more sympathetic to solar than any other state adminstration in the union. In 1980, the state energy plan, published by the Hawaii Department of Planning and Economic Development, held that it was technically feasible for Hawaii to become self-sufficient in energy by using renewable resources alone. Four years later, the same agency argued:

Our near total dependence upon imported petroleum makes the State especially vulnerable to economic and social disruptions which can arise from sudden shortages in supply or increases in price as we have experienced in the past. In addition, the large amount of capital which leaves the State to pay for oil also depresses the local economy and represents an attractive opportunity for economic development. A domestic renewable energy industry could provide us not only with a stable, abundant, and low-polluting energy source, but it can also provide benefits of increased local employment and additional tax revenues. There are few economic development opportunities in the State which are more promising."[37]

Despite federal cutbacks and the halving of the price of oil, the state government still maintains a commitment to solar energy. Businesspeople who invest in solar water heaters, photovoltaics, heat pumps, and ice storage units can earn a 50-percent state income-tax credit on their investment, which turns the state government and all its citizens into an investor in solar technology. However, that program has been insufficient to reduce the state's dependence on relatively cheap imported petroleum. Gone are the days when experts predicted that half of Hawaii's energy use would come from renewable sources by the year 2000.[38] Hawaii still generates 91 percent of its electricity with oil-fired power plants,[39] and the question is not whether but when a new coal-fired power plant will be built to meet increasing demand.[40] "Hawaii's current energy situation," wrote one consultant, [is like] "a wooden tub with millions of leaks. . . . From the utilities' perspective, putting in another hose and pumping in power looks like the most obvious answer to keep the wooden tub full. However, a more intelligent and far less costly solution is to plug . . . the millions and millions of leaks with reliable and readily available energy-efficient technologies and establish strong building standards to prevent more leaks from developing."[41]

The solar transition should be easier in Hawaii than in other states, because oil-fueled electricity, the only real alternative to solar, is relatively expensive, elevating the price consumers are willing to pay for solar.[42] The Hawaii Solar Energy Association, whose dozen or so members install solar water heaters, has been the most consistent and effective political lobby for solar solutions, aided by a pro-solar consciousness among the public at large. But it has been an uphill struggle. So far, developers and landlords allied to GasCo, the islands' natural-gas supplier, have successfully resisted attempts by the Hawaii Solar Energy Association to make solar water heaters a building-code requirement for new construction and building retrofits. Many homeowners also seem reluctant to believe in the efficacy of renewables, even though solar water heaters will pay for themselves in four or five years.[43]

Hawaiian Electric Industries, the politically powerful holding company that owns almost all of the islands' electric companies,[44] has decided to help promote solar water heaters because the devices are politically popular and would not cut too deeply into electricity sales, and because the promotion of the solar water heaters has been structured so that the utilities profit from every solar heater installed (see page 25). In order for solar water heating to succeed, argued Clyde Murley in 1993, then a Natural Resources Defense Council staffer in Honolulu, it was "of critical importance" to allow the utilities to make a profit from it, because compared to individual solar installers, utilities are large institutions with good credit that can be trusted to follow through on their commitments.[45] The twenty-year integrated resource plan of Oahu's Hawaiian Electric Company (HECO), submitted in July 1993, was supposed to address the questions of solar energy and fossil-fuel dependency. Under HECO's initial (and since revised) plan, about 105,000 of the 230,000 private homes in Oahu that lacked solar water heaters will have acquired them by 2013—replacing an estimated 50 megawatts of electrical capacity, explained Murley, who was working to encourage HECO to include solar energy in its twenty-year integrated resource plan. Under the 1993 plan, total generating capacity was expected to rise from 1,220 megawatts to about 1,560 megawatts by 2013, but neither solar thermal, photovoltaics, nor wind power was predicted to play a role in this expansion. By 2013, if things proceed as HECO intends, consumption of fossil-fueled electricity will have increased by 28 percent—95 percent of that generated by imported oil and coal.[46] In both relative and absolute terms, Hawaii's most populous island has been becoming more and not less dependent on imported fossil fuels.

Large real-estate developers have been indifferent to solar water heating systems and do not offer them as a routine option. Compared to electric heaters, developers regard solar water heaters as expensive and cumbersome to install, and they fear that malfunctioning solar heaters could trigger homebuyer lawsuits. Thus the new construction market for solar water heaters is confined almost exclusively to custom-built homes. Rolf Christ, president of the Hawaii Solar Energy Association and a wholesaler of solar water heating systems, estimates that solar water heaters have been installed ready-made in only 5 percent of new residences in recent years.[47] Keith Block, the knowledgeable and enthusiastic spokesperson for HECO's solar water-heating program, hopes that copayments to developers of $1,500 per unit will be enough to induce these developers to install solar water heating in half of the homes they build over the next five years. Despairing of making solar water heating a building-code requirement for all homes, Rolf Christ would like, at the very least, to require that developers offer solar water heaters as an option to all homebuyers.[48]

Landlords who own the buildings where 55 percent of Hawaii's inhabitants live[49] have additional objections to solar water heating. Rather than thinking of solar water heaters as devices that could lower their tenants' utility bills and make it easier for them to pay the rent, most landlords look at solar water heaters as an extra maintenance hassle. Since landlords don't pay their tenants' electric bills, they don't worry about the size of those bills.

Homeowners, as well, have expressed doubts about solar water heating. Some consider the devices unreliable, while others call them ugly. Instead of instructing their architects to hide the solar heaters or integrate them into the building design,[50] most homeowners just install electric heaters. In some Hawaiian communities, as in Florida, homeowners have even tried to ban solar water heaters from their neighborhoods altogether.

Like elsewhere in the United States, solar water heating in Hawaii flourished during the 1980s, when homeowners could claim a federal income-tax credit of 40 percent of all solar expenditures, with a maximum claim of up to $3,000. In 1985, the last year of the federal rebates,

6,500 solar water heaters were installed on Oahu, recalls Rolf Christ of the Hawaii Solar Energy Association. In 1986, the number of new installations crashed to 300. But Christ and his colleagues refused to give up. To make up for the loss of the federal tax rebate, they lobbied hard in the state legislature for increases in allowable state income-tax rebates for solar water heaters. From 10-percent tax rebates in 1985 (the last year of the federal solar rebates), the legislature increased the state rebate from 15 to 20 percent, and then in 1990 to 35 percent of the homeowner's cost of a water heater, where it stands today despite sporadic noises by the present governor, Democrat Benjamin Cayatano, to cut or eliminate the rebate.[51] Stimulated by the tax rebates, sales of new solar water-heating systems soldiered up from their 1986 collapse to over 1,000 in 1989 and to between 1,500 and 2,000 per year by the mid-1990s—about one third of all such sales in the entire United States.[52] Today, 15 to 20 percent of all residences in Hawaii have solar water heating, by far the highest proportion of any state in the Union.[53]

For years, the state's Department of Business, Economic Development, and Tourism[54] and Governor John Waihee III had argued for more attention to energy efficiency and solar. Finally, the Hawaii Public Utilities Commission ordered each of the state's private utility monopolies to come up with a plan that included energy efficiency and solar. The idea was to slow Hawaii's rapid growth in electricity consumption (3.2% in 1994 and 2.5% in 1995)[55] and postpone for three years the building of a proposed power plant, fired by imported coal, which would cost at least $500 million and trigger major rate increases.

In 1996, the Hawaii Public Utilities Commission (PUC) approved HECO proposals to encourage residential solar water heating and other more efficient water-heating devices for new[56] and existing[57] homes on Oahu, the island where four fifths of Hawaiians live. The plan is designed to replace 19.5 megawatts of fossil-fueled electrical-system demand by the year 2000. It also makes sure that HECO will be generously reimbursed for the electricity sales it will "lose" to solar water heaters. Under the terms of the program, the Hawaii PUC will allow HECO to spend $71.4 million of ratepayers' money to install 26,500 new water heating devices, of which 16,000 will be solar.[58] HECO's proposal promises to save 82.5 million kilowatt-hours of electricity consumption in five years through the year 2000. For those replacing existing conventional water heaters with the 16,000 solar units, the program makes clear financial sense. Under the program, as solar contractors tirelessly point out, a homeowner's investment in a typical $4,000 solar water heating system can be expected to save $518 per year in electricity bills, a payback time of 4 years when the $800 rebate from HECO and the 35-percent tax rebate on the remaining $3,200 are figured in. Over the expected 15-year life of a solar heater, a household can expect to save over $5,000 in electric bills.[59] The program should also be a financial and public-relations bonanza for HECO, and a winner for solar contractors and homeowner participants; but the benefits for nonparticipating residential ratepayers, particularly for the majority who do not own their own homes, are questionable.

The existing homes program presumes that HECO and its stock-holders have a right to the energy that the utility *did not produce* but *would have produced* if the Hawaii PUC had not approved the program.[60] Under the terms of the program, HECO gets to distribute $71.4 million from all residential ratepayers in the following manner:

1. $15.2 million in total state income-tax credits to 26,500 HECO customers who buy new water heaters, of which 16,000 will be solar;[61]
2. $30.9 million in administrative costs to run the program—advertising, water heater inspections by HECO personnel, evaluation of kilowatt-hours replaced, and so forth;[62]
3. $17 million in profits[63] that HECO *would have earned* if it sold the 82.5 million kilowatt-hours of electricity saved by the 26,500 new water heaters of all kinds;
4. $8.3 million in dividends[64] that *would have been earned* by HECO stockholders if the utility had sold the 82.5 million kilowatt-hours[65] of electricity replaced by the 26,500 new water heaters.

To these costs must be added $65.9 million in direct costs to the 26,500 households installing the new water heaters, for a total investment of $137.3 million in new water heaters, for an average of $5,181 per water heater—some solar and some not. If it seems unfair to ask nonparticipating renters to sustain a program that, in effect, bills renters to cut the utility bills of relatively affluent homeowners and that props up utility finances, those subsidies represent the political price that had to be paid for a solar water heating program in Hawaii in 1996.[66] While it is possible to imagine a cheaper and a more equitable way to do the sun's business in Hawaii, at least Hawaii could be proud that it would soon be installing 50 percent of the nation's new residential solar water heaters in a state with 0.5 percent of the nation's population.

On the Big Island of Hawaii, retired electrical engineer Jay Hansen is appalled by the inefficiency and backward thinking of the Hawaiian Electric Light Company (HELCO), which is, like HECO, a wholly owned subsidiary of Hawaiian Electric Industries, Inc. One of three public representatives on the thirty-three–member Integrated Resource Planning Committee that advises HELCO, Hansen believes that the utilities will convert to solar only if they are municipally owned and controlled. Hansen has vigorously fought a proposed new diesel-fired combustion turbine project for the Big Island, which will cost the company $32 million to install and cost consumers $463 million over its thirty-year life cycle, mostly for imported fuel.[67] Moanikeala Akaka, veteran community leader and elected trustee of the Office of Hawaiian Affairs, has heard rumors that Big Island brownouts can be blamed on HELCO's mismanaged maintenance policies and excessive promotion of electricity consumption. She noted that recent attempts to build power plants near the Kona airport and at Kawaihae have been defeated by local opposition, despite the brownouts, and recalled that Big Island activists in the late 1970s had briefly considered the idea of replacing HELCO with a municipally owned electrical system that could generate cheap geothermal electricity and benefit the entire island—until they discovered the true price of geothermal.[68]

The fiercest energy battles on the Big Island of Hawaii have been over a controversial proposal to build a 500-megawatt geothermal power plant. The proposed geothermal plant was to be sited in Wao Kele' o Puna, surrounded by the last native-growth rain forests in Hawaii, on the east rift of Kilauea Volcano, one of the two volcanoes considered sacred to the indigenous fire goddess Pele. A pamphlet published by the Pele Defense Fund protested this desecration of Kilauea: "Her person, her body/spirit, her power/mana (spiritual power) are the land of Hawaii. Native Hawaiians consider the volcanoes of Mauna Loa and Kilauea from their summits down to the ocean . . . to be her principle domain. Pele is manifest in magma, steam, vapor, heat, and lava. Her family of sisters and brothers are seen in kinolau (body forms) of forest plants and ocean life. Pele is the akua (goddess), aumakua (family goddess), and kupuna (wise elder) to thousands of Native Hawaiians."[69] Neighbors of a "pilot" 26-megawatt geothermal project have often complained about noise pollution and about the rotten-egg stench of hydrogen sulfide emanating from the project, and they feared the release of highly toxic mercury, lead, and arsenic into the air and the groundwater. Energy expert Amory Lovins claimed that the full-scale geothermal project, with its multibillion-dollar price tag, could "crash the state economy."[70]

The Pro-Geothermal Alliance was a Honolulu-based organization with five charter members: Hawaiian Electric Industries; Mid-Pacific Geothermal; True Geothermal (a powerful family-owned Wyoming firm involved in banking, ranching, and trucking); Campbell Estates (traditional missionary-descended Hawaiian landowners); and Puna Geothermal Venture (a subsidiary of Nevada-based, Israeli-owned Ormat Energy Systems). Other geothermal supporters included job-hungry carpenters, firefighters, machinists, electrical workers, longshore workers, laborers, operating engineers, and public-sector unions from the Big Island Labor Alliance, and the state department of business and economic development. Arrayed against this "state-federal-private" phalanx were the Pele Defense Fund, the Rainforest Action Network, the Big Island Rainforest Action Group, various Puna district community associations, and the Office of Hawaiian Affairs, which has supported some of the Pele Defense Fund legal interventions.

The proposed Wao Kele' o Puna geothermal plant was designed to supply electricity to Honolulu and other cities in Oahu through a risky undersea cable running under stormy Alenuihaha Channel to Honolulu and other cities in Oahu. The Pele Defense Fund and the Rainforest Action

Network charged that geothermal supporters also had a "hidden agenda" to provide power for the "seabed mining of manganese crust" near the Hawaiian Islands, which would foul the air and poison the sea nearby.[71]

The opposition to the proposed geothermal plant took many forms, from direct action to lawsuits. Palikapu Dedman, Dr. Noa Emmett Aluli, and other native Hawaiians formed the Pele Defense Fund and the Big Island Rainforest Action Group, and undertook a thirteen-year campaign of marches and sit-downs, with hundreds of arrests, in alliance with the Rainforest Action Network and the Sierra Club. These efforts helped to bring the issue to worldwide attention, and on March 4, 1994, the True Geothermal Energy Company announced its pullout from the project. At a victory news conference, Palikapu Dedman of the Pele Defense Fund warned his listeners that Wao Kele' o Puna was still in danger because it was located on land that had been illegally appropriated by private interests. In February 1995, the State of Hawaii officially gave up all attempts to promote geothermal development in the Wao Kelo' o Puna rain forest, ending one chapter in the struggle.[72] But Hawaiian Affairs trustee Moanikeala Akaka cautioned Pele's defenders to maintain their vigilance, reminding them that the 26-megawatt geothermal plant already in the area could serve as a prototype for new geothermal proposals.[73]

The good news comes from the village of Milolii on the south Kona coast of the Big Island of Hawaii, the "last Hawaiian fishing village in the world." The whole village runs on solar. About fifty fishing families live there, and the village has never been hooked up to the electric grid. In 1990, a foundation donated enough photovoltaic modules and batteries to equip about thirty houses with PV power, and a local engineer/electrician helped teach the people to hook them up. Shirley Casuga, one of the leaders in the project, said, "Once we all learned how to do it, we did all the work ourselves. Now we don't need generators or lanterns, and we can run the washing machine if we start at eleven when the sun gets high. All we have to do is check the water level in the batteries every month, and my batteries have run fine for six years, though some other people had to get new ones." She doesn't have a water heating system of any kind, "because it's so hot here." Shirley Casuga

TABLE 2-1: Solar Tax Rebates in California, 1980–1983

Family income	$50,000
Value of desired solar water heater and pool heater	$5,000
Federal tax liability	$10,000
California tax liability	$2,000
Federal solar rebate	$2,000 (40 percent of $5,000)
California solar rebate	$750 (55 percent of $5,000 minus the $2,000 federal rebate)

says the PV equipment costs about $8,000 per house, "and now everyone around here is doing it. . . . We don't need HELCO," she repeats, emphatically.[74]

California: Water Heating Redux

IN CALIFORNIA, solar hot water heating was reborn in the late 1970s thanks to ecological fervor, technical prowess, federal and state tax breaks, and the Arab oil embargo. But the person who really made it possible was Governor Jerry Brown, who collected a group of "whiz kids, Ph.D. progressive types" around him and made Sacramento "a hot-bed of renewable-energy ideas."[75] The most exciting institution he created was the Office of Appropriate Technology, headed by architect/philosopher Sim Van der Ryn, with whom Governor Brown created an energy policy that was "unique in the nation [and] consistently stressed indigenous, renewable, decentralized resources as best for California citizens."

Gigi Coe, who has worked in California government around energy issues for two decades, called the tax credits the "driving force underlying the rapid growth of the solar and wind industries" from 1976 until they were virtually extinguished in 1986 on both the federal and state levels.[76] From 1980 to 1983, a homeowner could claim a state income-tax credit of 55 percent of the value of a renewable home-energy application with a maximum claim of $3,000. California rebate claimants were required to subtract the federal rebate of 40 percent from the amount, so in most cases the real state-tax rebate came out to 15 percent of the value of the solar installation. For an upper middle-class family, the rebate might have worked out as follows during those years.

From the example in table 2-1, we see that a family could spend $2,250 of its own money for $5,000 in solar living improvements. If our model family had wanted to install $7,000 in solar improvements, it would have qualified for $2,800 in federal tax breaks (40 percent of $7,000 but $200 in state tax credits) because the maximum claim was $3,000. A different set of rules applied for commercial applications.[77] The 40-percent federal tax credit expired at the end of 1985, and Governor George Deukmejian, who succeeded Brown, terminated all state tax credits in a year or two. Until then the new solar industry, wind included, made quite a run of it in California.[78]

In San Francisco the Solar Center, a worker-owned co-op, installed hundreds of solar hot water heating systems. The Solar Center followed an egalitarian wage policy; democratically organized work and solar energy went hand-in-hand. Wherever the solar water heaters were installed, gas or electricity sales by Pacific Gas & Electric (PG&E), the country's largest utility monopoly, fell accordingly. During peak years business amounted to over $2 million, but Peter Barnes, the head of the Solar Center, never made more than $20,000 per year—less than some of the Center's top installers.[79]

The story of the Solar Center helps explain the reluctance unions have shown regarding solar energy. How could union workers be expected to support a new technology that cut into the market share of an electric and gas utility that hired union labor? The Solar Center was not a union shop, explained Barnes, "because as a pioneer operation, we just didn't have the kind of revenues to pay those kinds of wages." The book *Energy, Jobs, and the Economy,* by Richard Grossman and Gail Daneker, cites dozens of estimates of the abundance of jobs that would be created by mass application of solar technology—enough to put all of his unemployed members back to work, wrote the president of the Sheet Metal Workers' International Association in 1977. Other unions were equally generous in their estimates of the good jobs that could come from widespread solarization.[80] But unions representing workers in the coal, oil, and atomic industries, which were then highly unionized, kept their distance from solar entrepreneurs and tradespeople, reasoning that a well-paying job in the hand was worth two in the bush (see chapter 6).

Many of the well-publicized stories about quality-control problems caused by "tax-shelter developers" and fly-by-night solar hot water contractors were true. But good solar hot water systems installed by reputable organizations such as the Solar Center are still functioning. The property manager at the St. Francis Square Apartments—a cooperative 299-unit complex built at the initiative of the International Longshoremen and Warehousemen's Union, and completely equipped with solar water heaters by the Solar Center—reports that despite a four-and-a-half-day gas failure during a cloudy week in May 1991, the Solar Center system by itself was able to keep shower temperatures at a comfortable 105°F. In nearly fifteen years the system has functioned perfectly, he said, with only routine, twice-yearly maintenance to clean the solar panels, repair those broken by vandals, and check the electronic components.[81]

For some households, the potential savings could be quite substantial. The Epperson family of Oakdale, California, has propane bills of more than $50 per month for water heating for a household of six.[82] For T. Roy Epperson and his wife Esther, a $2,000 solar water heater could cut those propane bills to an average of only $10 a month. With a remodeling loan guaranteed by the Veterans Administration for ten years, the payback plus the supplementary propane bill would be less than $30 per month, a savings of more than $20 per month. After the loan was paid off, the Eppersons' cost of hot water would average $10 per month.

In California, the locus of solar activity since the early 1990s has been the municipally owned Sacramento Municipal Utility District (SMUD). SMUD is demonstrating how a vigorous combination of new loans and close supervision of solar contractors can simultaneously cut electric bills, guarantee solar equipment quality, and reduce the need to build new electric generating capacity (see chapter 4).

The Eclipse of Renewables

IT IS DIFFICULT to accept the reasoning that concern over business "efficiency" was the real reason for cutting R&D for renewables. Guaranteeing reliable solar hot water systems should be no more difficult than assuring trouble-free plumbing and electricity. Solar water heaters should be required by building code, as in Israel, and natural gas, propane, and electricity should be used merely as backups. Anxieties about reliability could be met with rigorous supervision of contractors' qualifications, reasonable performance guarantees, and repair facilities, just as they are with any other home technology. Rather than using the failures of the fast-buck artists to destroy the reputation of solar water heating, the political challenge should be to design systems of manufacturer warranties, contractor licensing, building inspection, regular maintenance, and citizen education—all of which strengthen the reliability of solar hot water heating as a technological system. Compared to the failures of nuclear power, the technical problems of solar water heating and wind power have been miniscule.

But solar never had a chance. The hoped-for cornucopia of clean energy and jobs was destroyed by president Ronald Reagan, whose most fervent supporters included the big energy corporations. David Stockman, Reagan's young budget director, was astonished at the avarice and cynicism of the dominant players during the tax-cut orgy that he initiated at the start of the Reagan administration in 1981: "Do you realize the greed that came to the forefront? The hogs were really feeding. The greed level, the level of opportunism, just got out of control. . . . What was new about the Reagan revolution in which oil royalty owners win and welfare mothers lose? Was the new philosophy so different when the federal subsidies for Boeing and Westinghouse and General Electric were protected while subsidies for unemployed black teenagers were 'zeroed out'? The power of these client groups turned out to be stronger than I realized . . . unorganized groups can't play in this game."[83]

By 1986, President Reagan and California governor George Deukmejian would carry out their election promises to end income-tax credits for renewables, and the result was predictable: from a $475-million industry employing 19,000 people full-time in California in 1984, solar water heating collapsed to $20 million in 1986.[84] Renewables were downplayed to such an extent that the Department of Energy's (DOE) Energy Information Administration even stopped keeping reliable statistics, so Public Citizen (the consumer watchdog group founded by Ralph Nader) had to step in to provide critical documentation.[85] Investment in new wind turbines in California tumbled from a billion dollars (supporting an industry with 12,000 employees) in 1984 to under $100 million in 1988.[86] Meanwhile, Colorado solar businesspeople were experiencing a similar crisis: Karen Jankowski of Alternative Energy Systems in Boulder recalls that her annual gross sales in solar hot water systems crashed from $9 million to under $1 million when the tax credits ended.

According to the Energy Information Administration, domestic shipments of American-made solar water heating panels fell from a peak of 12 million square feet to 0.7 million square feet in 1986, the year after the Reagan adminstration killed the federal energy tax credit.[87] Sales of solar hot water panels fell 91 percent between 1984 and 1986.[88] This collapse accelerated when deregulation and conservation caused the effective price of natural gas to fall by 78 percent.[89]

On the research and development front, Reagan budget cuts forced the Solar Energy Research Institute (SERI) to cut its workforce from 1,000 to under 500 in just a few months, with the biggest cuts in those engaged in outreach and education, and application of the new technologies. "It was an exercise in heartbreak," recalls one SERI survivor. "We had just gotten to the point where the organization was solid. . . . the administration singled us out and morale fell very fast."[90]

Other American-owned solar projects have stalled, as well. Arco Solar, the largest U.S. producer of photovoltaic cells, was sold to Siemens, a German multinational, in February 1990. Siemens Solar in turn sold the 6-megawatt Carrizo Plain photovoltaic array, which had sold solar electricity to PG&E, to an investment group from New Mexico that is dismantling and selling it, piece by piece. Mobil Solar Energy Corporation, a subsidiary of Mobil Oil Company in Billerica, Massachusetts, shut down its production of crystalline silica photovoltaic arrays in November 1993, claiming that the utility market it had hoped for had never materialized. Mobil Solar was sold to the German-owned ASE Americas, Inc., the next year.

The halving of oil prices, engineered in connivance with Saudi Arabia and facilitated by the efficiency-induced leveling-off of fuel sales, pulled all fossil-fuel prices down with it, especially natural gas.[91] As living standards fell during the 1980s, the only prices that fell faster were the cost of gasoline and the cost of flying. The price crash delivered the final *coup de grâce* to the solar hot water heating industry, which, outside of Hawaii and Sacramento, now consists largely of the upkeep of existing systems and a swimming-pools market in Florida and California. Wind-energy development likewise collapsed, despite dramatic decreases in startup costs.[92] New generating capacity installed in California wind plants went from 398 megawatts in 1985 to 19 megawatts in 1992. And some of the few surviving wind companies were driven overseas.[93]

Of the forty American companies making wind turbines in the early 1980s, only a few remain in business today, and Japan's Mitsubishi Heavy Industries has recently sold 660 new 250-kilowatt wind turbines to SeaWest, a San Diego–based company that operates wind turbines in California.[94] Though California's Independent Energy Producers Association likes to boast that "alternative" electrical generation capacity had risen from 2 to 18 percent between 1978 and 1993, most of that new capacity was in cogeneration, geothermal, and biomass capability. Truly pollution-free electricity from solar, wind, and small hydro sources now produce less than 2 percent of California's electricity.[95]

In June 1994, the California Public Utilities Commission announced that it would require the state's three biggest private utilities to purchase 14 percent of their new power needs from renewable sources, especially wind power, and the Kenetech Corporation, a San Francisco–based alternative energy company, told the press it had secured approval to sell a total of 945 megawatts of wind-power capacity to San Diego Gas and Electric, Southern California Edison, and Pacific Gas & Electric.[96] Kenetech's variable-speed wind turbines, which produce electricity for about 5 cents per kilowatt-hour, would be installed from 1997 through 1999.[97]

In the absence of a broad-based solar movement, the independents could not withstand the loss of tax incentives and government support.

Peter Barnes of the Solar Center, like tens of thousands of other entre-preneurs, left the solar arena, and most of the domestic solar industry shriveled up and died.[98]

The Solar Lobby and Its *Blueprint for a Solar America*

THE ORIGINAL SOLAR LOBBY of the 1970s, composed of impas-sioned people who valued principle over money, underestimated the ferocity of their opposition. The implicit theory of the Lobby's *Blueprint for a Solar America* was that when enough citizens and politicians under-stood that massive conversion to solar would be in the country's own best interests, a patriotic president and Congress would steer the ship of state away from the twin hazards of environmental disaster and foreign depen-dency. "We believe," they wrote, "that Americans will enthusiastically embrace solar alternatives if public policy merely allows them true free-dom of choice."[99] Having reasoned their way to the superiority of solar and presented their arguments to the American people in the nationwide Sun Day celebrations in 1978, they may have hoped that their opponents, like gentlemen patriots, would concede victory and fade quietly away once they had been bested in honest debate. But the desires of ordinary Americans didn't carry any weight, ultimately, because these citizens were excluded from the decision-making process. Why the energy cartels should be expected to voluntarily relinquish their historic dominance over energy, and wean themselves from the guaranteed markets and tax loopholes that had nourished them for so long in such royal luxury, was never adequately explained in the *Blueprint*.[100]

Under President George Bush, who had actually owned an oil com-pany, things went from bad to worse. "I put it this way," said Bush shortly before his inauguration in 1989. "They got a President of the United States that came out of the oil and gas industry, that knows it and knows it well."[101] The commonsense voice of ecological reason had been over-powered by the roar of oil money and the righteous ecstasy of a holy war for crude. "By God, we've kicked the Vietnam syndrome once and for all,"[102] crowed President Bush, after his victory over Saddam Hussein, stubbornly ignoring the fact that his administration had previously helped arm Iraq with billions in Department of Agriculture credits, on top of the weapons sold to Iraq by the U.S. during the 1980s in an effort to maintain "a regional balance of power" with Islamic revolutionary Iran.[103]

Richard Grossman, author and former executive director of Environmentalists for Full Employment and of Greenpeace U.S.A., has written that the Sun Day organizers relied too much on the Department of Energy and the government rather than on the grassroots citizen movements:

> The energy department was controlled by old [Atomic Energy Commission] boys, and even younger boys who were loyal to the fossil-nuke-chemical-utility corps. They openly ridiculed solar, wind, etc. So although sanity made a few inroads under Carter, resulting in new information, credibility, tax credits, etc., there was no fundamental change in who had authority over energy policy, technology, and investment. . . . Consequently, there was nothing solidly in place—laws, institutions, or a citizens' movement—to prevent Reagan from undoing all of the symbolic stuff and the limited machinery (like tax credits).
>
> The Sun Day initiators and organizers refused to make citizen control and authority an issue in their "educational and promotional" work. Not because they were evil, or secretly were against solar or sane energy policies. But because what did they know? In this country, especially since the late 1940s, it has not been common to talk about ownership and control as a public policy issue, as a variable. The activists from the '30s and '40s, who would have been leaders of unions, of farmers' groups, of civic groups, of the left, had been purged and silenced. The New Left never took energy or environmental stuff particularly seriously, and so had little influence from a political-economic standpoint on a new generation of largely self-taught grassroots anti-nukers and safe-energy organizers.[104]

Recall that the Port Huron Statement, written by Tom Hayden in 1960 to found Students for a Democratic Society, advocated mass use of nuclear power. Yet perhaps it is unfair to carp, given the weaknesses and divisions within and between the antinuclear movement and the solar advocates. If the solar movement had not been so potentially strong, the oil-infatuated Right would not have moved to destroy it with such premeditated fury.

Republican George Deukmejian won the California governor's race in 1982 on a platform that explicitly included the elimination of solar and energy conservation tax credits, and he quickly fulfilled the promise. Despite their conceptual and operational simplicity, solar technologies such as domestic solar water heating were vehemently characterized as unreliable, futuristic, and frivolous, setting the political stage for removal of the tax credits and sabotage of the budding solar industry. Mainstream environmental groups, including the Natural Resources Defense Council (NRDC) and the Environmental Defense Fund (EDF), made a virtue of *realpolitik* and left small solar businesspeople in the lurch, with the rationale that, for now, conservation was a cheaper alternative than solar. Solar Center founder Peter Barnes says David Roe of EDF was "the enemy, almost.... I met with him a couple of times and tried to get his support. He was friendly and charming, but was much more interested in pushing EDF's whole scheme of utility regulation. He had no interest in a decentralized, non-monopolized system. They just stood by and watched us get killed."[105] Rather than initiating serious programs to monitor and train solar hot

Stomping out Alternative Energy

In the mid-1980s, when big companies like Dow Chemical in Pittsburg, Union Oil in Rodeo, and the East Bay Municipal Utility District in Oakland, as well as a tiny one like the Bing Wong Wash Center ... in Berkeley, decided to install alternative-energy facilities, the PUC ordered PG&E to pay them a reasonable price for any excess energy they sent to PG&E's grid.

In 1985 the price PG&E paid was 7.2 cents per kilowatt-hour (kWh). But in 1987, after PG&E claimed the rate was too high, the PUC allowed it to drop its payment to 3.05 cents per kWh, according to Steve Adams, project manager for several cogeneration plants in Northern California. "It cut about one third of our revenue," Adams said. That move sent the projects to the brink of bankruptcy.... "The PUC and utilities were doing everything they could to keep alternative energy from going on line," said Jan Hamrin, a San Francisco-based energy consultant. "Ten years ago there were two hundred wind companies. Now there's half a dozen."

J. A. SAVAGE, 1991

water installers and drive the rip-off artists out of business, the opponents of solar water heating exploited the apparent abuses to abolish both federal and state tax credits, which killed the programs. Other solar initiatives of the early 1980s—such as the savings and loan association that financed $500,000 of the Solar Center hot water projects at low-interest rates in its first year[106]—were never institutionalized, and the proposed federal Solar Bank and mass federal and state government purchases of photovoltaics were never enacted.[107]

None of the major proposals of the 1979 *Blueprint for a Solar America* has ever been adopted. Today, the best-funded environmental groups, including the Natural Resources Defense Council, have deferred to utility claims that even solar hot water production is not "cost-effective," and such deference has been tolerated because it is not yet common knowledge among environmentalists that most people in Israel, Cyprus, Japan, Colombia, and Jordan already get their hot water from solar heaters, and that the Sacramento Municipal Utility District has successfully launched a multimillion-dollar campaign to replace thousands of electric water heaters with solar.[108] The California Solar Energy Industry Association reports that Southern California Gas fought the South Coast Air Quality Management District for the right to continue to heat swimming pools with gas and "strongly opposed" a 1991 bill that would mandate solar hot water heating in high-pollution areas like Los Angeles.[109]

As government subsidies for renewables dried up, the remnants of the solar industry and the major environmental groups decided to collaborate with their former antagonists (in a venture called the "collaborative process") in order to induce the corporations to make modest investments in conservation if not renewable sources of energy. In the early 1990s, the Natural Resources Defense Council and the Energy Foundation were especially persuasive in promoting demand "management" as a substitute for true reductions in energy use and increased production of renewably generated power (see chapter 5 for a discussion of the demand-side management or DSM strategy). The effect of this soporific atmosphere of compromise has been to suppress or delay adoption of those tested and proven solar applications that had intrigued millions and had begun to provide useful work to thousands of solar- and wind-power installers.

Growing Your Own Power:
Solar Culture in Northern California

B UT SOLAR TECHNOLOGY wasn't dead, it had just gone underground. Solar culture sank its deepest roots in communities and
homesteads in Northern California, whose citizens had their own
reasons, ideological or otherwise, for cherishing their freedom from the
grid, and avoiding the drain of dollars to outside institutions.

The city of Davis, California, home to a University of California
campus, pioneered the building of bicycle lanes and greenbelts beginning in the late 1960s. In the early 1970s, a small group of graduate students in ecology researched and wrote energy-efficiency revisions to the
local building code, which they proceeded to institute by getting three of
their number elected to the five-member city council. Builder-designer
Mike Corbett, who supported the revisions of the building code from
the beginning, went on to conceive and build Village Homes, a 60-acre
community of 240 passive solar homes with a community center and
swimming pool, interlaced with walking paths among communal gar-
dens and orchards. So famous did Village Homes become that French
president François Mitterand touched down in his official helicopter for
a tour in 1984.[110]

By the late 1980s Mendocino and Humboldt Counties, with their
annual Solar Energy Expo and Rally (SEER), had assumed the title of
solar and energy-conscious capital of the world.[111] For instance, in
Arcata, Larry Schlusser produces the handmade Sun Frost, one of the
world's most energy-efficient refrigerators.[112]

The Bank of Willits in Mendocino County is one of the few banks
in the United States to grant building loans for 100-percent solar homes,
even though federal mortgage guarantors like the FHA and the Veterans'
Administration are reluctant to insure mortgages on homes not connected to the grid. Keith Rutledge, of the Bank of Willits, notes that 50 of
the 5,000 loans the bank makes annually are for off-the-grid houses.[113]
Rutledge has also begun to compile a national database of banks and
appraisers familiar with off-the-grid photovoltaics and their costs and
benefits.[114] Along similar lines, the Real Goods Module Finance Program (sponsored by the Real Goods Trading Company of Hopland,

California) was announced, a program to offer finance packages for off-the-grid photovoltaic systems.[115] Rutledge himself lives off-the-grid, and stores some of his surplus solar-generated electricity in lead-acid batteries, using the rest of it to split water into oxygen and hydrogen. (The stored hydrogen is used as a cooking fuel.) He continues to drive a fossil-fueled car, but is designing one to run on hydrogen.[116]

David Katz was another of the talented hippie technologists baptized in the 1960s culture of resistance movements.[117] An electrical engineer, he worked for the Navy on warplanes for a few years as a well-paid civilian technician. In 1977, Katz quit and moved to Humboldt County. His career in alternative electricity began when he hooked up an extra alternator and battery to his Volkswagen Bug so that he and his girlfriend could have electric lights in their country cabin and avoid the smoke of kerosene lanterns. Two years later, he founded Alternative Energy Engineering (AEE) to help others light up their backwoods homesteads with automotive electricity. In 1980, Katz set himself up in the solar business by scavenging recycled photovoltaic panels from Exxon and Shell Oil solar subsidiaries that were shutting down. Under the slogan "Power to the People," AEE prospered in ecotopian Humboldt, buoyed at first by sales to self-reliant new settlers who lived beyond the electric grid in the backcountry, and by a hefty infusion of cash revenues from nouveau-riche marijuana-growers.[118] Today, AEE sports a hundred-page catalogue and a new headquarters built of recycled redwood and steel. The company claims to stock "the largest variety of equipment for making and efficiently using electricity anywhere," and ships its products around the world.[119] Katz concedes that homemade solar electricity is not yet "cost-effective" where utility-supplied power already exists, "unless you consider the actual cost of 'cheap power,' which everyone in the world is paying for in contaminated water, polluted air, and poisoned food."[120]

John Schaeffer founded Real Goods Trading Corporation in 1986 with $3,000 in capital, a mailing list of 2,000 customers, and a "strong commitment to educate Americans about the treacherous energy path we were following." For Real Goods, "the rollercoaster . . . has never stopped." In 1994, the company went public with a stock offering, and had 100 full-time employees by 1996, with another 60 hired on for the holiday season. Hundreds of new catalogue requests come in every day, and beyond sales, the company has continued to proselytize for clean energy. *Real Goods News,* providing a forum for discussion of energy and environmental issues, goes out to 75,000 to 100,000 subscribers four times a year, up from 10,000 in 1993.[121] An estimated 15,000 U.S. homes are now solar-powered—1,000 of them in and around Mendocino County—and the largest retailers and wholesalers of solar energy devices are located in the same region. Real Goods advertises hundreds of renewable energy products and produces the *Solar Living Sourcebook,* a combination Sears catalogue and encyclopedia of alternative energy.[122] Real Goods has recently celebrated the opening of an 11-acre Solar Living Center, the construction of which incorporates innovative ecological building techniques such as straw bales and rammed earth. The main building is equipped with photovoltaics and supplemental wind generators, with a utility intertie predicted to produce a monthly buyback check from PG&E.

Real Goods also cosponsors the Solar Energy Expo and Rally, one of the largest fairs of its kind in the world, with all-solar car races and every imaginable home solar device on display. Schaeffer maintains an aggressive rhetorical stance against fossil energy. Nevertheless, he welcomes the utility PG&E's cosponsorship of the SEER fair and says he would be happy to work as a PG&E subcontractor if the occasion arose. Both Schaeffer of Real Goods and Dave Katz of Alternative Energy Engineering report that business has doubled and redoubled in the last few years, and comparable expansion has no doubt been experienced by renewable-energy entrepreneurs in other regions of the country, including Jade Mountain in Boulder, Colorado; Backwoods Solar Electric Systems in Sandpoint, Idaho; and Fowler Solar Electric in Worthington, Massachusetts.

For practical visionaries like Katz and Schaeffer, the commitment to solar is a way to resist the "death culture" of cities, television, and government that people lit out for the hills to escape.[123] Katz helped set up a solar electric system in a school in Sandinista-controlled Nicaragua, refused to fill a large order for photovoltaic panels from the Guatemalan military, and tried to get a license to import $9,000 solar electric cars from Denmark. The January 1991 cover of Schaeffer's Real Goods catalogue featured a photo of U.S. troops exiting a C5A air transport plane in

Saudi Arabia, and was captioned "The Intolerable Price of Oil." In ensuing issues, a few letter-writers swore never to patronize Real Goods because of the company's anti–Gulf War position, but most correspondents praised the organization's stance. Real Goods has set up a traveling Institute for Independent Living for off-the-grid solar workshops, and to teach other techniques of energy independence, and the company claims that sales of Real Goods products saved enough electricity from its founding through March 1996 to run over 90,000 average American households for a year.[124]

The greatest achievement of the customers of firms like Alternative Energy Engineering and Real Goods has been to prove that solar works in everyday life and that it is possible to reduce electricity consumption by one half or three quarters and still live comfortably. In essence, solar consumerism has made it possible (for those people who are highly skilled and motivated, and for those who can afford it) to replace lots of dirty fossil energy with smaller amounts of clean energy and more efficient appliances. The magazine *Home Power*, bimonthly organ of the off-the-grid movement, is full of inspiring and technically sophisticated accounts of the obstacles these persistent pioneers have overcome.[125] Yet the real challenge for the solar movement is to enable the widespread adoption of energy-efficiency techniques and solar technologies by making available capital and credit to a much broader range of people. Without access to credit, few Americans will be able to afford the tools needed to generate electricity and heat with sunlight. Just imagine what would happen to new-car sales without automobile loans!

THE ELECTION OF Bill Clinton to the presidency in 1992 triggered an evanescent euphoria among environmentalists and other people normally skeptical about electoral promises and the powers of presidents. At least George Bush, the first president with his own personal oil company, had been removed from the White House. Candidate Clinton had been eloquent in his advocacy of an energy policy that "lets Americans control America's energy future" rather than "coddling special interests whose fortunes depend on America's addiction to foreign oil." The president-elect had enthusiastically advocated energy efficiency and conservation and expanded use of renewables, and had opposed "increased reliance on nuclear power" while providing extra support for solar research.[126] And the vice president–elect had just written a best-selling book called *Earth in the Balance* about "ecology and the human spirit," which meditated on the "dysfunctional" nature of modern technology, mentioned global warming two dozen times, lauded photovoltaics, and proposed a "strategic environmental initiative" (SEI) to jump-start new solar and other environmentally safe technologies.[127] Yet careful study of the fault lines of American energy politics suggests that alternation of the presidency from one party to another, and even such an escalation of rhetoric, should hardly be sufficient to alarm the customary masters of energy.

President Clinton could be commended for a 35-percent increase in federal R&D support for photovoltaics in his first two years in office, but most of the photovoltaic implementation projects were designed to be controlled by the electric utilities (see chapter 7 for further discussion of this point). Don Loweburg of Independent Power Providers, a group of almost 130 businesspeople involved in photovoltaic installation, points out that utilities actually install under 10 percent of photovoltaic modules in the United States.[128] Loweburg believes that the goal of the utilities is to drive the independents, who control over a third of the domestic PV market, out of business. "Once the public subsidy trough dries up," he writes, "expect the utilities to move away from PV and tell people it's not cost-effective. Just like they used to do before *they* got into PV. My goodness, we can't actually have people generating their own power!"[129] Of course, the Clinton administration demonstrated that it had no intention of replacing fossil fuels with alternative energy sources: the Department of Energy's blue-ribbon energy task-force seemed to blandly accept without question DOE estimates that fossil-fuel use in the United States would increase 19 percent by the year 2010, and likewise the assertion that in 2010, "fossil fuel is still expected to account for 85 percent of total consumption," as it had in 1995.[130] By 1995, new Republican majorities in the House and Senate had endangered the sharp increases the Clinton administration had managed to secure for photovoltaics and other research on renewable energy and energy efficiency.[131]

From the perspective of the oil companies, the loss of domestic oil and gas drilling jobs under Clinton, due to more stringent environmental enforcement, has presumably been balanced by passage of the North American Free Trade Agreement (NAFTA) and turning U.S. oil drillers loose in Mexico and the rest of the world, with typical tax breaks and government subsidies from U.S. taxpayers.[132] "What the oil majors really want," wrote *Business Week*, "is to explore for oil, take a percentage of the find, and help manage Mexico's vast fields." The Heritage Foundation went even farther, recommending that Mexico sell Pemex—the government-owned oil company—and all its oil deposits for $148 billion and use the proceeds to retire all of Mexico's foreign debt.[133] President Clinton's most assertive energy message was "more natural gas," which meant more drilling by the gas and oil companies. Clinton sealed and delivered this compact by appointing Arkansas gas man and kindergarten pal Thomas "Mac" McLarty as his first White House chief of staff. Clinton's repeated invocations of such mantras as "incentives" (the code word for "profits") and "least-cost energy planning" indicated that the administration intended to let the utilities make the final decisions about when and how energy efficiency and solar are to be deployed. Despite Clinton's expression of concern in January 1993 about "carbon taxes" and global warming, he couldn't even pass a 4.3-cent-per-gallon gasoline tax increase because he couldn't even persuade his own party to support stronger measures.[134] And despite White House lip service to the dangers of the greenhouse effect, on the policy level, support for solar energy has been insignificant compared to the efforts to stimulate "our domestic resources: coal, petroleum, and natural gas."[135]

Nevertheless, the renewable-energy insurgencies of the last two decades have proved several hypotheses:

1. Grassroots citizen-activists, if adequately informed and infuriated, could shut down multibillion-dollar energy projects, especially nuclear power plants.
2. Strict government energy standards for buildings, cars, and appliances could save massive amounts of energy and create hundreds of thousands of jobs.[136] According to the head of the California Energy Commission, the largest sources of electricity savings in California have been new state building-code standards and state appliance-efficiency standards, which saved

electricity equivalent to the production of seven large power plants between 1977 and 1990, despite a 20-percent increase in population.[137]

3. Low energy prices and deregulation stifle alternatives. As long as oil, gas, and coal supplies and services flow smoothly, and prices at the pump and meter remain more or less stable, the public at large has been content to leave ownership and control of energy to its customary masters in New York, Houston, Kuwait, and Saudi Arabia—threats of global warming, pollution, and transportation gridlock notwithstanding.

During the last twenty years, the energy establishment has lost only one major battle—the battle for nuclear power—to its organized citizen opponents.[138] The nuclear cartel has toiled in vain to overcome popular fear and skepticism.[139] No new orders for nuclear plants have been placed since 1978 in the United States, and nuclear plants are being shut down early for cost and safety reasons in California, Oregon, and Ohio.[140] The U.S. Council for Energy Awareness, the public-relations front for the nuclear lobby, still touts nuclear energy as the basis for a strategy of "energy independence" and a means to slow the greenhouse effect.[141] Nuclear advocates still hope to build a new reactor in Florida, and a Westinghouse/General Electric–utility consortium scored a last-minute $100-million R&D contract from the Bush administration to design a "foolproof" standardized light-water reactor, a program that may survive the Clinton cuts in nuclear power.[142] Clinton's secretary of energy Hazel O'Leary has called the nuclear option "difficult," the costs "non-competitive," and the waste-storage problems "daunting." The big question, according to one anti-nuclear lobbyist, is whether the president can make his "far-reaching" cuts stick under pressure from well-funded and influential pro-nuclear forces in Congress.[143]

By the 1990s, except among a small group of photovoltaic installers who fought to keep private utilities from invading the realm of off-the-grid installation, solar energy was moribund as a social movement, and the timid public-power movement seemed far more intent on defending its past gains than on imagining a new role for itself in the upcoming struggles over efficiency and solar in a world under stress. The solar policies of the American Public Power Association were indistinguishable from those of the industry-funded Electric Power Research Institute. Public-power advocates seemed to have forgotten that public power had won its greatest victories by courageously asserting that the hydroelectric power of the nation's rivers belonged to the people at large rather than private utility corporations, and by declaring that rural electrification was a public responsibility. In 1993–1994, the American Public Power Association did next to nothing to help the citizens' organization San Franciscans for Public Power and the *San Francisco Bay Guardian* to prevent the Park Service from handing over the Army-owned electrical system of San Francisco's magnificent Presidio to Pacific Gas & Electric. Most municipal utilities and rural electric co-ops seemed to take a position that the only reason to promote public ownership was the fact that publicly owned utilities tend to provide cheaper electricity for the average consumer.

Yet among publicly controlled municipal utilities there were signs of a change, like shafts of sunlight after a thunderstorm. The voters of Sacramento had elected to shut down the Rancho Seco nuclear power plant, which had been built and operated by the Sacramento Municipal Utility District. To make up for the loss in generating capacity, SMUD aggressively championed efficiency and solar power—and bought supplementary electricity from Pacific Gas & Electric (see chapter 4 for a more complete description of SMUD's innovative programs). SMUD has been careful, however, not to let its leadership in solar and conservation evolve into an open critique of the legitimacy of private ownership per se. Perhaps that caution was justified: though over 20 percent of the nation's electric capacity is publicly owned, only 3.5 percent of the country's electric power is actually produced by municipally owned systems governed by voter-elected boards.[144] Overall, it proved far easier to say no to nukes than yes to solar. Emboldened by the apparent weakness of public-power advocates, House Republicans in 1994 moved to end the historic requirement that municipal utilities and rural electric co-ops get preferential access to inexpensive federal hydropower. Under the Republican plan, such power would have to be sold to the highest bidder.[145] After the 1994 elections, which placed Republican majorities in the U.S. Senate and House, such attacks quickly multiplied.

And as for the question of independently generated electricity, the proof is in the rates: in all states but Minnesota and Wisconsin, private electric utilities buy low and sell high to customers who own grid-connected small windmills or stand-alone photovoltaic solar modules. Widespread installation of solar water heaters and rooftop photovoltaics would obviously (and permanently) remove notable sectors of demand from the utilities' markets. In the technological maneuvers of the energy corporations, the issues of centralized control and market demand are always paramount. Pacific Gas & Electric owns large natural gas reserves and so quite logically favors gas over all renewables.

In alliance with environmental groups, private utilities have moved to assert their dominance over energy efficiency and solar electricity programs, and to transform both into a revenue source without destroying their main line of business: selling increasing amounts of electricity and gas at a profit. As new demand growth slumped at home, subsidiaries of U.S. electric utilities began to actively build new fossil-fuel power plants overseas. At a stockholders' meeting in April 1994, Pacific Gas & Electric announced that U.S. Generating Company—its 50–50 subsidiary with Bechtel—was "evaluating" two promising plant projects in the Far East. If consummated, the deals would represent the giant utility's first overseas ventures.

By the mid-1990s, the energy question had become a slow burn, like a subterranean coal fire in an abandoned mine, threatening to burst through to the surface. A new generation of citizen leaders had begun to ask an array of old-new questions about energy: How much was enough? How should the costs of the transformation to a green economy be shared?[146] Could solar energy guarantee more jobs and community control? Could the energy cartels be trusted? Was public ownership with local control a necessary if not a sufficient condition of a solar economy? In local pockets of common sense all over America, ordinary people once again began demanding a voice in the nation's energy future and trying to steer America toward the sun.

CHAPTER 3

The Empire of Oil

IN 1854, A GROUP OF BUSINESSMEN led by George
Bissell hired Professor Benjamin Silliman Jr. of Yale
University to analyze a sample of Pennsylvania
rock oil. Professor Silliman's $526.08 report proved that at least one
distillate would make an excellent illuminating oil, and that report
became the basis for attracting investors to the newly formed Pennsylvania Rock Oil Company. Using salt-drilling technology, "Colonel"
Edwin L. Drake and his drillers brought in a producing well at a depth of
69 feet in Titusville, Pennsylvania, just as their funds were running out.
Almost at once the first oil rush was on, and a drilling and real-estate
fever was underway.

At first, almost all of the new product was refined and sold as illuminating oil, also known as kerosene. Soon, however, production and
refining had risen so quickly that oil producers began to market a heavy
petroleum fraction as boiler fuel, in direct competition with coal. Meanwhile, an oil man named John D. Rockefeller had begun buying up crude
under the Standard Oil name and refining it himself. As he became the
nation's biggest refiner, he set out to control all refining, systematically
buying up his competitors or driving them out of business. Rockefeller's

primary weapons were the special rebates he received from the railroads, who were so avid for business from their chief oil shipper that they would even pay secret "drawbacks" on oil shipped by Standard's competitors. A drawback worked as follows: A competing shipper might pay a dollar a barrel to send his oil by rail to New York. The railroad would turn around and pay 25 cents of that dollar back, not to the shipper, but to the shipper's rival, Standard Oil! Standard, in effect, was being subsidized by its own competitors. Then, using surreptitious secret affiliates, Standard began forcing rival refineries out of business through regional below-cost pricing, which Rockefeller subsidized with sales revenues from markets where Standard had a monopoly. By 1879, Rockefeller controlled almost all of U.S. refining capacity. The remaining competitors tried to break out of the Standard stranglehold by secretly building a pipeline to a Reading Railroad railhead in order to ship their oil to Philadelphia and New York. But Rockefeller soon bought out that operation as well. As production rose, Standard aggressively sought to create a market for heavy fuel oil as a boiler fuel to replace coal. By 1889, one third of all oil was sold for this use.

The automobile added a giant new market to the business of Standard Oil, especially when Henry Ford began mass production of the Model T. Gasoline, which had previously been discarded as too volatile and explosive, became the principal fuel of the internal combustion engine. By the time the Standard "octopus" had been broken up by court order in 1911, fuel oil had replaced kerosene as the main petroleum product, to be replaced in turn by gasoline shortly after World War I.[1]

The famous (or infamous) oil depletion allowance was originally set at 5 percent in 1913. The depletion allowance was based on the legal fiction that subterranean oil constitutes a "capital asset," which is used up as oil is extracted, and thus should not be taxable as income. The allowance originally made sense because of the recognized financial risk of wildcat drilling. As the power of the industry increased, so did the size of the oil depletion allowance, and soon even the pretense that it had some relation to invested capital was abandoned. By 1926, the depletion allowance had been hiked to 27½ percent, thanks to the influence of Secretary of the Treasury Andrew Mellon, banker and former president of Gulf Oil. The provision worked as follows: the owner of wells in which

$50,000 had been invested and which produced $1 million worth of oil each year could deduct $275,000 per year from taxable income, or $2,750,000 over ten years . . . thus the origin of thousands of Texas oil millionaires (and a few billionaires).[2] Today, the oil depletion allowance stands at 15 percent.

During the the 1920s, oil-field developers tried to waste as little oil as possible during the extraction process, since geologists believed the world was running out of oil. In 1930, Dad Joiner's wildcat strike brought the huge East Texas oil field into production for the first time, and wreaked havoc with petroleum prices already softened by the Depression. The major companies were caught off guard, and the independents—the "poor boys"—bought leases and found oil in great quantities. Between early 1930 and the end of 1931, the price of crude fell to as low as 4 cents per barrel, much of this refined in "coffeepot" refineries slapped together to peddle cut-rate gasoline in the Midwest and South.[3] The independents were perceived as out of control and accused of destroying the market for the majors.

With an abundance of new crude flooding the market, the call for "conservation" became a weapon with which the big companies and their allies in government drove the "poor boys" out of business. To solve the problem of low prices, FDR's secretary of the interior Harold Ickes urged the governors of the principal oil-producing states of Texas, Oklahoma, and Kansas to prorate oil-production quotas and dam the flood of new crude oil until the price came up. Now in the name of conservation, the "poor boys" were accused of drilling so many holes so close together that they bled off the sometimes gusher-producing "bottom hole" gas pressure needed to drive the oil to the surface without costly pumping. The accusation was valid enough, but the accused had merely been playing by the big boys' rules.[4] Some independents refused to go along with prorating, fighting these measures with lawsuits, illegal "hot oil" shipments, and the mysterious dynamiting of major-owned oil pipelines.[5] In August 1931, Texas governor Sterling Ross, a founder and former chairman of Humble Oil, a Standard Oil of New Jersey subsidiary, declared East Texas to be in a "state of insurrection," and dispatched thousands of National Guardsmen and Texas Rangers on horseback through unseasonably heavy mud to enforce prorating, and to shut down the recalcitrants. The governor's small civil war ended shipments of hot oil almost entirely, and soon "an eerie quiet" settled over East Texas.[6] By the spring of 1932—aided by a stiff federal tariff on oil imports—the price of crude had risen to a dollar per barrel and commercial "order" was reestablished.

Under pressure from the major oil companies, the governors of Texas and Oklahoma agreed to follow the lead of the Texas Railroad Commission and restrain production to remain within marketable limits. Those production limits were enforced by federal administrative order under Franklin Roosevelt. Thus, both the federal and state governments—coordinated behind the scenes by the "private government of oil" had become accessories to price-fixing, and all but three of the dozens of coffeepot refineries were driven out of business.[7]

Oil was a critical component of the Allied victories over Germany and Japan in World War II. In the summer of 1941 the United States stopped shipping oil to Japan in retaliation for the relentless Japanese advance through China and Indochina. This embargo was the major factor in Japan's decision to launch a sneak attack on Pearl Harbor. The inability of Japanese tankers to make it through the American submarine net from the East Indies oil fields kept the Japanese fleet in its home ports and Japan's planes on the ground.[8]

The Devil and Mexican Oil

The Mexican poet López Velarde described Mexico's situation best by saying "The devil has given Mexico oil." Why? Because we have a two-thousand-mile border with the U.S., which is a very powerful nation that consumes a lot of energy. Before the expropriation [of U.S. oil companies in 1938], the U.S. and U.K. both consumed a great deal of Mexican oil. They took everything they could from Mexico. They dried up our most important reserves in the course of a few years. When the Mexican government nationalized the oil industry, it did so as a matter of preservation. Mexico produces about two and a half million barrels of oil a day more than it consumes. Most of this excess is sold to other nations. We have about twenty years' worth of proven reserves. However, once these reserves are exhausted we will have nothing, and the U.S. oil will be worth more. We would be trade partners with the U.S. but, because the U.S. is so powerful, we are not. We cannot fix the price for oil; we cannot impose any conditions on the U.S. This is why we say it was the devil who gave oil to Mexico.

GUILLERMO LÓPEZ PORTILLO, 1992

In Germany, the I. G. Farben chemical trust allied itself with Adolf Hitler shortly before he took over as chancellor, on the promise that he would support Farben's experiments to produce high-quality synthetic gasoline from coal, even though this would cost ten times as much as gasoline refined from petroleum. By 1943, with the use of countless slave laborers whom the Nazis systematically worked to death, synthetic gasoline production had been cranked up to 142,000 barrels per day, which supplied over half of the German army's fuel supply and more than 90 percent of the Luftwaffe's aviation gasoline.[9] The turning point in Hitler's ability to fuel his tanks and airplanes came in the spring and summer of 1944, when Allied bombers smashed the vast Leuna complex and reduced production to almost zero. The first of these raids, wrote Albert Speer, Hitler's production chief, was "the day the technological war was decided."[10]

Over half of the military tonnage shipped overseas by the United States during World War II consisted of fuel and other petroleum derivatives.[11] Indeed, 1943 proved to be the last year in which the United States shipped more oil than it imported. Today, by contrast, roughly half of the oil consumed in America is imported. The chief backups to domestic production are Saudi Arabia and Kuwait, both de facto U.S. protectorates, which sit smack in the middle of half of the world's known oil deposits.[12]

These oil fields were one of the biggest prizes of World War II, because Europe's coal mines, which had supplied over 90 percent of Western Europe's energy, lacked the capital and the stable labor relations to rebuild. The point was driven home during the bitterly cold winter of 1946, when the Thames froze at Westminster, and British industry effectively shut down for three weeks. "It is not a joyful thing, but it is a national necessity to import more oil," Hugh Dalton, Britain's chancellor of the Exchequer, told U.S. Secretary of State George Marshall. By 1948, writes Daniel Yergin, it was already clear to American planners that one fifth of Marshall Plan aid was likely to be spent importing oil and oil equipment. In 1950, oil supplied 8.5 percent of Western Europe's energy needs; by 1970, its portion was 60 percent.[13]

Replacing coal with oil had the collateral advantage of destroying the power bases of traditionally militant mineworkers unions, "where in many countries, communists occupied leading roles."[14] Environmental considerations also helped spur the switch to oil. London's "killer fogs," heavy with sulfur dioxide and soot, induced Parliament to pass the 1957 Clean Air Act, which banned the home burning of coal. Oil, and later natural gas, took the place of coal for industrial uses, as well. Not only was oil easier to ship and burn; it used less labor, generated no dust, and was relatively smog-free, and by the late 1950s, oil cost less than coal.[15]

As the citizens of Europe and Japan bought cars, the demand for oil increased exponentially. In Germany, for instance, there were 50 people per car in the 1930s and only 2.3 people per car in 1985.[16] The increase in Japanese car ownership was even more dramatic. As European and Japanese oil consumption soared, the major U.S. oil companies got half of the business. The European and Japanese oil markets were the demand side of 1945's spoils of war: a permanent, almost tax-free contribution to the coffers of oil companies and transnational investment banks in the United States and Britain. Corporate income-tax obligations on oil produced in tax havens such as Saudi Arabia were assumed to have been paid in the host country, so they didn't count as income for U.S. tax purposes.

The Arab oil embargo, and the emergence of the Organization of Petroleum Exporting Countries (OPEC) as a strong producers' cartel in the aftermath of the Arab-Israeli war of 1973, tripled the international price of oil, and some alarmists believed that OPEC might boost the price to $60 or $100 per barrel. President Richard Nixon, wounded by the Watergate scandal, announced Project Independence in 1973 with the following words: "Let us set as our national goal, in the spirit of Apollo, with the determination of the Manhattan Project, that by the end of this decade we will have developed the potential to meet our own energy needs without depending on any foreign energy source."[17] But Nixon's pontifications now ring as hollow as an empty oil drum, as today the United States imports more petroleum than ever.

For a few years, with multibillion-dollar federal subsidies, the major oil companies hedged their energy bets with huge purchases of coal, oil shale, and nuclear properties. There has been fierce lobbying to open Indian reservations, federal property (particularly in Alaska), and the continental shelf to exploration for oil and uranium.[18] A related project, instituted under heavy utility and investment bank pressure, was intended to induce the federal government to subsidize the mining of oil shale and the generation of electricity from nuclear reactors and cleaned-up coal-fired plants. Spurred by the oil price hikes of 1973 and 1979, the federal government also spent over $2 billion in a vain attempt to extract synthetic fuels from coal on a commercially viable basis.[19] "The entire synfuels program had a quality of madness to it," wrote Cohen and Noll in *The Technology Pork Barrel*. "Project after project failed. . . . Goals were unattainable from the start. Official cost-benefit studies estimated net benefits in the minus billions of dollars. . . . the dogged continuation of the research and development program seems incredible."[20]

Some oil companies invested heavily in solar as well: Arco Solar became the world's largest producer of crystalline silica photovoltaic cells, and Mobil created a subsidiary called Mobil Solar Energy Corporation. When Arco decided that producing solar cells wasn't profitable, it sold the subsidiary to Siemens, the German electrical/electronics giant. Likewise, Mobil Solar Energy Corporation, based in Billerica, Massachusetts, shut down its production of crystalline silica photovoltaic cells in late 1993, and sold out to a German combine. Mobil Solar cells had been used successfully in a rooftop solar experiment sponsored by the New England Electric System on a moderate-income housing development in Gardner, Massachusetts, but projected bulk sales to utilities never materialized, according to two laid-off employees, because the participating corporations never really committed to this marketing program.

Drowning the Sun in Oil

IN 1981 RONALD REAGAN, the energy barons' white knight, rode triumphant into the White House. For twelve years, the theme would be back to the future, as domestic oil production declined and petroleum imports and the trade deficit soared, while the price of oil came down. Federally funded solar research was cut to the bone, and solar tax credits for homeowners abolished. One of President Reagan's first official acts was to rip President Carter's solar hot water heater off the White House roof. Oil was sacrosanct. President Reagan's budget director David Stockman recounts the catalyzing effect his attempts to eliminate tax breaks for big oil companies had on the man in the Oval Office:

> Up to now, the President had remained almost entirely passive. At a meeting that day we brought up the . . . tax . . . proposals, starting with the oil depletion allowance. All of a sudden, the President became animated. Our proposal unleashed a pent-up catechism on the virtues of the oil depletion allowance, followed by a lecture on how the whole idea of "tax expenditures" was a liberal myth.
>
> "The idea implies that the government owns all your income and has the right to decide what you can keep," said the President. "Well, we're not going to have any of that kind of thinking 'round here."[21]

Reagan and Bush worked closely with Saudi Arabia and the oil companies to keep the price of oil around $20 per barrel. The initial task was to pressure Saudi Arabia to flood the market with new production, to bring the price down from a peak of $35 per barrel in the early 1980s.[22] By early 1983, the Treasury Department had completed a secret study that concluded that the United States would profit if oil dropped from $33 to $20 per barrel, and the administration worked to get Saudi Arabia to cooperate with this scenario by raising its oil output by several million barrels a day. After a meeting between President Reagan and King Fahd in February 1985, Saudi Arabia's four major oil customers—Exxon, Chevron, Texaco, and Mobil—proposed a "netback pricing" scheme that would guarantee the companies $2 per barrel on all the Saudi oil they sold. With this new cost-plus incentive, U.S. oil refinery production and imports soared and the price of oil fell (as Treasury had predicted) to $10 to $15 per barrel. Gasoline prices quickly dropped 45 percent, helping create the sensation that was the ostensible Reagan economic boom.

When the price of oil fell to $10 per barrel on the spot market in early 1986, Texas economists began to "worry" about the adverse effect of low prices on an "already battered [domestic] oil industry and on the segment of the banking industry that has large energy portfolios," reported

Robert D. Hershey Jr. in the *New York Times*. "Much of Louisiana and the Southwest have been economically crippled [and] the Texas unemployment rate [is] now above the national average." So the administration sent Vice President Bush to Riyadh to persuade Saudi Arabia to limit production and raise the oil price, and, lo and behold, the price did rise, to $20 per barrel. U.S. oil drillers and producers survived, and Saudi Arabia continued to prosper. During this maneuvering to raise the world oil price, White House spokesman Larry Speakes continued to mouth a familiar catechism: "The way to address price stability is to let the free market work"—as if politics had nothing to do with the price of oil. The idea was this: keep oil prices high enough to ensure the profits of the Saudi-American oil complex and to keep the most efficient domestic producers in business, and low enough to keep the world hooked on oil and to hamstring the development of conservation and solar alternatives in the U.S. and overseas.[23]

By the mid-1980s, OPEC had been tamed. Though Saudi Arabia and Kuwait had nationalized their oil wells, they continued to market through established channels and to reinvest their surplus petrodollars through U.S. banks. In fact, the largely American major oil companies continued to exercise joint control with the Saudis over world oil sales and prices, especially in the United States. The net effect on the world economy was to tax all oil-importing countries, for the benefit of the sheiks of Saudi Arabia and the oil companies of the United States, and to cycle the lion's share of tens of billions of surplus petrodollars through American and British banks for reinvestment. Nominal nationalization notwithstanding, the sheik-ruled oil-producing countries had become part of a vast network of producers and consumers still marketing through U.S.-based oil companies and still dependent on the U.S. military for protection against nearby, much more populous Muslim countries such as Iran and Iraq.

"When 'our' oil in the Middle East is in jeopardy," observed scholar Michael Parenti, "*our* sons and daughters must go and fight, and *our* taxes are needed to finance these ventures," though few of us own oil stock. "To protect one dollar of their money, they will spend five dollars of our money . . . monopolizing the private returns of empire while carrying little of the public cost."[24]

The War for Oil

WHEN IRAQI DICTATOR Saddam Hussein launched his murderous and foolhardy invasion of Kuwait and refused to withdraw without preconditions, the military reaction was reminiscent of East Texas in 1931, only this time involving the whole country. Saddam Hussein's Stalinesque moustache made him the perfect mediagenic patsy in a didactic exercise to teach the world who was really number one when it came to oil. The Iraqi army refused to stand up and fight, and thousands of Iraqis died, some of them buried alive in their trenches by armored bulldozers. In an unforeseen aftermath, thousands of U.S. soldiers complained of chemical-poisoning symptoms and reproductive- and immune-system damage, perhaps as a consequence of exposure to Kuwaiti oil-field fires, radioactivity, Iraqi poison gas, or anti–poison-gas medications administered by Army medical personnel.[25]

World Rent-a-Cops

Saudi Arabia has paid more than half of the $600 million cost of last month's U.S.-led military deployment in the Persian Gulf, Arab diplomats in Riyadh said. The diplomats said the Saudis made a $330 million payment to the U.S. Kuwait is believed to be paying about 40% of the cost.

Wall Street Journal, November 7, 1994

The Gulf War "was for oil, the umbilical cord that feeds the world's industrialized nations, not for democracy," said Robert Krueger, chairman of the Texas Railroad Commission, the entity that regulates oil production in Texas.[26] Operation Desert Storm proved that the real program for "energy independence" was military control of the Mideast rather than the intelligent exercise of American know-how to harness the power of the sun. A few obscure environmental publications asked "Whatever Happened to Solar?" but most of the U.S. media lined up and joyfully saluted the military pyrotechnics, recalling President Lyndon Johnson's famous quip, "Reporters are puppets."[27] The ease with which the Allied forces rolled over the Iraqis suggests that the prewar scare talk about Saddam's "awesome" military machine was mostly a spectacle orchestrated at the Pentagon and parroted by a compliant press.[28]

With the war won, Kuwait's ruling Sabah family returned home to face public accusations of having squandered over $4 billion of a $100 billion Kuwait Investment Office portfolio meant to supplant oil revenues when the oil runs out.[29] Three months after he left office, former President Bush paid a triumphant three-day visit to Kuwait, accompanied by his sons Marvin and Neil, ex–Secretary of State James Baker, and former White House Chief of Staff John Sununu. A couple of weeks later Baker, Sununu, and the Bush sons returned, representing firms such as Westinghouse and the gas-pipeline giant Enron on a fishing expedition for Kuwaiti business contracts that promised to total billions of dollars. (General Norman Schwartzkopf, by contrast, stayed away from deals that could have netted him "hundreds of millions in commissions.")[30]

Likewise, in recent years, the extravagance and corruption of Saudi ruling circles has begun to outrun the supply of oil money. Saudi Arabia's King Fahd agreed to grant American Telephone & Telegraph a $4 billion contract to re-equip Saudi Arabia's phone system, even though European bids had supposedly come in at half that amount.[31] The ruling House of Saud, already divided over the question of succession to the ailing King Fahd, has now begun to face principled opposition from the Committee for the Defense of Legitimate Rights.[32] More and more, it seems, the fate of the ruling dynasty has become dependent on its special relationship to the United States. To defend Saudi rule and keep neighboring states at bay, the United States has been maintaining a permanent military presence on Saudi soil, which has generated a violent underground resistance.[33]

The permanent mobilization of armed forces in the Middle East has also had an invigorating effect on the market for American arms. "Saudi Arabia is our best customer," remarked a spokesperson for the Pentagon's Defense Security Assistance Agency.[34] Between 1985 and 1992, Saudi Arabia bought $18.3 billion worth of American arms, according to the Congressional Research Service, and the desert monarchy was scheduled to buy another $30 billion, plus $6 billion in commercial airliners from U.S. aerospace manufacturers.[35] In the next decade, McDonnell-Douglas alone expects to reap $5 billion from a Saudi purchase of seventy-two F-15 jet fighters.[36]

By 1994, the world price of oil had stabilized at $20 per barrel, and crude cost as little, in constant dollar terms, as it had before before the first oil shock of 1973. If the cheap oil of the cherished "American way of life" required permanent military hegemony in the Persian Gulf, most Americans didn't mind, so long as the inevitable wars were quick and victorious. No one expressed the prevailing attitude better than the person who hung a sheet reading KICK THEIR ASS AND GET THEIR GAS from an interstate overpass to salute the 82nd Airborne as it rumbled through Georgia on its way to fight Iraq in late 1990. With gasoline prices low again, thirsty pickups and four-wheel-drive sport/utility trucks became the new vehicles of choice. Solar energy and efficient vehicles remained a hard sell. The voices that argued for a stiff tax on imported oil, to keep the nation's "energy future" from being held hostage by the "whims of the international market," were few and far between.[37]

In 1992, as President Bush's term of office was winding down, skeptics smelled oil behind the "humanitarian" intervention he dispatched to Somalia. Four major oil companies had exclusive exploration rights in Somalia, and Conoco's compound in Mogadishu was serving as the de facto U.S. Embassy when the Marines landed.[38]

Today, the Reagan-Bush energy legacy is evident everywhere.[39] Oil imports are higher than ever, surpassing in dollars the trade deficit with Japan, which is mostly an automobile deficit.[40] Electricity, more than ever, is being generated by fossil fuels and nuclear plants.[41] It was no coincidence that the "landmark" energy legislation of 1992 neglected to set limits on cars and oil, and that utilities had won the right to invest anywhere.[42] The energy policies of Presidents Reagan and Bush could justly be summed up in four words: "Arabia, Kuwait, coal, and nuclear." The Clinton administration would add "natural gas" to the formula.

Natural Gas: Bridge to Solar Energy or Just Another Fossil Fuel?

BILL CLINTON'S CAMPAIGN promise to "increase U.S. reliance on natural gas" was a cornerstone of his campaign platform for placating oil companies, oil drillers, and environmentalists as well, so we can expect a green light for natural gas from Washington through 1997 at the very least.[43] Today, the Clinton administration touts natural gas as the "environmentally safe" fuel for cars and trucks, and plans to stimulate its expanded production and use as a way to reduce dependency on imported oil. Vice President Al Gore clinched the argument: "There's no question that we have to shift, as rapidly as we can, to alternative fuels. The automobile is essential to our future. What is not essential is the 400 million tons of carbon emissions that gasoline vehicles send into the atmosphere each year."[44]

The illustrious biologist, author, and third-party presidential candidate Barry Commoner has argued for twenty years that natural gas—a fossil fuel composed mostly of methane (CH_4)—is a suitable fuel as a transition to solar, and he seems to imply that utilities might help lead this transition.[45] Gradually, argues Commoner, fossil-fuel methane could be replaced by "solar" methane generated from readily available biomass: "garbage and sewerage in cities; crops, manure, agricultural residues, and cannery wastes in farming areas; wood in forest areas; seaweed along the coast." This biomethane, supplemented increasingly by renewable hydrogen, produced by electrolysis with photovoltaic or wind-generated electricity, could be converted into methane, and would be piped through existing gas mains so smoothly that the transition would hardly be noticed (see chapter 8 for more discussion of hydrogen). Commoner cited the case of a cattle feedlot in Guymon, Oklahoma, which has piped methane from steer manure to Peoples Gas in Chicago. And yet, despite assertions that gas companies are "greener" than oil companies, there is, unfortunately, no evidence that the natural-gas industry has thought seriously about future conversions to solar methane or renewable hydrogen.[46]

The political side of the argument for methane is that the electric and gas utilities—regulated by public utilities commissions—could form a natural counterweight to oil and automobile interests. Utility advocacy of either electric- or natural-gas–powered cars in the face of auto industry indifference to non-gasoline cars is used as evidence that there is a real conflict of interest that environmentalists could exploit between regulated utilities and unregulated oil and auto companies. But the conflict is more apparent than real.

Those who are hopeful about this basic conflict of interest between petroleum/automobile interests and electric and/or gas utilities might be surprised to know that the six top producers of natural gas—according to the American Petroleum Institute—are Chevron, Amoco, Texaco,

Exxon, Arco, and Shell, and that 60 percent of natural gas is produced as a spinoff of oil production.[47] In August 1993, the *Oil and Gas Journal* projected that the dollar value of domestic natural gas would exceed the value of the oil produced that year in the lower forty-eight states.[48] The fact that one tenth of U.S. electricity is generated using natural-gas fuel also suggests that there is no inherent conflict between gas producers and electric utilities.[49] What makes money for utilities makes money for the oil industry, as natural gas becomes one of its main products.

Some have argued that natural gas would be a particularly appropriate "transition fuel" to a solar hydrogen economy because much of its infrastructure could be adapted to utilize hydrogen, and because gas-fired power plants can be rapidly turned on or off to mesh with intermittent solar energy sources such as photovoltaics and windpower.[50] Environmentalist advocates of methane correctly note that it produces fewer pollutants and less carbon dioxide per unit of heat generated than either coal or petroleum. Southern California Gas (SoCal Gas) wants environmentalist support for a proposed $250 million annual R&D subsidy from the Department of Energy. SoCal Gas was quick to remind environmentalists that it was the first large company to support the controversial South Coast 1989 Air Quality Management Plan.[51] Natural gas is cleaner-burning than other fossil fuels commonly available, argued SoCal Gas, and furthermore, it is a fuel that generates one third less carbon dioxide gas per unit of energy than carbon-based fossil fuels such as coal and petroleum. The problem is that natural-gas vehicles generate almost as much smog as gasoline vehicles.[52] Conveniently omitted by the natural gas lobby, charges Greenpeace, is the information that methane, the chief component of natural-gas, is twenty times more potent by weight than carbon dioxide as a greenhouse gas. Inevitably, as production and use occur, large quantities of methane either escape unburned into the atmosphere, or are transformed into carbon dioxide itself.

Another argument for natural gas, say the gas utilities, is the fact that at present prices and consumption levels, about sixty-two years of proven conventional reserves of domestic natural gas remain to be extracted, compared to less than ten years for domestic oil. Yet gas use in recent years has been increasing at a rate of 4 percent per year, and it seems likely that reserves will run out quickly and its price will soar if natural gas is used to replace gasoline and coal on a massive scale. Today the United States has less than 4 percent of world natural-gas reserves, from which it pumps almost one quarter of the gas presently used. And the whole of North America contains only 8 percent of the world's known gas reserves.[53] If U.S. consumption of natural-gas usage continues to double every twenty years or faster, North American supplies (including those of Canada and Mexico) will be burned up in two or three decades. Even so, cautions Ralph Nader, critics shouldn't "fall into the trap" of believing that the world is running out of oil and natural gas, as many 1970s experts did, because shale oil and dissolved natural gas in the Gulf of Mexico can always be exploited—if the price is right and the political will is there.

In the transportation arena, the utilities, both gas and electric, cooperate in a whole series of projects with automobile and oil interests. Big Three auto companies and the Electric Power Research Institute are partners in a DOE-led consortium called the United States Advanced Battery Consortium (USABC), which aims to develop a stronger, longer-lasting battery for electric-powered cars, and the gas utilities have their own plans for natural-gas trucks and automobiles.[54] Thus, the dispute over whether "clean" vehicles in California should be powered by gas or electricity turns out to be an argument over increments from the oil companies' point of view, because in either scenario, increased electricity use for transportation will increase the use of natural gas, which in California accounts for one third of the electricity supply and one half of "dependable" capacity.[55] A California Air Resources Board proposal that 2 percent of cars sold in 1998 (increasing to 5 percent in 2003 and 10 percent in 2008) had to be "zero-emissions vehicles" was a short-term victory for the electric utilities, because it is impossible to build zero-emissions vehicles that run directly on natural gas, but automobile company pressure forced the Air Resources Board to withdraw the proposal in early 1996. In Northern California, where Pacific Gas & Electric sells most of the electricity as well as gas, the same car/oil/utility coalition is working together to design natural-gas–powered cars and refueling stations. (So far, no one has figured out what to do about the wave of lead-acid battery waste and industrial lead poisoning that will result if the USABC is unsuccessful in developing a cleaner alternative battery.)

Without exception, all major energy constituencies—electric and gas utilities, oil companies, auto companies, and coal corporations—are determined to promote automobile transportation, whether the power to keep cars running comes from natural-gas–fired or coal-fired electricity, or involves vehicles fueled by petroleum or natural gas. Carl Weinberg, former research director at Pacific Gas & Electric, guesses that at present consumption rates there is only a forty-year supply of natural gas in the world.[56] Increased demand would no doubt create a shortage, leading to higher prices for gas producers and higher income for oil companies. The flaws in the gas strategy will come when unavoidable shortages commence—a process that has already begun in Texas—and prices and utility bills escalate rapidly.[57]

In terms of new initiatives to support development of renewable fuels and energy sources, the deregulation of the natural gas and coal trade under the North American Free Trade Agreement and the General Agreement on Trades and Tariffs (GATT)—passed in 1993 and 1994, respectively—could completely undermine national sovereignty by prohibiting special provisions in support of solar, for example, as "non-tariff barriers to trade." The effect of unencumbered "free trade," according to Greenpeace, is to "foster co-dependency on dirty energy supplies." When Mexico and Canada and Saudi Arabia are drained of oil in a generation or two, according to one nightmare scenario, those nations will be forced to turn to U.S. coal for their energy supplies.[58] When North America is sucked dry of natural gas, the only way to maintain a cheap supply at a reasonable price will be to continue to impose our military hegemony on those regions that have the gas. There will also be unrelenting political pressure to ship natural gas around the world on expensive and ultra-hazardous liquefied natural-gas tankers, which could explode with the force of thousands of tons of TNT.[59]

Today deregulation, cheap electricity, and natural gas are all the rage, but few people are paying attention to what will happen when the price of natural gas and oil go up, as they most surely will, after falling by 75 percent in the last decade.[60] What will happen when the new, unregulated "independent power producers" of cheap electric power fired by combined-cycle gas turbines pass on whopping rate increases to the public as the price of natural gas soars? Will big industry come weeping to the public, hat in hand, as the savings-and-loan investors did? Are the energy corporations crippling American industry by reinforcing an addiction to cheap fossil fuels and electricity?[61] Will there be a massive ratepayers' revolt when utilities try to stick consumers with doubled and even quadrupled utility bills?

A fossil-fuel civilization is a dinosaur devouring its own tail: it will eat itself to death; the only question is when. The price of oil, still determined by the delicate machinations of the Washington-Riyadh political axis, is a rudder for the world economy. As hundreds of billions of electronically traded petrodollars slosh around the world economy, governments and central banks have found that the traditional Keynesian instruments of fiscal and monetary policy have lost their power to steer national economies. Compared to the infinite abundance of sunlight, wind, and growing plants, large fossil-fuel deposits—especially oil—are extremely rare. The way for governments and citizens to regain the power to steer their own national and local economies is to reduce dependence on imported fossil fuels and cut back on greenhouse gas emissions. The real solutions to the energy problem are more democracy, more efficiency, more use of renewable energy, and more local control. To hold otherwise is to help rearrange the deck chairs on the Titanic.

CHAPTER 4

Public Power

*I believe in municipal ownership of these
monopolies because if you do not own them,
they will in time own you. They will destroy
your politics, corrupt your institutions, and,
finally, destroy your liberties.*[1]

—Tom Johnson, Mayor of Cleveland, 1901–1909

MANY OF THE UPCOMING BATTLES over the production and control of electrical power will reprise the struggles for public ownership of electricity during the past century and a half. But those previous contests are largely unknown to the general public, because the corporate winners have dominated the historiography of electric power and because public-power advocates have lost their missionary zeal.

Today the electric power industry is changing rapidly. Because of energy-efficiency regulations, new motors and appliances use less electricity, so the growth in electricity demand has slowed dramatically. Big power users want to lower high electricity prices and are pushing to open up regional electricity markets to outside competition—a concept known as "retail wheeling."[2] Nuclear power plants are no longer being built in the United States, though nuclear advocates continue to organize for a comeback. The main dispute over American nuclear power today is about who will pay for the mess these power plants have already created; in many cases, the cleanup costs associated with decommissioning nuclear plants will amount to twice the original cost of constructing the plants.

In California, Governor Pete Wilson and both houses of the legislature decided to charge ordinary ratepayers rather than utility shareholders for $30 billion in runaway nuclear costs through 2001.

Meanwhile, the price of wind power and solar energy has continued to fall. Technology already exists today that would enable us to supply all the energy the world needs from solar thermal collectors, wind turbines, and photovoltaic (PV) cells that can be built into the roofs and walls of existing structures.[3] As the cost of transforming sunshine into electricity continues to decline by half with every decade, a vast battle is looming over whether the sun and the wind should be considered inherently public resources like hydropower, or whether the solar technologies and electricity itself should be under the private control of monopoly utilities. By relatively early in the next century, electricity might be produced with a few thousand dollars worth of photovoltaic shingles on the roof of a typical house, and the electric company's role might be limited to maintaining the transmission grid and storing and distributing the surplus generated during sunny hours. As the burden of electricity generation shifts to renewable sources of energy and comes under local control, and as electrical machines and appliances became ever more efficient, electricity production and distribution could decline as a proportion of total economic activity. The significant revenue thereby saved could be reinvested in local communities.

As "customer-located generation" becomes easier and more commonplace, increasing numbers of citizen activists are beginning to argue for community ownership of electricity, as a necessary condition for establishing a genuine solar economy. Under public ownership, solar electricity would be subject to local control just as schools and water districts now are. And yet, as rooftop solar electricity becomes a prime source of electric power, the utilities will undoubtedly attempt to control it. To avoid becoming modern-day "solar sharecroppers," ordinary people will have to make solar electricity the object of intense political thought and action. And unfortunately, in recent years neither the moribund public-power movement, nor the weakened labor movement, nor an increasingly fractious environmental movement, has thought seriously about the democratic implications of wind and solar energy.

The Genesis of the Electric Power Industry

ONE HUNDRED and ten years ago, when electricity was in its infancy, no one was certain how electric power would be sold and who would buy it. At first, the "dynamo" was thought of "as a kind of - appliance, like a hot-water heater, to be installed at any gold mine or grand hotel that wanted to generate current for electric light. . . . No one . . . had thought of the [electric current] coming out of the dynamo as a commodity in its own right, a product that could be made in one central place and then sold to the general public—though gas for gas lamps had been made and piped and sold that way in San Francisco for a quarter of a century."[4] The first large-scale use of electricity was for streetlighting. The new lights were demonstrated and sold at great public spectacles, much like today's manned rocket launches. Charles Brush, an inventor and entrepreneur, promised to make downtown Cleveland "as bright as day" on the evening of April 29, 1879, with twelve of his own electric arc lights powered by a dynamo of his own design. As a crowd of thousands gathered, Brush motioned for his assistant to throw the switch: "The first globe flickered with purplish light. Thundering cheers gave way to astonishment as the other lights came on. In the awed quiet the Cleveland Grays band struck up its brass, and artillery boomed along the shorefront. Cleveland became the first city to light its public square with electricity." A year later in Wabash, Indiana, Brush turned on four 3,000-candlepower lamps atop a 200-foot-high courthouse dome on a moonless night. According to an eyewitness, " 'the strange weird light, exceeded in power only by the sun, rendered the square as light as midday. Men fell on their knees, groans were uttered at the sight, and many were dumb with amazement.' "[5] Despite the success of Brush's light show, Wabash city leaders, repelled by what they knew about the rate-gouging of private power companies, refused to issue a franchise for a privately owned concern controlled by Brush, and eventually established a city-owned system.[6]

Many cities wanted electric streetlights, but it wasn't clear at first whether electricity could be generated and sold for a profit. Thomas Edison, financed at first by J. P. Morgan, the nation's top investment banker, built New York City's first generating plant in 1882, but Morgan was reluctant to rush ahead in building generating capacity. Competition from gaslights was still strong, and the cost of buying and maintaining electrical generating equipment and transmission wires—which remained idle during the day when the lights were turned off—was extraordinarily high. Another problem was that the direct current then in use could not easily be transmitted much farther than a mile. For the time being, Morgan thought, there was more money to be made in buying up electrical equipment patents and manufacturing the equipment.

What changed this picture was the introduction of alternating current (which could be transmitted dozens of miles) and the construction in 1887 of an electric streetcar in Richmond, Virginia, which created a demand for electric power during the day as well. This trolley car (which got its name because it "trolled" for power from an overhead line) proved an instant hit. By 1895, ten thousand miles of track had been laid in over eight hundred cities. Now financiers such as J. P. Morgan and Samuel Insull bought into the generation business in a big way, and Morgan founded General Electric to gobble up electrical equipment manufacturers. Giant financial combines led by Morgan and Insull also battled against municipal utilities over who was to collect the vast sums that industrialists and consumers would soon be spending on electrically powered transportation, lighting, motors, and appliances.

People who believed that electricity, like roads, should be a public responsibility fought bitterly against the power trusts, who sought to keep control of electricity production as a "dividend machine" for private monopolies. In most cases the private monopolies won, yet these private combines were later brought to a stalemate during the New Deal era, when the Tennessee Valley Authority and other federal hydroelectric complexes were built. Promoters fighting for city streetcar and electrical franchises paid off mayors by the hundreds and aldermen by the thousands. In this welter of headlong expansion and corruption, the electric power business was born, and gigantic fortunes were made and lost.

In San Francisco, the financial battle between gas and electric interests lasted only a decade. Rather than slug it out over lucrative municipal lighting contracts coming up for renewal, the San Francisco Gas Light Company and the Edison Light and Power Company merged into the

San Francisco Gas and Electric Company in 1896, capitalized at $20 million. In 1905, SFG&E spearheaded the creation of Pacific Gas & Electric from what were once over five hundred independent companies, and the question of gas versus electricity became a technical rather than a financial issue. Gas continued to be the system of choice for cooking and heating, though some gas streetlamps lingered on until 1930. Today, PG&E is the largest private utility in the United States.[7]

Even so, the substitution of electric for gas streetlighting was neither foreordained nor instantaneous. In 1884, the efficiency of gas lighting had been increased six times by the Austrian Carl Auer von Welsbach. The patent for his Welsbach gas mantle, a "lacy asbestos hood which became incandescent when attached over a burning gas jet," was bought by the United Gas Improvement Company of Philadelphia, which also owned the widely licensed Lowe patent for producing coal gas. A United Gas manufacturing subsidiary in Gloucester, Massachusetts, supplied the whole country with patented Welsbach mantles and serviced its streetlamps. But electric streetlights continued to gain ground, and by 1940, the Welsbach Incandescent Light factory had closed its doors forever.

The Birth of the Public-Power Movement

THE MAIN THREAT to what promised to be unlimited profits came from those who believed that electric power and mass transit—including the railroads—were too important to franchise away to private monopolists. To legitimize private utility monopolies and to forestall the demand for public ownership, Samuel Insull, the administrative genius of Chicago's Commonwealth Edison, told an astounded 1898 convention of utility executives that competition was "economically wrong" because it led to duplication of power capacity and electrical lines and drove the cost of electricity too high. To sell the idea that electricity was a "natural monopoly," yet also temper public opposition to the granting of monopoly charters, Insull proposed state-by-state regulation of private electric monopolies as a substitute for unmitigated private competition and complete public ownership. The idea of a regulatory commission for electrical utilities was borrowed from the experience of the Interstate

Commerce Commission (ICC), which had served the interests of the very railroads the commission was theoretically supposed to regulate. Former U.S. Attorney General Richard Olney had quickly put to rest the notion that the regulatory commission could be a mortal threat to business:

> The [Interstate Commerce Commission] ... can be ... of great use to the railroads. It satisfies the popular clamor for government supervision of the railroads at the same time that that supervision is almost entirely nominal.... It thus becomes a ... barrier against hasty and crude legislation hostile to railroad interests.[8]

Samuel Insull counseled utilities to lobby, state by state, for electric monopolies that would be supervised by the states under a system of uniform records and accounting, and urged them to avoid taking a public stand on whether private or municipal ownership should take precedence.[9] Insull was certain that the state public utilities commissions (PUCs) he proposed would be at least as pliable as the Interstate Commerce Commission had proven to be. The utilities would end up regulating themselves, in effect, and foil the cries for public ownership provoked by popular revulsion at the arrogance of the railroads and capitalist monopolies such as Standard Oil.

As Insull predicted, the public utilities commissions dealt a "stacked deck" in favor of utilities. "The inequities in state regulation soon became apparent ...," wrote Richard Rudolph and Scott Ridley in *Power Struggle.* "The sky was the limit as to how much a company could spend on a case, and all expenses would eventually be charged to consumers through their rates. In the early days the costs of carrying through with a [consumer] complaint normally ran anywhere from $10,000 to hundreds of thousands of dollars. In these proceedings city attorneys proved to be no match for specialized utility counsel brought in from other parts of the country, [and] the utilities were able to secure and pay technical experts who were reluctant to testify for the public side because, as one consulting economist candidly explained, 'Engineers and accountants experienced in electric lighting ... are in the employ of private companies or expect to be.'" Members of public utilities commissions— usually appointed by governors—were reluctant to oppose wealthy private utilities. Between 1907 and 1912, in Progressive-dominated Wisconsin, fifty of fifty-two utility requests for rate increases were approved, and requests for reductions by Wisconsin cities and towns were granted in only eleven out of thirty-nine cases.[10] Rather than providing a ceiling, the effect of PUC regulation was to set a guaranteed floor under utility rates and profits.

Yet the momentum toward public power could not be stopped, only slowed. Tom Johnson, who served as mayor of Cleveland, Ohio, from 1901 to 1909, built his political career on the struggle for municipal ownership of the electric utilities and the street railways, which he attacked for bilking the city's citizens out of millions of dollars. Johnson had worked his way out of poverty as the inventor of an automatic fare-collection box and as the owner of street railway companies, and he knew how the political game was played by the utilities: "They make a daily, hourly business of politics, raising up men in this ward or that, identifying them with their machines, promoting them from delegates to city convention to city offices. They are always at work protecting and building up a business interest that lives only through its political strength."[11]

After an exhausting struggle, including successful referendums and a dramatic public hearing in which he accused two city councilors of accepting bribes from the Cleveland Electric Light Company (CELC), Johnson and his allies managed to annex the nearby city of Brooklyn and its municipal power plant. The city of Cleveland finished its own power plant by 1914, and sold its electricity for 3 cents per kilowatt-hour, as opposed to the 10 cents CELC charged. It was a stupendous political achievement, "the first example of a major city selling its own power and providing reliable service at a drastically reduced rate"—especially since this campaign was carried out against the CELC, a subsidiary of J. P. Morgan's General Electric.[12]

Tom Johnson's victory was not an isolated instance: by 1912, fully one third of the 5,396 power companies nationwide were publicly owned, and most of them produced their own power. Struggles similar to Cleveland's, not always successful, occurred in countless towns and cities around the country, including Los Angeles, Chicago, Detroit, and San Francisco. To encourage the national movement for public power,

Republican Senator George Norris of Nebraska and early conservationists such as Gifford Pinchot (who was appointed as the first chief of the U.S. Forest Service by President Theodore Roosevelt) formed the National Popular Government League to fight for the right of local governments to run electrical systems and to keep private corporations from taking control of the nation's rivers for power generation. Senator Norris would become the major advocate of public power and economic democracy in Congress and would be roundly excoriated as "socialistic" and "communistic" by the electric power industry. To bolster local public-power organizing, Pinchot, Wisconsin Senator Robert LaFollette, Scott Nearing, and Wisconsin state legislator Carl D. Thompson formed the Public Ownership League in 1916 to promote the concept of public power at a grassroots level. The basic idea of the League was that "those public utilities which are natural and essential monopolies—water, light, power, and transportation"—should be publicly owned and developed and ultimately controlled by locally elected boards.

From an initial membership of one hundred, the Public Ownership League grew to five thousand strong by 1922. Carl Thompson became the head of the organization and would devote the next thirty-three years of his life to the promotion of public power. To counteract the isolation of small municipal power systems, Norris, Pinchot, and Thompson dreamed of establishing a national "superpower" generation and transmission system to supply isolated municipal utilities and relieve them of the necessity of depending on private utility conglomerates for their electricity supply. Thompson organized a National Public Ownership Conference in 1924 with support from the American Federation of Labor, the Electrical Union, the National Grange, and the League for Industrial Democracy. The conference proposed a bill for a nationwide public-power system, which was introduced the next year into Congress by Republicans George Norris in the Senate and Oscar Keller of Minnesota in the House. That proposal was the forerunner of what became the Rural Electrification Administration, the Tennessee Valley Authority (TVA), and other federal projects founded during the New Deal. To this day, federal authorities such as TVA and the Western Area Power Administration sell inexpensive hydroelectric power on a preferential basis to publicly owned utilities.[13]

Despite the work of Norris and Thompson, the big story of the 1920s was the consolidation of local utility organizations into fifteen giant regional power trusts that controlled 85 percent of the nation's electricity. The model for the holding-company tradition of stock watering and fraud was provided in Chicago by Sam Insull, who became the main challenger to J. P. Morgan as the nation's preeminent utility magnate. After convincing the Illinois legislature to institute a system of state regulation in 1913, Insull created Middle West Utilities, a parent holding company, and inflated by a factor of ten the value of the operating companies so he could market the holding-company stock at ten times its real worth. Electric rates paid by consumers at the bottom of this financial pyramid could be elevated, since those rates were set by public utilities commissions based on the nominal value of the operating companies.

To undermine and buy out small private and municipal utilities, Insull and other utility magnates would offer preferentially low electric rates to the largest customers of the takeover target to undermine their rate base, a tactic that threatens to reappear today if retail-wheeling becomes widespread. The monopolists also prevented municipal utilities from rebuilding their own plants by interfering with the municipals' efforts to raise investment capital. One strategy frequently used was power dumping: offering below-cost wholesale power rates subsidized by higher rates in neighboring communities. Such manipulations proved extremely effective. Between 1909 and 1923, the proportion of municipal utilities wholly dependent on private utilities for their electricity supply rose from 7 to 33 percent. But although Senator Norris and the Public Ownership League were ultimately unable to expand the realm of public power during the 1920s, they were able to stop the takeover of many prime hydroelectric sites—such as the unfinished Muscle Shoals dam on the Tennessee River—by private power interests.[14]

To cover their financial shenanigans, the utility trusts set up speakers' bureaus and press-relations offices and actively courted individual editors and reporters and the news services such as United Press, Associated Press, and Reuters. Their main lobby was the National Electric Light Association, which had affiliates in most states. Ernest Gruening, later Alaska's first senator, wrote his famous exposé *The Public Pays* (1931) to recount the results of three years of investigations and hearings into the utility industry's "propaganda campaign . . . designed to subvert public opinion so that these companies might maintain their monopolisitc status quo, as well as conceal and further their excessive profiteering and unsound financial practices." J. B. Sheridan, head of the Missouri Committee on Public Utility Information was a typical lobbyist. He served as the sole manager of the annual meeting of the Missouri Press Association, in 1923 and again in 1925. In the latter year, 320 local editors and their families attended the meeting, where Sheridan felt the entertainment budget was well spent, according to a letter he wrote to a utility executive:

> Gee Mr. Buck, what the country press is worth to people who are honest and use it honestly is beyond calculation. I have spent as much as $300 in three years entertaining editors, etc. Some of them do enjoy a little drink. All of them are "God's fools," grateful for the smallest and most insignificant service or courtesy.[15]

The utilities also spent thousands of dollars criticizing university and high school texts that they considered unfavorable to private utilities, and hired sympathetic academics from Harvard, Yale, MIT, and the University of Toronto (among others) to write books and reports which promoted the private-utility point of view. Altogether, note Rudolph and Ridley in *Power Struggle,* the industry spent $28 million of its annual $33 million publicity budget to get favorable stories into print and to keep out the stories it didn't like.

Samuel Insull was riding high in 1929. At a time when a new Ford cost $700, his personal fortune had exploded from $5 million to $150 million, and he maintained a "palatial" home on Chicago's Gold Coast and a 4,300-acre spread in nearby Libertyville, Illinois. But all the good publicity and watered stock certificates in the world couldn't paper over the truth when Insull couldn't pay off the bond notes for Middle West Utilities, his personal holding company. In his battle with the Morgan interests, Insull had overbuilt and underbilled to keep his large industrial customers, but the Depression cut heavily into that market. The bubble finally burst for Insull and thousands of shareholders on April 8, 1932, when five Morgan representatives demanded evidence that Insull could make good on a $10-million note that was due on June 1. When Middle

West Utilities collapsed, the holding company left 73,447 stockholders in the lurch; the eventual stock market losses on all nineteen Insull companies totalled between $500 million and $2 billion. It was the largest bankruptcy in American history up to that point, and it happened as the Depression reached its nadir in a presidential election year.[16]

FDR and the Tennessee Valley Authority

PRIVATE-POWER INTERESTS tried to block Franklin Delano Roosevelt as the Democratic nominee for president in 1932 because as governor of New York he had opposed a takeover of the St. Lawrence River's hydropower resources by the Niagara Hudson Power Company, a subsidiary of the Morgan-controlled United Corporation. Instead, Governor Roosevelt had created the Power Authority of the State of New York, which recommended building a publicly owned dam across the St. Lawrence River. By this time, it was common knowledge that for a typical month's consumption of 250 kilowatt-hours of electricity a family paid $19.50 in Albany, $17.50 in New York City, $7.80 in Buffalo, $6.93 in Dunkirk (which had a city-owned power plant), and only $2.79 across the border in Canada, where electric power was supplied by publicly owned Ontario Hydro.[17]

Seven weeks before the 1932 election and three months after Insull's final collapse, Roosevelt delivered a major speech in Portland, Oregon, lambasting the "fraudulent monstrosities" that had led to Insull's "ultimate ruin." The Democratic presidential candidate asserted that the government had the "undeniable right" to set up "government-owned and -operated" utility services "as a national yardstick to prevent extortion against the public and encourage wider use of electricity." In his campaign speech, Roosevelt proposed building massive public hydroelectric systems in the four corners of the United States, and implored the people to "judge me by the enemies I have made. Judge me by the selfish principles of these utility leaders who have talked of radicalism while they were selling watered stock to the people and using our schools to deceive the coming generation."[18]

In a series of bitter struggles during his presidency, Roosevelt created the Tennessee Valley Authority (TVA), moved to expose and curb the power of utility conglomerates, and fought for almost universal rural electrification. The power companies almost immediately filed court challenges to the legitimacy of TVA and refused to allow their lines to be utilized for power transmission. To break this monopoly, TVA encouraged farm communities to organize rural electric cooperatives to which it could deliver wholesale electricity—at half of the price the private utilities charged. TVA's staying power was demonstrated when director David Lilienthal helped organize the Alcorn County Electric Power Association in 1934 in the back room of a furniture store in Corinth, Mississippi. By 1938, seventy-two successful rural electric cooperatives had been set up in close working relationship with TVA.[19]

Meanwhile, Roosevelt and his public-power allies went on the offensive against the electric power companies. Accusing the private utilities of trying to maintain a form of "private socialism," a Public Utilities Holding Company Act was proposed with a "death sentence" clause, which would have required that all utility holding companies be dissolved after 1940.[20] The "death sentence" clause never made it out of committee, though, partly because the Edison Electric Institute (successor to the now-discredited National Electric Light Association) launched a $1.5 million publicity campaign to convince Congress and the American public that the Roosevelt administration was engaged in a plot to destroy the free-enterprise system. Even so, as it was finally passed, the Public Utility Holding Company Act of 1935 required utilities to file detailed accounts of their financial activities and plans with the Securities and Exchange Commission, which was authorized to approve or reject any proposed expansion or conglomeration of utility interests. This is another safeguard to the public that is under assault by utility lobbyists today.

That same year, President Roosevelt, encouraged by the success of the new rural electric cooperatives in the Tennessee Valley, signed an executive order establishing the Rural Electrification Administration. The Emergency Relief Act of 1935, passed with heavy support from farmers' groups, allocated $100 million to build rural electric lines. Early in 1936, Senator Norris and Representative Sam Rayburn sponsored and passed a law that gave the Rural Electrification Administration permanent statutory authority. Under the legislation, farmers were encouraged to form cooperatives and apply for credit to bring electric power to their farms.

Farm families were ecstatic. In one county, a group of families "staged a mock funeral for a kerosene lantern," and a lineman for a Mississippi co-op recalled that turning the power on for the first time "was the greatest thing in the business. We wired the houses, brought out the appliances, put in the meter with the family crowded around waiting. When the first switch was turned on, they literally cried and shouted with joy."[21] Because of the New Deal, the proportion of farms with electricity soared from 11 percent to 43 percent between 1935 and 1944. Though the private-utility industry managed to halt the advance of public power when the New Deal was over, the public-power sector—composed of federal electric projects such as TVA, rural electric co-ops, and municipal power authorities—still controls about 20 percent of electricity generation and distribution in the United States. Many older people in the countryside fondly remember rural electrification as the most successful program of Roosevelt's New Deal.

Hetch Hetchy and the Public-Power Movement in San Francisco

A T THE TURN OF the century, the City of San Francisco bought the water rights for the Tuolumne River, 130 miles away, and proposed building a dam and reservoir in the narrow Hetch Hetchy Valley to capture the Tuolumne's water and generate electric power. Since Hetch Hetchy Valley is right in the middle of Yosemite National Park, the city had to lobby for specific congressional authorization to build the dam. The Raker Act, which passed Congress in 1913, guaranteed to San Franciscans the right to be supplied with Hetch Hetchy power and water. Excess water and power were supposed to be supplied to public agencies in nearby districts. To prevent exploitation by private-utility interests, the Raker Act provided that those rights would "revert to the Government of the United States" in case of any attempt to "sell, assign, transfer [or] convert" those rights to private interests. The floor debate in Congress proves that the explicit intent of Representative Raker was to have the city supply power directly to its inhabitants on its own municipal power lines:

> MR. SUMMERS: Is it the purpose of this bill to have San Francisco supply electric power and water to its own people?
> MR. RAKER: Yes.
> MR. SUMMERS: Or to supply these corporations, which will in turn supply the people?
> MR. RAKER: Under this bill, it is to supply its own inhabitants first....[22]

The Sierra Club and its founder, John Muir, bitterly opposed the Hetch Hetchy dam. Muir swore that San Franciscans might as well turn their cathedrals and churches into water tanks if they dammed Hetch Hetchy, "for no holier temple has ever been consecrated to the heart of man." It would not be the last time that the Sierra Club opposed a dam.[23] (In 1989, when the secretary of the interior proposed, half seriously, that the dam be dismantled and Hetch Hetchy be returned to its natural state, the Sierra Club immediately supported the proposal.)

San Francisco completed its Hetch Hetchy powerhouse in 1925, and proceeded to string transmission wire ninety-nine miles from the dam towards San Francisco. At Newark, southeast of the city, the Hetch Hetchy cable "conveniently" met a PG&E substation that was already "conveniently" connected to the San Francisco power grid, thirty-five miles away. Although San Francisco, wrote J.B. Neilland, "had purchased enough copper wire to complete the Hetch Hetchy line, word suddenly rocketed from city hall that further construction funds were exhausted. San Francisco's two power companies, Great Western and PG&E, refused to sell their systems to the city, and the board, instead of using eminent domain to acquire them, approved a contract on July 1, 1925, to hand over Hetch Hetchy power to PG&E at Newark. The copper wire was stored quietly in a San Francisco warehouse and ten years later sold for scrap." Thus, San Francisco residents had paid $50 million ($1 billion-plus at 1991 prices) to build a dam, power station, and transmission lines—only to see that cheap power sold to PG&E wholesale for resale to San Franciscans at exorbitant retail rates. In the next election, voters defeated every San Francisco supervisor who had favored the sale of

Hetch Hetchy power to PG&E. Through 1941, PG&E spent a total of $200,000 to defeat eight San Francisco bond issues designed to buy up PG&E's local transmission and distribution cables.

From the beginning, most of the local newspapers were hostile to any attempt to reverse PG&E's raid on the municipal purse and the pockets of individual citizens. In addition to normal business-class solidarity, there were personal ties that bound PG&E to the *San Francisco Chronicle*: Joseph O. Tobin, nephew of PG&E director Joseph S. Tobin, became a *Chronicle* owner by marrying Constance de Young, daughter of one of the de Young brothers who had founded the newspaper.

The consequence for San Franciscans, of course, has been inflated electric bills for over half a century. According to one 1941 congressional estimate, PG&E bought Hetch Hetchy power for $2 million per year and resold the electricity in San Francisco for $9 million per year. When the capacity of the Hetch Hetchy power plant was increased by 2 billion kilowatt-hours annually, the excess power was made available, cheaply, to other municipal utility districts and to big PG&E accounts such as Shell Chemical, Dow Chemical, Hercules Chemical, and others.

In 1982, Public Power Advocates made a serious attempt to place public power on the San Francisco agenda. The group collected enough signatures to place Proposition K on the ballot, which mandated a study of the costs and benefits of municipalization. To defeat it, PG&E spent $680,000 (versus $20,000 by Public Power Advocates), the largest sum disbursed up to that time to defeat a municipal proposition, and Prop. K went down to defeat, with only 37 percent of the votes cast in its favor.

In recent decades, all the major Bay Area media, except the weekly *San Francisco Bay Guardian*, have ignored the issue of public power. The *Bay Guardian* tirelessly points out that the city's electric rates are higher than those of eighteen of the twenty publicly owned power systems in California: thus a typical electric bill of 500 kilowatt-hours in Palo Alto cost $24.43 in October 1990—half of the $51.97 for a comparable residential charge in San Francisco. Palo Alto's city-owned electric company buys back domestic photovoltaic electricity at the retail rate.[24]

Pacific Gas & Electric works assiduously to keep the issue of public power off the agenda and almost always manages to bring politicians of the major parties around to its point of view. A study prepared by the office of San Francisco mayor Art Agnos conceded that Hetch Hetchy profiteering costs city residents at least $15 million extra per year. Yet neither the mayor, nor the Natural Resources Defense Council, nor Turn Toward Utility Rate Normalization (TURN, the 50,000-member statewide consumers' watchdog group), will openly advocate municipalization, and a recent proposal to study the subject was narrowly defeated by the city's board of supervisors. PG&E beseeched its San Francisco stockholders to oppose such a study on the grounds that, in other cities, "millions" had been spent "to study takeovers that were not then pursued."

Even so, the municipalization issue never quite dies in San Francisco. In the 1991 mayoral campaign, one of the candidates proposed that San Francisco build its own transmission wires, block by block, to replace PG&E's.[25] But Pacific Gas & Electric is forever, whereas mayors come and go. Mayor Art Agnos, considered the more "progressive" candidate in a bitterly fought runoff election in 1991, promised to appoint Nancy Walker, a popular public-power advocate and former chair of the board of supervisors, to the public utilities commission. Once again PG&E seemed to get the last laugh. When Agnos lost in a close election, he delivered on his promise, in what the Bay Guardian called "Agnos' last dirty trick": rather than appointing Walker to a three- or four-year term as he could have, he "quietly arranged for her to serve fewer than two months, for a term which expired a week after the new mayor took office."[26]

The attempt to undermine San Francisco's control of Hetch Hetchy proceeds apace. In the summer of 1993, Representative Wayne Allard, a Republican of Colorado, introduced an obscure amendment to a public lands bill that would have obliged San Francisco to pay fees of $20 million per year for the right to operate Hetch Hetchy. After frenzied behind-the-scenes lobbying by California's congressional delegation and San Francisco mayor Frank Jordan, the amendment was killed. Since their takeover of both houses of Congress in 1994, Republicans are once again threatening to take back Hetch Hetchy from the city of San Francisco. "Selling electricity or health insurance . . . always makes lots of money," wrote investigative reporter Tim Redmond of the Bay Guardian, "so whenever the government tries to get a piece of the action, private business fights like hell to stop it. The whole concept seems pretty simple to me."[27]

Meanwhile, Pacific Gas & Electric has launched an attempt to take over ownership of the electric power grid at the Presidio, a defunct military base in the heart of San Francisco that the Army has turned over to the National Park Service for civilian use. The official "technology evaluation criteria" for the Presidio required that newly constructed infrastructure should be designed to "save energy, water, and other scarce resources" and "to enhance sustainability," among other criteria, by powering the new Presidio National Park with renewable technologies such as photovoltaics and wind power. A study by a National Park Service staffer showed that a takeover by PG&E would cost the Park Service $7.7 to $14.8 million in upfront transfer costs, and make it impossible for the proposed Presidio National Park to take advantage of hydropower from the Western Area Power Administration at 3.8 cents per kilowatt-hour (a municipal utility would be granted preferential rates). The ultimate difference for the consumer at the new national park between buying PG&E power at 14.3 cents retail and power from the Western Area Power Administration would amount to over $5.5 million per year in electricity bills—in addition to the proposed transfer fee.[28] Under attack by San Franciscans for Public Power and the *Bay Guardian* in 1994, the National Park Service was forced to open up the bidding to all comers, including the publicly owned San Francisco Public Utilities Commission (which runs Hetch Hetchy and the electrical systems for the San Francisco International Airport and the city's streetcar system), and San Franciscans for Public Power itself. When the Park Service again awarded the Presidio contract to PG&E, Joel Ventresca, chair of San Franciscans for Public Power, and the *Bay Guardian* again made a great noise, obliging the city's public utilities commission to sue to rescind the contract, on the grounds that the bidding process had been flawed. At the end of July 1996, the issue of who would supply electricity to the Presidio was still in litigation.[29]

The success in halting PG&E's Presidio gambit seemed to jolt the public-power movement into action. San Francisco supervisor Angela Alioto, chair of the Select Committee on Municipal Public Power of the board of supervisors and head of its Standing Committee on Municipalization, initiated the passage of an ordinance that appropriated $150,000 for a feasibility study of municipalization of San Francisco's electrical system, over the objections of then-mayor Frank Jordan. "I think PG&E will get the feasibility study they want," observed Ventresca, who claimed that the personnel of the firm that won the study contract had too many long-term ties to PG&E.[30] The feasibility study was set for delivery in October 1996.

In Los Angeles, sixty years ago, municipalization was supported by the voters (and even by the chamber of commerce) as a pragmatic, piecemeal reform, because the competing and fragmented private utilities charged too much and could not maintain uninterrupted service. Thus to the hypothetical question, "Do you believe in municipal ownership?" the progressive Municipal League of Los Angeles noted that utility stock-holders would answer "no" and socialists would unequivocally say "yes," but "the average, sane, disinterested businesslike citizen" would say "Yes, at some times, in some places. . . ." Bolstered by cheap Boulder Dam hydropower and loans from the federal government's Reconstruction Finance Corporation, the Los Angeles Department of Water and Power took over most of the electricity business in Los Angeles during 1936.[31]

In Cleveland, Ohio, the banks never gave up the struggle for the reprivatization of city-owned Muny Light. In 1973, M. Brock Weir, president of the Cleveland Trust Company (and a former PG&E director), threatened to deny debt refinancing to the city of Cleveland unless it sold Muny Light. Mayor Dennis Kucinich refused to do so, and called for a referendum. Bucking huge campaign contributions from the power industry nationwide, the Cleveland electorate voted 2–1 to retain public ownership. Kucinich explained his reasons: "We gave the people a choice between a duly elected government and an unduly elected shadow government. . . . If I had cooperated with them and sold Muny Light to the private utility . . . everyone's electric rates would have automatically gone up. It would have set the stage for never-ending increases."[32]

"Sixty Percent of the Underwriting Business": Wall Street and the Utilities

WHAT IS THE BANKS' role in utility financing? Oil companies are largely self-financing because of the oil depletion allowance (see chapter 3) and other de facto government subsidies and tax breaks, but utilities must raise their funds from private capital markets. In their book *Power Struggle*, Rudolph and Ridley argue that Wall Street investment banks, which carry three quarters of utility paper, are the prime determinants of electric utility policy. Financing charges for construction of power plants amount to 40 percent of utility costs. Thus, financing charges are about equal to total fuel costs, according to Eugene Meyer, vice president of Kidder Peabody: "Fees are collected at every stage—for advising a company on its financial plan, for selling its stock and placing its debt, and from dividends on power company securities the brokerage house or bank might own."[33] Bank influence is reflected on the utilities' boards of directors: in 1990, three of the thirteen outside directors of PG&E, the nation's biggest utility, came from banks or savings-and-loan associations.[34]

And the dependence is mutual: utility investments constitute a third of investment bank industrial financing. If the utilities don't build, the banks don't make money. Pennsylvania utilities commissioner Michael Johnson alleges that the utilities serve as "handmaidens for the banking industry to provide them with lucrative investment fields. In the utility industry the chief thrust today is how many loans can be floated, how many bonds can be sold and at what kind of interest. . . ." [35] Johnson sees the banks as major players in pushing utility overexpansion, because the utilities are principal vehicles of Wall Street investment bankers.

Utility-industry construction plans in the mid-1970s had been based on projections of a 7-percent annual increase in energy use, which would have led to a doubling of generating capacity every ten years—as had occurred in the 1950s and 1960s, in tune with the slogan "Better Living Electrically." But the reaction of most energy buyers to the OPEC-induced price hikes of the 1970s was to conserve energy. The utilities were hit with a triple whammy: their fuel prices quadrupled and demand flattened just as they got stuck with a huge overhang of existing and planned new generating capacity—much of this, nuclear—for electricity that they couldn't sell. To deal with the revenue shortfall, the utilities tried to shove doubled and redoubled electricity rate increases through state public utilities commissions. Normally compliant PUCs, with an outraged public snapping at their heels, began to balk at rate increases, and the utilities were in big trouble.

At first, the banks were loath to believe that the milk in their utility cash cow was drying up: "The thing you've got to remember," said one Wall Street utilities analyst, "is that 60 percent of the entire underwriting business in this country comes from the public utilities industry. Nobody wants to rock that boat."[36] Yet by 1979, the boats had begun to rock furiously, as one environmentalist recalls:

> Interest rates were breaking records every month, along with oil prices driven up by the fall of the Shah [of Iran] and by the war between Iran and Iraq, which had produced an oil crisis more economically serious than the one in 1973. Nearly all utility companies had redoubled their commitments to power-plant construction in response to the first oil crisis; and, as interest rates reached 15 to 16 percent, they were discovering that they couldn't afford to keep borrowing the capital that their projects demanded. Going ahead on construction could take more than a company's entire annual revenue. Sixteen-percent interest also did ghastly things to the target cost of a nuclear plant with several more years to completion, when the planning assumption had been that the interest rate would stay below 5 percent.[37]

The major New York investment banks finally became convinced that the financial crisis of the utilities was serious by the bankruptcy of the Washington Public Power Supply System (WPPSS) on July 24, 1983. Suddenly, $7 billion of WPPSS paper—borrowed to build five nuclear plants—was worthless, in the biggest municipal-bond default in the nation's history. In New York, Chemical Bank rushed to organize 75,000 WPPSS bondholders into a political lobby in order to sue hundreds of public officials for fraud, but the damage had been done. What had happened to the nuclear plants? What had happened to the money?

Fifteen years earlier WPPSS, a consortium of 118 public and private rural cooperative power systems and three private electric companies, had begun to plan two nuclear power plants near the Satsop River and three more near Richland, Washington, as part of a gigantic program intended to comprise forty coal plants and twenty nuclear plants throughout the Pacific Northwest. According to the "Net Billing Agreement," which Washington Senator Henry "Scoop" Jackson pushed through Congress, the ratepayers from seven Northwest states would have to pay all building costs, even if no electricity was ever produced. Bonds with that kind of guarantee earned the top AAA rating, and sold easily. Energy hogs such as the aluminum industry, whose twenty-year contracts for cheap hydro-generated electricity were running out, were reputed to have pressured local WPPSS consortium members "to build ahead of their needs." Donald Hodel, head of the Bonneville Power Authority, a crucial WPPSS associate, and later secretary of the interior under Ronald Reagan, threatened recalcitrant towns with brownouts and economic strangulation if they refused to sign on the dotted line for the new utility bonds.[38]

Construction finally began in 1976 at the height of the rainy season in Satsop, "one of the wettest places on earth." Total confusion reigned:

> Half the cleared site went sliding downhill in huge mud streams. Contractors responded by covering fifteen and a half acres with plastic. As work schedules slipped steadily behind, claims emerged that contractors were milking the project for all it was worth. Reports also revealed that contractors jockeyed for position with their equipment while sometimes waiting weeks for material to arrive. Crews arrived to do work eighty feet up only to find that scaffolding had been taken down by other workers busy with another job. . . . As the stories of boondoggles at the construction sites continued to surface, the acronym for the project . . . changed from WPPSS to "Whoops."[39]

In the desert near Richland, where a three-reactor WPPSS project was in the works, only the topography differed: workers called it the "construction gig that never ends," and a few workers pulled their boys out of high school and put them to work welding at $750 a week.[40] By January 1982, the bills for two of the unfinished reactors had already soared to $7 billion, and electricity demand in the region had fallen so far that these two plants had to be cancelled. Estimates for finishing all five reactors would soon top $24 billion.

Among the hardest hit were the six hundred inhabitants of Drain, Oregon, who faced a collective bill of $4.5 million. Bills for electricity that would never be delivered often exceeded home mortgage payments. The public's patience was wearing thin, indeed.[41]

The revolt began in February 1982, in Grays Harbor County, Washington, where the county utility commissioners had told a dozen citizens that nothing could be done about a six-fold increase in electricity rates. Two weeks later, Dorothy Lindsey called a meeting to confront the commissioners, and 3,000 county residents showed up. "For years Washington [State] residents have sat back and let others run their public power," Lindsey told them. "Those days are over. We're well aware that we're taking on the greedy contractors, the greedy bankers, the nuclear industry, and all the entrenched power management with their $125,000 salaries. But we can win!" The two commissioners present were stunned. To the cheers of the multitude, Lindsey swore that they would organize a night of protest that would black out the entire state. Dozens of people came to the microphone to pledge not to pay their bills so that Wall Street would have to "eat the debt." That meeting in the Hoquiam High School gym was the beginning of protests that spread throughout the WPPSS moneyshed in seven states and led directly to the notification of default on July 24, 1983.

The New York State Power Authority was another large and autocratic public entity that had been captured by corporate interests, in this case heavy industry. According to the *Buffalo News*, the authority was squandering 695 megawatts of the cheapest power in the country, selling electricity to more than seventy companies in the Niagara Falls area at 0.5 cents per kilowatt-hour, a ten-year subsidy worth over $2.1 billion. The next cheapest power in the country cost almost 5 cents.[42] This cheap power was basically used to keep obsolete heavy chemical and steel plants alive, while their parent firms reinvested the profits elsewhere. The Oil, Chemical and Atomic Workers International Union (OCAW), which first called these sweetheart deals to the attention of the *News*, was concerned because employment among its members had fallen 60 percent in the twenty years beginning in 1970. The biggest beneficiary of this public largesse was Occidental Chemical, which saved $38,479,935 in 1990 over what it would have paid for its allocation of 111,400 kilowatts if it had paid the going rate of at least 5 cents. Rather than propping up dying industries at an average cost of $22,580 per job every year, the union recommended that the power authority use its cheap power to produce electrolytic hydrogen. For Richard Miller, policy analyst to the president of the OCAW union, the issue was about "how undemocratic public power leads to socialism for the rich, costs workers their jobs, and wastes energy." Miller suggested that continuing power subsidies be conditioned on continued capital investment in the Niagara Falls region, and that displaced workers be given four-year scholarships to retrain themselves for other work. Rather than waste the cheap power, Miller argued, continued access to it should be dependent on "maximum feasible energy conservation" so the low-cost power could be freed up for other uses.[43]

To reinforce public control of the authority, OCAW's Miller recommended that a public board made up of industry, labor, environmental, and government representatives decide on allocation of low-cost power. And he further recommended that the power authority investigate the possibility of using some of the cheap hydroelectric power to produce hydrogen and establish an R&D center for what was expected to be a growing market in clean transportation and heating fuel. A local representative proposed a bill in the New York State Assembly that would have mandated that all official New York State vehicles be powered by propane, natural gas, or hydrogen, so long as anticipated life-cycle costs of these "clean" vehicles were held to no more than 25 percent higher than comparable vehicles running on gasoline or diesel.[44]

The head of the New York State Power Authority rejected the union's accusations of power squandering, beyond carrying out a vigorous defense of its record.[45] Eventually, industrial rates were raised substantially, but the power authority refused to allow a public board to decide on the allocation of low-cost power. Obviously, the argument was just beginning.

The Sacramento Municipal Utility District: A Public-Power Success Story

PUBLIC POWER in Sacramento, California, owes its existence to the McClatchy family, for its fiery and relentless advocacy in the *Sacramento Bee* for municipal control of natural resources, and for the Sacramento Municipal Utility District (SMUD), in particular. James McClatchy, founding publisher of the *Bee,* proudly "fought monopoly and utility pirates wherever he found them," and his son C. K. McClatchy "fought as unyieldingly, with the same uncompromising passion." In 1928, five years after SMUD was founded, C. K editorialized, "So thoroughly have the utilities captured the body social, the body politic, the body commercial, the body club womanly, the body editorial, and the body ministerial, that all over the nation it becomes an occasion for sneer when any man or any woman dares to stand out in the sunlight and battle for the sovereign rights of a sovereign people against the arrogant dominations of public utility corporations, which should be the servants of the people, and not their throttling czars."[46]

In 1923, "after years of agitation and organization," Sacramento's electorate voted to create the Sacramento Municipal Utility District, to be gòverned by a five-person board elected by the people.[47] Voters were influenced by the fact that the Federal Power Act of 1920 had given public utilities preference over corporations or individuals in filing claims for hydroelectric development. Pacific Gas & Electric fought the transfer mightily in court, but in 1946—twenty-three years later—the buyout of PG&E's generation, transmission, and distribution facilities was consummated for a price of $10,632,000.[48] According to the agreement, SMUD would at first buy all its bulk power from PG&E, and then develop its own power sources. The transition was effected through $23 million in Rural Electrification Administration loans, which were available because SMUD was expanding rapidly into rural areas of Sacramento County. SMUD's second major power source was the Upper American River Project, an eleven-dam "Stairway of Power" completed in 1971. But SMUD still needed another source of power.

The "Atoms for Peace" program, inaugurated by President Eisenhower in 1953, seemed to have all the answers and more. Harold E. Stassen, Ike's Special Assistant on Disarmament, told *Ladies' Home Journal* readers that the "world of the future that nuclear energy can create for us" would be a world "in which there is no disease . . . where hunger is unknown . . . where food never rots and crops never spoil . . . where 'dirt' is an old-fashioned word, and routine household tasks are just a matter of pushing a few buttons . . . a world where no one stokes a furnace or curses the smog, where the air is everywhere as fresh as on a mountaintop and the breeze from a factory is as sweet as from a rose."[49] The atomic propaganda barrage was aimed at every sector of society. In *Our Friend the Atom*, a 1956 Walt Disney "Tomorrowland Adventure," elementary school children were told that "An atomic airplane could circle the earth many times without ever landing for fuel," and that "a rocket ship powered by the atom would not have to carry the enormous weight of fuel required by present-day high-altitude rockets . . . and will help us to cast off the shackles of gravity and fly freely through space." These schoolchildren were sermonized with the nuclear gospel, "Atomic power has come just in time, for coal and oil are far too valuable to burn! The magic power of atomic energy is beginning to work for mankind throughout the world. Atomic energy will give more food, better health—the many benefits of science—to everyone."

McClatchy's "Sacrifice"

The history of [the federal] government, in so far as public property and public utilities are concerned, is one of sacrifice of the interests of the people for the pocketbooks of the exploiters. That has been true of public lands that have been given away to the railroads; with timber lands; with coal lands; with water power from the navigable streams of the country; now with oil possessions. Uncle Sam has proceeded on the theory that the citizenry of the United States have no right to anything that any syndicate or corporation desires. What is true of the national government is true as well of state and municipal governments.

C. K. MCCLATCHY,
The Sacramento Bee, 1924

Like scores of electric power producers around the country, SMUD was swept up in the enthusiasm for nuclear power incited by the president and the nuclear lobby. When General Electric sold a nuclear power plant to Jersey Central Power and Light Company in 1962, everybody began "to look at nuclear power from an economic standpoint," especially since the word was out that "conventional fuel sources" (coal in the East and oil and gas in the West) "would not be available for future power generation." Expected to cost 0.45 cents per kilowatt-hour to generate, nuclear electricity was promoted as "far more economical than that generated from other fuels"—a prediction that would be laughable now if it hadn't cost so much money and human suffering. SMUD's decision to build its Rancho Seco nuclear plant was triggered, says one insider, when "the Atomic Energy Commission conducted a symposium for the people who might ... wish to look at nuclear power as a commercial reality. I attended the meetings, where specialists indicated various problems had been overcome, and the big firms such as General Electric, Westinghouse, and Babcock & Wilcox were present and indicated they were ready to receive orders for power plants." Ultimately, SMUD bought a Babcock & Wilcox design, which was then built by Bechtel at Rancho Seco. The plant was a twin of the Three Mile Island plant that experienced an almost complete meltdown in 1979, and whose cleanup was finally completed in 1993—fourteen years and tens of millions of dollars later.[50]

Rancho Seco was in trouble from the start. Construction costs, originally budgeted at $180 million, soon doubled to $350 million.[51] Commencing generation in 1975, the plant's design and construction were so slipshod that operators managed to keep it on-line only 40 percent of the time. One engineer who worked on the project commented that SMUD didn't seem to realize until it was too late that a nuclear plant had to be much more tightly engineered than a dam. "In one case," reported the New York Times, "a worker dropped a light bulb while replacing it in a control panel. That caused a short that tripped a fuse, cutting power to much of the control room. An automatic safety system then flooded the hot reactor vessel with cold water so fast that metallurgists feared there was damage."[52] This near-meltdown at a Three-Mile-Island clone catalyzed activists into organizing Citizens for Safe Energy in 1979 and calling for a shutdown.[53] "The Ranch" was temporarily closed in 1985 for what ultimately proved to be a $400-million, three-year reconstruction.

"It was a terrible time for the utility," recalls Ed Smeloff, later president of SMUD's Board of Directors. "SMUD nearly collapsed financially in the 1980s under the burden of a failing nuclear plant. In three years, from 1985 to 1988, electric rates doubled while the utility's credit rating plunged. Ratepayers were angry, employees were demoralized, and top managers were hired and fired routinely. The press had a field day as one problem after another with the troubled Rancho Seco nuclear power plant was exposed."[54]

In 1987, a broad-based grassroots group called Sacramentans for Safe Energy—troubled by the rate hikes and reminded by the 1986 Chernobyl catastrophe of the dangers and costs of nuclear power—collected enough signatures to put a shutdown referendum on the June 1988 ballot. Ed Smeloff, the lone board member skeptical about Rancho Seco, called for a permanent shutdown of the plant in January 1988, and a month later General Manager Richard K. Byrne made a similar recommendation. Byrne's act of political courage caused the board to fire him almost immediately. The June 1988 referendum to close Rancho Seco was defeated by a small margin under the slogan "Give the Ranch a Chance," with the understanding that voters could review their decision after a one-year trial run. During the next year Rancho Seco experienced three lengthy outages, and on June 8, 1989—despite furious opposition from 300 Rancho Seco employees and 1,600 skilled trades contract workers, the voters elected to shut down Rancho Seco by a 53- to 47-percent margin. (In both elections, the *Sacramento Bee* favored closure.) Rancho Seco thus became the first nuclear plant in the world to be closed by popular decision.[55]

The worker reaction to the shutdown was less bitter than expected, because Sacramento was in the midst of a roaring construction boom; meanwhile, the high-level nuclear executives and technicians—having proved their worth in Sacramento—were eagerly recruited to other nuclear plants around the country. Business agent Gary Mai of Local 1245 of the International Brotherhood of Electrical Workers represented the three hundred permanent workers at Rancho Seco. Mai still swears by nuclear power and believes that the plant should have been saved, but he found SMUD "as fair as they could have been, given the political atmosphere at the time. To my knowledge," he recalls, "there were no actual layoffs, and many of the permanent workers were reabsorbed into other positions."[56] The permanent employees who were not asked to stay got up to twenty-six weeks of severance pay, and the 1,300 contract workers got thirty days of severance pay. Most evidently found other jobs easily. "If the shutdown had happened now, it would have been a lot nastier," said Smeloff in 1993, "because unemployment is so high now."[57]

The new board of directors, elected in 1988, decided to hire practical visionary S. David Freeman, an advocate of conservation and solar for a quarter of a century, as SMUD's managing director.[58] Freeman's career had begun almost forty years earlier with the Tennessee Valley Authority. An engineer and a lawyer, he had been a U.S. Senate staffer and headed up the controversial Energy Policy Project of the Ford Foundation in the early 1970s. Later, Freeman worked in the White House for President Carter, who appointed him managing director and chairman of the board at TVA, the nation's largest public-power system.

Faced with a 900-megawatt hole in generating capacity at SMUD, Freeman led SMUD to diversify its power sources with purchases of electricity from other utilities and the construction of 1,000 megawatts of new capacity—two thirds natural gas and one third a mixture of wind, solar hot water, solar thermal, and photovoltaics. SMUD is also trying to bypass PG&E and buy its natural gas directly from the Mojave Pipeline. According to the Natural Resources Defense Council, SMUD spends 6.4 percent of its revenues on energy efficiency, the highest in the United States, compared to 2.0 percent for PG&E and 1.4 percent for Southern California Edison.[59] And the simple act of planting trees is one of the utility's most popular energy-saving measures.

In 1990, SMUD helped organize a partnership with the Sacramento Tree Foundation, whose goal was to blunt the city's infernally hot summer afternoons by planting 500,000 shade trades. When mature, these "air conditioners with leaves" can cut home air-conditioning costs by up to 40 percent. SMUD pays for the trees, which participants select from a list of appropriate species. Participants must take part in a course to teach them how to plant and look after the trees. Each new tree is registered on a computer database so that its progress can be monitored. Twenty-eight thousand trees were planted in 1991, and the program peaked at 50,000 trees in 1994. SMUD's contribution to the Foundation was $1.1 million in 1995, when 40,000 trees were planted. Joanna Julienn, the Sacramento Tree Foundation staffer who serves as liaison for the program with SMUD, supervises twenty-seven full-time employees. "This city has a long tradition of tree-planting and care going back to the middle of the last century," says Julienn. "A century ago the *Sacramento Bee* ran dozens of editorials and articles promoting trees. Today the *Bee* has been crucial in preparing the ground and informing the public about the present program. Working with SMUD has been a real pleasure; the fact that the utility is owned by the customers creates a whole different set of community expectations. I wonder whether a partnership like ours would have been possible with a privately owned utility, whose primary responsibility is to its stockholders rather than to its customers."[60]

By contrast to SMUD's program, the tree-planting program of Pacific Gas & Electric looks positively anemic, despite the fact that PG&E has ten times as many customers as SMUD. PG&E used to provide $5 rebates for up to two trees per customer in some areas, and has supported three small tree-planting pilot projects. The big utility always appeared less interested in promoting tree-planting than in the public-relations benefits of claiming support for planting trees. *Plant an Acorn . . . Grow an Oak,* a PG&E flyer for schoolchildren, contained *no* follow-up addresses or phone numbers. The 1992 PG&E *Trees and Energy Conservation Calendar,* mailed out free to anyone who calls PG&E about trees, constituted a marvelously illustrated primer on appropriate shade trees for California, but also neglected to mention PG&E's rebate program. The organization California Releaf was mentioned without explanation, in tiny print at the bottom of the calendar above a toll-free telephone number. A call to 1-800-TREE-GEO elicited an ad pushing the "fuel-efficient" Chevrolet Geo from General Motors and an option to request literature regarding California Releaf. There was no mention of the $5 PG&E rebate program, and at no point did the telephone message offer to connect the caller with a local chapter of California Releaf.[61]

Sharon Dezurick, the head of PG&E's program, conceded that most of the emphasis had been on printing up informational literature. With a couple of exceptions, the utility has steered clear of organizing communities for tree-planting. In the first full year of the program, Dezurick claimed that 16,000 trees were planted as a result of the individual incentives. As Dezurick stated: "Trees are a great way to get your foot in the door. They have a lot of PR value, and people like them. For some people, trees may be the wedge that gets them to look at the whole issue of energy conservation."[62] By 1993 or 1994, PG&E's shade-tree program had been cancelled, and by August 1996, it was hard to find anyone inside the company who even remembered the program. Heidi Lian, an employee in Vegetation Management Services, whose mission is to oversee 1,200 contract employees who keep power line rights-of-way free of overhanging branches, remembered that Sharon Dezurick had been laid off in 1993 or 1994 when PG&E abolished the shade-tree program "for lack of funds." Lian says that the utility still tries to plant a few trees at public occasions such as Arbor Day and Earth Day. In her spare time, Heidi Lian is studying to become a certified arborist, and she helps organize a volunteer effort very popular with PG&E employees that is called the Phoenix Project, in which they plant several thousand trees in national and state forests every year.[63]

Guided by the solar slogan of "sustained, orderly development" (a term coined by Don Aitken of the Union of Concerned Scientists), SMUD also began a program to replace the electric water heaters in the district with solar water heaters. In the first four months of the program, $2.3 million was freed up for solar conversions, financed by rebates and low-cost loans to homeowners. Solar water heaters are great energy savers even if they are not so important in reducing peak energy demand, since the thermostatically-controlled heating element is turned on only one twelfth of the time. In Sacramento, the solar heaters save homeowners an average of 2,600 kilowatt-hours per year. Each solar heater shaves 0.4 kilowatts from peak demand during summer consumption peaks, replacing $2,000 in construction and operations costs to SMUD over the fifteen-year life of the unit.[64]

Sacramento contractors have been pleased with the new solar water heater programs, though they have grumbled about SMUD's strict quality controls and attempts to select the contractors allowed to participate in the program. Terry Parks of Harvest Sun Energy reported that gross receipts from his solar hot water heating business quadrupled from $50,000 in 1990 to $200,000 in 1991. Parks expected to show a gross of $750,000 for 1992, with 230 solar hot water heaters installed, despite the fact that SMUD's tough inspectors required him to retrofit 135 heaters that had caused certain problems, at the expense of the (Japanese) manufacturer. Harvest Sun picked up fifteen new employees in 1992, up from two in 1990—mostly people with building-trades experience in plumbing and solar. Parks pays his two-person crews $300 for one-day installation jobs that take from six to ten hours to complete.[65] As the program has developed, SMUD's "Operating Guidelines for Solar Domestic Water

Heaters" (OG-300) have become the de facto industry standard. Eventually, under pressure from the solar water heating contractors, SMUD agreed to go along with a vendor-driven program, and sales and installation of solar water heaters expanded somewhat. Between May 1992 and August 1994, over 3,000 SMUD customers installed solar water heaters.[66]

But the momentum of the program has slowed from 1,100 new heaters installed in 1994 to a predicted 300 in 1996, says Cliff Murley, SMUD's solar water heating expert. One problem, says Murley, is that a family has to use a great deal of hot water to make the systems pay off quickly. Forty-one dollars per month is the typical portion of an electrical bill attributable to hot water for a four-person household with an electric water heater in the Sacramento area, and SMUD's studies show that solar water heaters replace 60 percent of that demand. A solar water heater with a predicted life span of 15 years sells for $3,000, financed by an $800 rebate and a $2,200 loan from SMUD, repayable at $29 per month for 10 years. Backup electric water-heating costs after the solar heater has been installed might be expected to average $16 a month, which means that homeowners are paying up to $4 more per month than they would pay with an entirely electric system during the 10 years while the loan is being paid off. And despite improvements in the 1990s, solar water heaters still suffer from a lingering reputation for unreliability, so many homeowners have adopted a wait-and-see attitude. At first, SMUD had estimated that there were about 35,000 households with electric water heating, but it scaled down that estimate to 28,000, meaning a potential market too small to benefit from economies of scale.[67] Because electricity and natural gas are cheap and plentiful at present, adopting the Israeli practice of requiring solar water heaters in the building code has been politically impossible. The result of all these factors is that only 5,000 new solar water heaters are now installed each year in the United States, compared to about seven million conventional heaters, split equally between gas or oil and electric.[68]

Similar programs with comparable incentives are under development at SMUD in solar cooling, building design, and photovoltaics. In the summer of 1992, SMUD volunteers built the first photovoltaic recharging station on the West Coast for electric auto commuters.

SMUD has also been installing rooftop PV modules on more than one hundred roofs per year. Participating homeowners are volunteers who have agreed to help study the performance of the grid-connected modules, which are installed and owned by SMUD. Residential customers who want to personally promote solar energy can sign up as "PV Pioneers": for a contribution of $4 per month, they agree to allow SMUD to install a 2-kilowatt photovoltaic array on their houses and to help the utility monitor its functioning. From the beginning, the PV Pioneers program has had far more applicants than the utility can accommodate. By the end of 1995, 350 arrays had been installed, 105 of these in 1995 alone. [69] So many homeowners have signed up that the program is oversubscribed. "Realistically," guesses Smeloff, "we may be a generation away from the widespread use of photovoltaics," but if the enthusiasm of these SMUD customers is any indication of the potential mass popularity of renewable sources, despite their current price premiums, that day may come sooner.[70]

Under the leadership of Jan Schori, who succeeded David Freeman as general manager, SMUD continues to be a world leader in the utilization of solar technologies. The utility operates an experimental 5-megawatt wind plant in nearby Solano County, which takes advantage of the sea breezes that usually blow hardest at those times when electricity demand for air conditioning peaks: on summer afternoons and evenings. If SMUD's wind plant continues to be successful, the utility plans to open bids for another 45 megawatts of wind capacity, enough to serve about 13,000 customers. SMUD also operates 4.3 megawatts of photovoltaic capacity near the site of its shut-down Rancho Seco nuclear plant, and is accepting bids for an additional 10 megawatts of renewable generation capacity to come on-line between 1998 and 2002.

Air conditioning accounts for some 70 percent of summer demand peaks in Sacramento. To reduce that demand, SMUD is sponsoring a Solar Advantage House project "to incorporate solar design and low-energy cooling strategies" into new construction and retrofits. Overall, SMUD claims that its comprehensive efficiency and solar program has managed to reduce its annual increase in baseload demand from 2.2 percent a year to 0.6 percent, and to halt increases in peak power demand.[71] Financially, former General Manager S. David Freeman has boasted, SMUD has been able to raise itself from junk-bond purgatory to a senior bond rating of A from Moody's and A- from Standard & Poor's since the 1989 voter shutdown of Rancho Seco.[72] Though electric rates are expected to rise in coming years, at 8.18 cents per kilowatt-hour, SMUD's residential rates are still 40-percent lower than the rates of neighboring PG&E.[73] Jan Schori, Freeman's successor as general manager at SMUD, has continued to support solar water heating and photovoltaics, despite the pressures of cheap natural gas and the move toward deregulation.

For SMUD, demand-side *reduction* pays—and so does solar. Once a crisis had been created by nuclear mismanagement and cost overruns, the people of Sacramento invoked the democratic decision-making process that had been established for their benefit—a process that enabled them to reorient their power system toward the sun. With a quarter of American electric-generating capacity due to become obsolete in the next decade—even without an increase in demand—the time has come to seize the day and make the "sustained, orderly development" of solar energy a reality throughout the country.[74]

The Public-Power Option

MOST OF THE 1,750 municipal utilities in place today were founded during the movement in support of "municipal socialism" for basic services that rose up between 1890 and 1925, in reaction to the abuses of private power by the railroads, trolley companies, electric and gas utilities, and insurance companies. In Ohio, part of the progressive legacy is the exclusion of private insurance companies from workers' compensation insurance, which are still primarily the exclusive responsibility of a state fund though large businesses, as usual, can self-insure to fulfill their required workers' compensation coverage.[75]

The McClatchys' public-power campaign in the *Sacramento Bee* was part of that movement, though the McClatchys were certainly not socialists. The McClatchy family's fervor for publicly owned and controlled enterprises was multiplied in states and cities and towns across the whole United States during the flowering of the public-power movement. Publicly owned and locally controlled electric companies continue to prosper. Despite their small size, many have been competent enough and responsive enough to the needs of citizens that they are hard for private utilities to dislodge.

In recent years, municipalization has become a serious political option only when supply interruptions or rate shocks have created a crisis. Activists have begun talking about electricity-buying co-ops modeled after rural electric co-ops, which would negotiate special rates with independent suppliers. As nuclear plants age and face high shut-down and decommissioning costs, nuclear safety and reliability problems will intensify and cost more to remedy. As with SMUD's Rancho Seco nuclear plant and Portland General Electric's Trojan plant, more rate shocks and safety and reliability problems can be expected in the near future from nuclear-dependent utilities—particularly those with obsolete reactors (see chapter 4).[76]

Public power has many facets, and only a naive person would view any one of those facets as a panacea for the energy problems threatening the nation. For example, in Nebraska—the only state where private utilities are publicly owned and controlled—nearly all electricity is generated from coal and nuclear plants. From the customer/owner's perspective, only two major crises can occur with non-nuclear power plants: the power can go off and the bills can skyrocket. The American Public Power Association likes to remind people that "Public ownership

---

TABLE 4–1: Electricity Rates (cents per kWh), 1994

	Residential	Commercial	Industrial
Public	6.7	6.7	4.9
Private	8.8	7.9	4.9

Source: Diane Moody, "Public Power Costs Less," *Public Power* (January–February 1996, p. 40). Moody's figures are based on reports to Energy Information Administration (EIA), U.S. Dept. of Energy, on Form EIA-861, *Annual Electric Utility Report.* Residential and commercial rates for rural electric cooperatives fall between rates of publically-owned and privately-owned utilities; rural cooperative rates are slightly lower.

---

of a utility is like owning your own home. Private ownership is like renting it."[77] The strongest argument for public power is that the utility is locally owned and controlled, and that its rates are often approximately 10 to 25 percent lower than private-power rates, especially for residential and commercial customers (see table 4–1). With public power, there is no incentive to oversell electricity consumption in order to raise profits and maintain stock prices for the benefit of absentee investors and banks. And yet, although low electricity rates are an economic stimulus to local commerce and to the public at large, they also spur higher electricity consumption. In 1989, the average public-power residential customer consumed 10,443 kilowatt-hours per year, 15 percent more than the average private-power customer (see table 4–1).[78]

The beauty of municipal power is that this arrangement permits local citizens to supervise their own electrical system. Large cities such as Los Angeles and Seattle and thousands of smaller towns have entrusted the transmission and distribution and even the generation of electricity to municipal utilities run by professional managers and governed by a publicly selected board. According to a 1988 Gallup poll, a majority of Americans believe that it is more important to use energy efficiently than to build new sources of power supply, and over half prefer the development of solar energy supplies above all others. (Only 12 percent thought nuclear energy should be developed first.)[79] When the people in Sacramento, for example, opposed nuclear power and favored solar, they could pressure the publicly controlled SMUD board to hire enlightened managers and introduce solar into the power mix. On the other hand, if the citizens are indifferent and their leadership is unenlightened, a municipal utility will ignore solar and energy efficiency and build or contract for big new coal and gas-turbine plants to meet anticipated demand.

Nationally, only 20 percent of electricity is produced by publicly owned companies. Ten percent of electrical capacity is federally owned, and 6 percent is produced by rural electric co-ops; only 3 percent is produced by municipally owned systems. Obviously, some public-power systems are more democratic than others. The New York State Power Authority is governed by a board of trustees appointed by the governor and subject to confirmation by the New York State Senate. The most democratic public utilities have governing boards chosen by direct election of the people. Sacramento's municipal utility is governed by a five-person board elected by the citizens of the district SMUD serves; if the constituents are dissatisfied with outages and rate hikes, or frightened about a nuclear plant, they have a mechanism for debating solutions and choosing leaders who will implement those decisions.

The Tennessee Valley Authority (TVA), on the other hand, is governed by a three-person board appointed by the U.S. president and approved by the Senate. Because it was isolated—and so much inexpensive TVA hydropower was available there—Oak Ridge, Tennessee, was the site of the gaseous-diffusion plant that separated the uranium-235 isotope for the first atomic bombs. TVA's involvement with the bomb is part of the reason for this utility's importance to the nuclear lobby, which has continued to try to expand TVA's nuclear program despite a nuclear debt of $14 billion, says Michelle Neal, organizer for the Tennessee Valley Energy Reform Coalition (TVERC). Originally, TVA planned to build seventeen nuclear power plants at seven sites in Tennessee, Alabama, and Mississippi. Eight of these were cancelled in 1982 and 1984 after expenditures of $5 billion,[80] as construction costs kept rising and demand for electricity levelled off.[81] The reputation of nuclear power had been further damaged by a dreadful 1975 electrical-insulation fire at TVA's Brown's Ferry plant and the 1979 near-meltdown of the Three Mile Island plant in Pennsylvania. The U.S. General Accounting Office (GAO) reported in 1995 that "TVA's nine remaining nuclear units have had a long history of operating and construction problems."[82] As of August 1996, five of the remaining TVA plants were in operation, and two were in "mothballed" status, probably never to generate electricity again. Two other non-operating plants may be converted to coal gasification or to combined-cycle natural-gas. In January 1996, the 1,250-megawatt Watts Bar II nuclear plant came on-line after twenty-three years of construction, at a cost of $7 billion.[83] The eighteen local, regional, and national groups that constitute TVERC want to end the use of nuclear power at TVA, and establish a democratic mechanism for citizen oversight of the utility. TVA, Ralph Nader has noted, is a "dream demolished" since its original purposes of developing the Tennessee River Valley and supplying cheap power were achieved.[84]

Municipal utilities, though they generate only 3 percent of total electric power, function as a moral and financial yardstick to the rest of the utility industry. As if to blur the basic distinction, private utilities now refer to themselves not as "private," but as "investor-owned."[85] The goal of citizens should be to come as close as possible to direct election of utility boards. Members of state PUCs should be chosen by direct election rather than gubernatorial or presidential appointment, to assure maximum democratic oversight of electrical monopolies. The further removed that such monopolies are from democratic control, the greater the possibility for abuses, whether they are publicly or privately owned.

All private utilities are chartered by the states and governed by self-perpetuating boards of directors. Richard Grossman advocates withdrawing the charter of utilities and other corporations that violate their charters and the public interest.[86] Another way to extract concessions from a private utility is to build a political movement to municipalize it. Such a campaign, even if unsuccessful, will chasten or humble the utility under siege and may extract concessions on rates and governance.

Where there is widespread public interest, pioneering and precedent-setting solar- and wind-power systems could be developed within municipal systems. However, in Nebraska—the only state that excludes private utilities—the cause of renewables has been difficult to advance. If the Lincoln Electric System (LES) in Lincoln, Nebraska, is typical, municipal systems are proudest of the fact that they are able to provide cheap electricity to their customers. Members of the nine-person administrative board of the Lincoln Electric System are nominated by the mayor and confirmed by the city council, and must live in the LES service area. The Lincoln City Council must approve rate changes and long-term financing. In 1985, according to Terry L. Bundy, manager of the Power Supply Division, electricity rates per kilowatt-hour were actually reduced 10 percent, and system-wide power rates have remained lower than 5 cents per kilowatt-hour. Though LES has provided energy-efficiency audits, total electricity sales continued to grow at a rate of 2.5 percent a year, well exceeding the population growth rate.[87] In 1990, Lincoln's electricity mix was 70.3 percent coal-generated, 22.4 percent nuclear, and 5 percent hydroelectric.[88] Nebraska municipal utilities should be encouraged to investigate solar and wind technologies with a view, at first, to lowering the summer air-conditioning peaks, although the largest and most accessible coal deposits in the upper Midwest are located adjacent to the country's strongest and most consistent winds.[89]

Since 1980, almost thirty towns and counties have established new publicly owned utilities, none of them with more than 16,000 meters.[90] The most active public-power campaign in a large city in recent years has been in Toledo, Ohio. Half of Toledo Edison's electricity is generated by three costly and unreliable nuclear plants—one of these is an unstable clone of Rancho Seco—which have driven up the average residential electric bill to $58.19 per month, compared to $37.54 in Cincinnati and $47.93 statewide.[91]

In 1989, Mike Ferner won a Toledo City Council seat as an independent. Ferner's most important campaign plank was a promise to establish an Electric Franchise Review Committee (EFRC). Working with Citizens Action, he helped gather 27,000 signatures in a petition drive to establish the EFRC and examine "alternatives" to Toledo Edison, including a municipal utility. The city council passed EFRC before the 1989 elections. Director Paula Ross of Citizens' Action said, "The citizens of Toledo are outraged at the high rates they pay for electricity . . . . Coming from a region which will have a surplus of wholesale power for the next twenty years, we could buy all the energy we need at reasonable prices without having to generate any ourselves."

In 1990, Citizen Action exposed the fact that Toledo Edison had founded and almost entirely funded a "grassroots" group called Citizens Against the Costly Government Takeover to fight municipalization. In May 1992, EFRC released a report calling for rate reductions of 17 to 20 percent, which would still keep Toledo Edison out of bankruptcy. Under pressure from the city council, the mayor appointed a committee to seek negotiated rate reductions, but the talks adjourned without any rate reductions by Toledo Edison. In response, the city council voted to charge a reconvened EFRC with a Phase II study, to examine the feasibility of a city-owned electric system that would buy and distribute power. In the summer of 1993, the *Toledo Blade*'s Gallup poll showed that 79 percent of Toledans polled thought that the city should put pressure on Toledo Electric to lower rates, and that by a margin of 47 to 43 percent people favored municipalization.[92] According to Ross, no one at Toledo

Edison has been able to find a document that gives the company an exclusive electric franchise, and she claims that Toledo's city charter prohibits the city from granting one in any case.[93]

Mike Ferner's idea was to create a small municipal utility that would expose Toledo Edison to direct competition, which is permitted by state law in Ohio and at least a dozen other states. "We got where we are today through a lot of hard work," Paula Ross said. "Our people go door to door and talk with the people about the issues of cost and feasibility and keep challenging Toledo Edison to come up with options and answers. Our immediate goal [was] to get the city council to approve a $401,000 appropriation for Phase II of the Electric Franchise Review Committee. The job of Phase II would be to figure out the best place in Toledo to start competing directly with Edison in service and price."[94] Then Councilman Mike Ferner ran for mayor, and municipal power became an important plank in his campaign.

"A municipal system," argued Ferner, "would encourage independent power producers, cogeneration, and demand-side management programs to help keep rates down. These actions will stimulate economic development, protect the environment, and save our citizens money."[95] But Ferner lost the election by only 672 votes out of 92,470 votes cast, and a conservative majority took over the city council. As Ferner's supporters struggled to regroup, Paula Ross, who had served as Ferner's campaign manager, conceded that municipal power was a stalled issue in Toledo for the time being.[96]

In the final analysis, the public ownership and local control of utilities, per se, is hardly a guarantee against the plunder of the public legacy or the squandering of publicly owned resources. Only when combined with farsighted vision, vigorous stewardship, and democratic accountability can public power live up to its full promise of providing abundant and affordable (and eventually solar-generated) energy for all the people it serves. The further that public power entities stray from democratic control, the greater the temptation for abuses.

CHAPTER 5

Green Capitalism and Wall Street Environmentalism

The Yale group was suspicious at first of us stuffy
Wall Street lawyers, and we were, if not suspicious,
at least cautious about these young students with
stars in their eyes. But we made a go of working
together. . . .[1]

—STEPHEN P. DUGGAN, cofounder,
Natural Resources Defense Council

F
OR THE PAST TWENTY YEARS, mainstream
environmental leaders have argued that there is
no unresolvable contradiction between energy
conservation and the fundamental goal of the utilities: producing electricity for a profit. And San Francisco—the city that is headquarters for
Pacific Gas & Electric, the Sierra Club, the Energy Foundation, and
branches of the Natural Resources Defense Council (NRDC) and the
Environmental Defense Fund (EDF)[2]—has served as a kind of crucible
as these and other industry and environmental groups have sometimes
fought, sometimes found common ground—all the while developing
new ways of looking at energy.

"To be fair," argued Zach Willey, a utility analyst for the Environmental Defense Fund in the mid-1970s, "[conservation] ought to be
treated exactly like an investment in a power plant. If PG&E makes the
investment, they should be able to put it into the rate base and earn a
profit on it, just like any other investment they make. Otherwise they'll
have no incentive to invest in it. Even if conservation is a better deal than

a coal plant, it won't look the same to them, because it won't pay off the same way."[3] According to the EDF formulation, the utilities wouldn't have to do any of the energy conservation work themselves; they would simply mobilize the capital to pay for insulation, solar water heaters, or photovoltaic devices. The EDF energy theorists envisaged a situation in which the utility would function like a bank: "Ban them from the [energy conservation or renewable energy] business, and let competition and innovation flourish among the small entrepreneurs; but when the homeowner got the bill [for purchase or installation of energy-efficient, home-scale technologies] . . . send it on to PG&E. The utility would pay, and then it would put the amount into rate base and . . . earn a profit on it just as though the investment had been made in a power plant."[4] In that way "utilities could finance enormous amounts of conservation without interfering with the small entrepreneurs, and it would be to their own economic advantage."[5] What made the EDF arguments so disarming was the way in which they were couched in the free-market language that political conservatives adore.

The idea was "to make conservation as attractive as dynamos" to utilities who, with their access to Wall Street capital markets, would transform America's electrical system. Revitalized state public utilities commissions would be the political carrot-and-stick with which to entice and browbeat the utilities onto the path of energy righteousness. Customers would be satisfied so long as their lights stayed on and their utility bills didn't rise too quickly. For utilities, the new energy-efficiency strategy would hopefully shore up sales and stock prices in an era of stagnant growth. What environmental utility theorists were advocating was an industrial development strategy in which the utilities would steer investment capital away from wasteful purchases of fuel and new generating capacity, and toward the development and purchase of more energy-efficient electric motors, windows, insulation—and even shade-tree plantings to reduce peak electricity demand in the summer.

Furthermore, in what seemed to some like an implausible conflict of interest, utilities would serve as bankers and venture capitalists for the technologies that would replace purchases of their electricity and natural gas, without reducing utility profits. Rather than building new power plants to increase electricity *supply*, new-style utility management would focus on managing *demand* to slow down the need for new power plants. This strategy has come to be known as "demand-side management," or DSM in modern utility jargon. Some analysts considered modern DSM to be completely new; others viewed DSM as a contemporary permutation of century-old load-management practices, which occasionally required conservation but usually demanded expansion.

"What exactly is DSM?" asks an article in the *EPRI Journal,* published by the industry-funded Electric Power Research Institute. Promotion of "energy efficiency" is just one aspect of the DSM picture. "In its broadest definition, it incorporates all kinds of actions utilities take to modify their customers' demand for electricity. These actions can include . . . programs that aim to reduce electricity use, to redistribute electricity demand . . . more evenly throughout . . . the day, and even programs that encourage strategic load growth. Utility efforts to influence customer demand date back to . . . Thomas Edison's Pearl Street facility . . . in the 1890s. . . .When nighttime lighting was the only load, Edison hired people to promote electric motors and other daytime uses of electricity."[6] Perhaps demand-side management was just a fancy phrase for "load management . . . decreasing on-peak sales but also increasing off-peak electricity use."[7]

A few examples will give a sense of how utility-sponsored DSM is supposed to work. In San Francisco, the Bank of America earned a $235,000 DSM rebate from Pacific Gas & Electric on a $1.1 million energy retrofit of its headquarters building, which saved the bank $400,000 a year in energy costs. It took the Bank of America only twenty-six months—slightly over two years—to earn back the $865,000 it had paid out of pocket for the retrofit, an annual rate-of-return of 46 percent per year on the capital invested. In another case cited by *Business Week,* Wisconsin Electric Power Company paid for half of the cost of a new $1.5 million system for melting steel at the Charter Manufacturing Company in Milwaukee. Over ten years, the change will save an average of $300,00 per year in electricity expenses for the steelmaker, and allow Wisconsin Electric to avoid building $2 million in new electric capacity—clearly a win-win situation for both parties. In New York City, the National Audubon Society incorporated a myriad of commonsense innovations (double-paned windows, new wall and roof insulation, ultra-efficient office equipment, and specially configured fluorescent lights) as part of its $14 million remodel of the century-old Schermerhorn building. Building manager Ken Hamilton reports that annual energy costs are about $86,000 per year, or 39 percent of "normal," under the estimated $140,000 per year for typical buildings of this size and type. The new fluorescent lights were partially funded by a modest utility DSM rebate. The greatest cost savings, says Hamilton, are attributed to a new gas-fired heater-chiller, which replaced the use of electricity for summer air-conditioning. According to New York architect Randall L. Croxton, who supervised a similar renovation of the headquarters of the Natural Resources Defense Council, "It's the interaction of simple decisions that leads to dramatic change."[8]

The Rise and Fall of Conservation

UNDER GOVERNOR Jerry Brown, California became the first state to try out the new theories of energy regulation on a grand scale. When he took office in 1975, Brown appointed energy reformers to the California Public Utilities Commission. The most notable appointee was John Bryson, a cofounder of the Natural Resources Defense Council, a major environmental law firm and lobbying group.

The Natural Resources Defense Council had been founded in 1970 as a legal defense fund for the environment by Stephen P. Duggan and Whitney North Seymour Jr., both Wall Street lawyers who had fought the Storm King power plant and reservoir in the Hudson River valley. At the urging of the Ford Foundation, the founders incorporated a group of recent Yale Law School graduates—Gus Speth, John Bryson, Richard Ayres, and Ed Strohbehn—who had formed a similar organization. The Ford Foundation agreed to provide ample support to the new group on the condition that it accept a conservative board of trustees, with Duggan himself as the first chairman, and Laurence Rockefeller, along with other wealthy conservatives, as its members. The Ford Foundation also required that all NRDC litigation be cleared with a board of five "gurus"—all past presidents of the American Bar Association.[9] Chairman Duggan was a partner in the New York law firm of Simpson, Thatcher & Bartlett, at the center of the world of high finance and industry.

Over the years, his firm's clients have included Lehman Brothers, Kuhn Loeb, Manufacturers Hanover Trust, American Electric Power, Pennsylvania Power & Light, Burmah Oil, Chrysler, Ford, General Motors, and General Electric. Historically, according to the *American Lawyer Guide to Leading Law Firms*, "utilities comprise a significant portion of the firm's corporate work."

In an informal sketch of NRDC's first twenty years, Duggan wrote that he and Seymour founded NRDC in order to create a stable professional organization of lawyers and scientists, which could overcome the disadvantages of a "disorganized . . . finger-in-the-dike approach to protecting the environment." In some areas of litigation, NRDC quickly became the dominant environmental force. Richard Ayres, part of the original Yale Law School group, in 1991 recalled that he had helped file thirty-five of the first forty lawsuits under the newly passed Clean Air Act of 1970, just one of an impressive array of issues in which NRDC has played a major role: clean water, beach pollution, western water, lead in gasoline and in drinking water, pesticides on apples, nuclear arms testing, and dozens of others. By 1993, NRDC's annual budget had grown to over $17 million, with a staff of more than 150, and a membership of more than 170,000 people, over a thousand of whom were contributing more than $1,000 annually. "Over time," wrote Robert Gottlieb in *Forcing the Spring*, "NRDC became the environmental organization most identified with the technical expertise needed to draft legislation, issue reports, and use litigation as a tool in the policy process. . . . Applauded for its continuing emphasis on expertise while simultaneously expressing a lack of interest in the tactics and strategies of mobilization, [NRDC] had secured for itself a central place in the organizational culture of contemporary mainstream environmentalism."

Upon being appointed to the California PUC in 1975, Bryson soon made his mark. "We regard conservation," opined this PUC under Bryson's direction, "as the most important task facing utilities today," and they promised to regulate utility rates according to the "vigor, imagination, and effectiveness" of their conservation efforts.[10] Even Pacific Gas & Electric, which had planned to spend more than $13 billion for power plant construction in the coming decade without spending a nickel on conservation, finally promised to make available zero-interest loans for home weatherization.[11] And the Abalone Alliance and other antinuclear groups had organized major occupations of the Diablo Canyon nuclear plant site. To the chagrin of utility nuclear engineers and construction giants such as Bechtel, nuclear engineers were being laid off and attic-insulation specialists were being hired.

Political activist Tom Hayden's Santa Monica–based Campaign for Economic Democracy initially tried to bar utilities from involvement in small-scale, solar energy sources, on the theory that the giants would overwhelm the small entrepreneur, drive out competition, and then strangle the new technologies in order to maintain fossil-fuel and nuclear predominance.

But David Roe of the Environmental Defense Fund argued that there was no inherent contradiction between utility profits and the promotion of conservation and renewable energy. Roe contended that it was imperative to cut the utilities into the demand-side deal by offering them a fixed profit share on their expenditures to *reduce* demand, expenses which ran the gamut from rebates on energy-efficient refrigerators to conservation commercials aired on TV. It may have seemed "too good to be true," wrote Roe, but under utility-controlled demand-side management, all that the utilities would have to relinquish would be their energy-promotion advertising, which had historically overshadowed research and conservation efforts.[12]

By the time Jerry Brown left the governor's office in 1983, electricity use had stopped expanding and the price of natural gas and coal used in electricity generation had begun to fall. The state was awash in excess electricity capacity; the power of environmentalists and solar advocates was waning. Voters elected conservative Republican George Deukmejian to succeed Brown, and Deukmejian set out to reshape the California Public Utilities Commission in his own image. In a short time, California utilities mysteriously lost interest in their highly successful conservation programs and either cashiered their energy-saving experts or transferred them into energy promotion. (Most chose to quit rather than switch.)[13] *The Decline of Conservation at California Utilities*, a special report by the Natural Resources Defense Council, details the results of the utilities' silent sabotage during the 1980s.[14] By 1988, the utilities had cut their conservation spending by more than half. Southern California

Edison's conservation spending collapsed dramatically, from $138 million in 1985 to $36 million by 1989.[15] As state and federal tax rebates and utilities' loan programs for solar hot water heaters lapsed, the utilities simultaneously beefed up their incentives and advertising to encourage customers to buy *more* electricity. PG&E's Architectural and Security Light program, wrote NRDC's Chris Calwell, "pays commercial customers $200 per kilowatt of 'exterior night lighting' that operates at least 2,000 hours a year" and $300 per kilowatt to installers of refrigerator cases." The utility also agreed to subsidize up to $100,000 of the initial cost of such projects.[16] Though energy conservation reduced pollution and saved money for the public, PG&E grumbled to NRDC that conservation programs constituted "unselling our product," and tried to quietly scuttle them.[17]

NRDC and the "Collaborative Process"

To cope with the unrestrained pro-business political environment of the late 1980s, the Natural Resources Defense Council proposed to use the carrot of profits, rather than the stick of government regulations, to "reward" utilities for investments in customer energy efficiency. Under certain conditions, utilities could even "decouple" utility revenue from electricity sales. Ralph Cavanagh, senior staff attorney for NRDC, initiated the California Collaborative Process—part negotiation, part encounter group—to carry out these changes. With the implied threat of NRDC-initiated regulatory and legal actions in the background, the California PUC instituted a "statewide collaborative process" in 1990 which involved all the major energy "stakeholders": utilities, organizations of both industrial and citizen consumers, and NRDC. But Cavanagh had no intention of using the Collaborative Process to threaten utilities and the building industry with new building-code regulations and other "command and control" measures that they abhorred. Instead, Cavanagh urged the electricity stakeholders "to focus on the idea of utility shareholder incentive [a.k.a. profits] for increased investment in energy efficiency."[18] As a result of this "collaborative process" the California PUC, urged on by the utilities and NRDC, passed a regulation to sweeten the conservation pot by allowing utilities to earn a profit on the energy demand they had *replaced* by investing in conservation.

The California Public Utilities Commission agreed to stand aside while the leading utility and consumer stakeholders (everybody except the utility unions, builders, and residents of neighborhoods where power plants were built) came to a consensus on how to grant "shareholder incentives" for utilities to conserve energy and promote efficiency. For the first time in history, utility monopolies would be guaranteed a profit both for selling and for *not* selling electricity and gas.

As governments in the 1980s cut their expenditures on grassroots energy programs to the bone, the Energy Foundation stepped into the funding gap. Founded in 1991 by the Pew Charitable Trusts and the Rockefeller and MacArthur Foundations, the Energy Foundation aimed "to assist in the nation's transition to a sustainable energy future by promoting energy efficiency and renewable energy." In its first four years of operation, it disbursed $30.9 million and made NRDC its premier grantee.[19] Hal Harvey, an engineer who hails from Amory and Hunter Lovins's Rocky Mountain Institute, was chosen to head the Energy Foundation. Substantial Energy Foundation grants have gone to organizations such as the Union of Concerned Scientists to track the price of and performance advances in wind, solar, and biomass technologies, and to publicize those findings widely. But the main purpose of the Foundation's work, wrote Harvey in 1991, was to *"influence energy decision-makers at the point of decision"* (Harvey's emphasis). However, one "social question" that does concern the Energy Foundation is, How can the energy awareness of decision-makers be raised during times of falling oil prices?

The Collaborative Process was sold by the Energy Foundation and NRDC as a model that would promote efficiency and slow the increase

---

"Collaborative Principles of Practice" According to NRDC

Participants agreed at the outset:
- *to make decision by consensus*
- *to see the process through*
- *to adhere to a set meeting schedule*
- *to speak with one voice*
- *not to use the process as a lever in other proceedings to undercut other participants' interests.*

RALPH CAVANAGH, 1992

---

in electricity demand, not only for the entire U.S., but for the world at large. Suddenly, NRDC became an ardent public defender of PG&E, whether the issue was high electric rates or PG&E's environmental credentials.[20] The beauty of this "new" demand-side management strategy, from the utilities' perspective, is that it would take the environmental heat off the utilities without greatly reducing—let alone stopping—the increases in power use and fossil-fuel consumption. In areas where there is an excess of unused generating capacity during certain seasons and times of day, DSM can be used to promote electricity sales without needing to add expensive generating capacity. More fuel is burned, more greenhouse gases and pollution are generated, and more money flows to utility coffers without the necessity of major new capital investments. When the rising demand can no longer be "smoothed" into the system, then it is time to build new power plants. When demand is steady or falling, the utilities can use their DSM premiums to buy increased advertising, thereby attempting to pump up demand, especially among big customers. By grabbing hold of the steering wheel of demand-side management—with NRDC and Energy Foundation support—the utilities hoped to make sure that demand-side *management* never became demand-side *reduction*.

The alliance between Pacific Gas & Electric and the Natural Resources Defense Council was announced to the public in a full-page PG&E newspaper ad headlined "This Award Belongs to All of Us," celebrating the fact that President Bush had bestowed on PG&E the President's 1991 Environment and Conservation Challenge Award. NRDC's Ralph Cavanagh was quoted in a panegyric to the nation's largest private utility: "PG&E programs benefit every sector of the economy. The farmer, the factory owner, or the family of four can save money and improve the environment through PG&E's various energy-efficiency efforts." The ad also praised General Motors, Shell, and Chevron for helping "develop the first generation of cleaner-burning natural gas vehicles and the stations to refuel them," while completely ignoring the fact that those companies had done much of their "environmentalist" research for the environment with taxpayers' money, and that PG&E and Southern California Edison were buying and building fossil-fuel generation plants around the globe.[21] Other contradictions were likewise ignored. Natural gas, after all, is a fossil fuel that will be exhausted eventually; what's more, methane (the major component of natural gas) is a more potent greenhouse gas than carbon dioxide. The environmental group Greenpeace has claimed that a switch from coal to natural gas for electricity generation consitutes a "dangerous detour on the road to a clean energy future."[22]

It might have seemed disingenuous for Cavanagh and NRDC, in such a public manner, to praise the auto and oil companies (which had adamantly and successfully opposed more stringent mileage standards) on the occasion of an award from President Bush, whose primary efforts on the energy issue had been military: the liberation of Kuwait from Iraq, and the invasion of Somalia, where Conoco, Chevron, Amoco, and Phillips had exclusive oil exploration contracts.[23] Domestically, Bush opposed any initiatives that might limit the freedom of the oil industry or the automobile manufacturers.[24]

An Unholy Alliance

MOST OF THE MONEY for NRDC's work with electrical utilities comes from the Energy Foundation, which has made NRDC its largest grants recipient. Tracing back the money trail raises the next question: Who are the benefactors of the Energy Foundation? The answer is, three giant foundations: the John D. and Catherine T. MacArthur Foundation, the Rockefeller Foundation, and the Pew Family Trusts, which together fund the Energy Foundation to the tune of $10 million per year. [25]

The endowment of the Pew Charitable Trusts is now worth about $34 billion, reported journalists Alexander Cockburn and Ken Silverstein in their book *Washington Babylon*, and together the seven Pew Trusts invest about $20 million per year in the environmental movement. Even more important, Joshua Reichert, the chief of the Pew Trusts' environmental sector, helps coordinate the Environmental Grantmakers Association, which altogether donates about $350 million per year. "Pew never goes it alone," write Cockburn and Silverstein, "which means no radical opposition to its environmental policies can get any money.[26] Just one of the seven Pew trust funds, the Pew Memorial Trust, earned

$205 million in "investment income" in 1993 from holdings in Weyer-hauser, forest products and paper ($16 million); Phelps Dodge, mining ($3.7 million); International Paper ($4.6 million); and Atlantic Richfield, which is eager to open the Alaskan arctic to oil drilling ($6.1 million). Cockburn and Silverstein claim that the income received by the Pew Memorial Trust from these "rapacious" companies is twice as large as the trust's disbursements. If either the Natural Resources Defense Council or the Energy Foundation has taken any positions that would jeopardize the value of their ultimate benefactors' endowments, they have kept these objections well hidden.

Evidently, the working assumption of NRDC has been that private utilities and the investment banks behind them are powerful, immortal institutions that must be placated rather than confronted. The strategic alliance between the utilities, NRDC, and the Energy Foundation was supposed to make the utilities as bankable in the new sphere of conserva-tion and efficiency as they had been in the palmy, pre–oil-shock years of 7-percent annual increases in electricity consumption. The alliance pre-sumed that private utilities would remain in control of the production of electricity, and that the relationships between investment banks and utili-ties—dating back a century to the days of J. P. Morgan and Samuel Insull—would remain intact. On the world stage, the Collaborative Process was conceived of as an important step in demonstrating that cap-italism could be both green and profitable. The secret to success would be in "managing" the ecosphere by "producing more with less," in the words of *Our Common Future*, the influential report of the World Commission on Environment and Development (WCED), chaired by Norwegian prime minister Gro Harlem Brundtland.[27]

Yet, historically, the energy reform movement's biggest successes have come from public political pressure against the utilities and other energy corporations, in order to limit their activities by statute and regu-lation. Nationwide, tighter safety and air pollution standards as well as neighborhood resistance have driven up the cost of new power plants and extended construction lead-times by years, delaying or halting the building of unnecessary plants.[28] In California, increased public interest since the late 1970s in conservation and efficiency has "avoided the need to build 8,000 megawatts of new capacity."[29] A California Energy Commission study indicated that most of the slowdown in the growth of energy consumption could be credited to the steep rise in fuel costs and electricity rates, which had forced consumers and industries to conserve electricity, and to the political movement to reform building codes and electrical-appliance and motor-efficiency standards.[30]

The utility-sponsored Electric Power Research Institute concedes that cost-effective programs exist that could actually *reduce* electricity usage. But to do so, concedes EPRI, "California would have to 'tighten' building standards, adopt new standards for buildings that are remodeled, establish state-of-the-art standards for commercial and residential appliances (particularly those not currently covered), require more aggressive utility DSM programs, and expand research and development programs for efficiency technologies and policies."[31] Professor Arthur Rosenfeld of the Lawrence Berkeley Laboratory (and now with the U.S. Department of Energy) believes that with the maximum feasible energy efficiency programs in place, it should be possible to cut total electricity consumption in California by 40 percent in twenty years and still maintain economic growth and living standards.[32] But such a reduction in brute electricity consumption would require a major change in building practices and would, by definition, cut heavily into the sales of electric power, the main product of electric utilities. Despite the enthusiastic talk by NRDC's Ralph Cavanagh about "transforming a mega-utility," and "building attractive models for sustainable development," it soon became apparent that the Collaborative Process agreement would at best slow the rise in the consumption of fossil-fueled electricity and "delay

Paying for Collaboration

Philanthropy and its purposes remain the same as when John D. [Rockefeller] dispensed millions to winch the family name out of the mud. Today the environmental movement receives about $40 million a year from three oil companies which operate through front groups politely described as private foundations. The top two are the Sun Oil Company (Sunoco) and Oryx Energy. (The latter has vast holdings of natural gas in Arkansas, and throughout the oil patch.) The Pew family once entirely controlled the two companies and still has large holdings in both of them; Oryx shareholders recently sued the Pew operation for insider trading.

ALEXANDER COCKBURN AND KEN SILVERSTEIN, 1996

the time when a new plant is needed."[33] Without an alarming rate shock or a crisis in the natural-gas supply, both the building industry and the utilities can be counted on to fight against such "command and control" programs, no matter how much energy these might save or how many jobs they might create. Serious energy-saving programs will come about only if the public understands what is at stake and builds a movement to implement and enforce them.

The spectacular career of John E. Bryson, cofounder of the Natural Resources Defense Council, and now the president and CEO of Edison International and its unregulated subsidiary, Mission Energy, exemplifies the *entente cordiale* between the utilities and the wealthiest environmental groups. Bryson, fresh out of Yale Law School, came back to California to open NRDC's West Coast operation. He was tapped by Governor Jerry Brown to head the California Public Utilities Commission at the age of 34, and then jumped to the private sector as the chairman and president of SCEcorp, now known as Edison International. Perhaps some mainstream environmentalists believe that Bryson's ascent demonstrates the movement's success in infiltrating an environmental point of view into big business, while keeping profits flowing. For some of the more ambitious of NRDC's lawyers, John Bryson's $1.4 million annual salary may also represent comforting proof that personal poverty is not a prerequisite for being an environmental do-gooder.[34]

For the utilities, the California Collaborative Process was an immediate public-relations success. One utility executive observed, "If someone had told me a couple of years ago we would have a group like NRDC supporting shareholder incentives, I wouldn't have believed it. They are willing to listen—to understand." The attitude of NRDC's Cavanagh, who guided the meetings, received high marks from one PG&E official, who praised Cavanagh for "portraying the decline in spending for conservation by the utilities as a 'crime without villains' [which] enabled the utilities to save face."[35]

While the Collaborative Process in California legitimized the concept of energy efficiency as a tool and goal of energy management, it turned the control and profits of that tool over to monopoly utilities that have an interest in expanding electricity consumption, while meanwhile stilling public criticism. And barred from expansion locally, California utilities are now vigorously seeking markets elsewhere—in areas free from PUC regulation. Despite boasts by PG&E and Southern California Edison that they are stabilizing emissions of carbon dioxide in California-produced electricity, both companies have created unregulated subsidiaries to build and operate coal- and gas-fueled power plants throughout the world. Their goal is to grab market share in Latin America and Asia, where electricity consumption is expected to double every six to fifteen years.

U.S. Generating Company, a PG&E subsidiary owned 50–50 with construction giant Bechtel, owns or is building at least seven coal-fired and four gas-fired power plants in the United States. In 1994, U.S. Generating became the nation's second-largest non-utility electricity generator by acquiring J. Makowski Company, Inc., a Boston-based firm that specializes in producing both natural gas and electricity, and which maintains offices in Canada, Germany, India, and Scotland. Investment analysts surmised that the new acquisition was aimed at bolstering PG&E's presence in the rapidly expanding Asian market.[36]

Mission Energy, an Edison International subsidiary, bought a 32.5 percent interest in Indonesia's $1.8 billion Paiton coal-fired 1,230-megawatt power plant, in partnership with Mitsui, General Electric, and P. T. Batu Haitam Perkasa of Indonesia, a coal supplier.[37] Mission had also invested over $300 million in Carbon II, a 1,400-megawatt coal-burning plant in Mexico. Energy economist and consultant Charles Komanoff called the Carbon II project "disgraceful," and eventually pressure from environmentalists and national-park advocates forced Edison International to pull out of the project (see chapter 6 for a more detailed account of the Mission Energy/Carbon II story).[38]

Long forgotten are the days when Edison International CEO and president John Bryson, as head of the California Public Utilities Commission, published articles with titles such as "California's Best Energy Supply Investment: Interest-Free Loans for Conservation."[39]

Can Private Utilities Voluntarily Reduce Electricity Sales?

PRIVATE UTILITIES have never intended for demand-side management to result in demand-side *reduction* of energy sales. To claim otherwise is to promote a grand illusion. The reason is simple and goes to the heart of private monopoly enterprise: utilities will not support a strategy that "unsells" their main product. In a private corporation, the primary legal and financial duty of management is to serve its stockholders. To raise the stock price, management must increase sales. An increase in sales will raise demand for more electric power, which can only be met by new capital investments, which in turn triggers new earnings growth, higher dividends, more buyers for the stock, and higher stock prices. Even if rewarded with a premium for encouraging demand-side reduction, a capitalist organization preoccupied with its own profits will not and cannot embrace a goal of *reduced* sales.

According to California utility projections, if utility-sponsored DSM operates as planned, electricity use will increase at 1.4 percent per year instead of the 1.9-percent annual increase that would occur with no DSM programs in place. Most of this increase will be due to increased commercial usage of air-conditioning and office equipment. So, instead of doubling electricity production every thirty-seven years, these utility-sponsored DSM programs could be expected to slow the doubling time to every fifty years.[40]

The profit-sharing assumptions of demand-side management as practiced by utilities mask some other tricky assumptions. Under the DSM agreement reached with the California PUC, the commission calculated the dollar amount that the utilities would have otherwise had to spend to meet anticipated demand if no utility-funded efficiency investments had taken place. In 1991, for example, the commission estimated that PG&E would have had to spend $347 million to meet the demand that was avoided by investing $147 million in conservation. PG&E's profit, or "utility shareholder incentive," for participating in the conservation programs totalled $30 million, or 15 percent of the $200 million difference between $347 million and $147 million. In the new parlance, that $200 million is called the "net life-cycle benefit."[41] The giant utility claimed it would invest $2 billion in conservation programs by the year 2000, which would reduce by three fourths the anticipated increase in electricity demand.[42]

Critics of these procedures frequently point out that estimates of "anticipated demand," as well as of the true effect of utility conservation expenditures in slowing that demand, are nothing more than educated guesses. Rather than crediting utility DSM programs for reduced electricity consumption in a stagnant California economy, it would make more sense to assume that many households and businesses were doing everything possible to cut their utility bills as California's economy faltered in the early 1990s.

The utility's incentive to exaggerate "avoided demand" projections is obvious: the larger the avoided demand, the larger the utility's profit; moreover, the less spent to bring about a given quantum of avoided demand, the larger the utility's profit. Dale Sartor, president of a local chapter of the Association of Energy Engineers, has noted that to achieve the quickest possible results, Pacific Gas & Electric was putting a huge emphasis on promoting fluorescent light bulbs while ignoring medium- and long-term opportunities for reduction in energy use.[43]

Ultimately, said Eugene Coyle, a utility economist with TURN (Turn Toward Utility Rate Normalization), a San Francisco-based consumer group, "utility activities which support growth in earnings per share by promoting sales growth" could be expected to "swamp" the decrease in electricity consumption caused by DSM programs. Rather than paying utilities to unsell electricity, Coyle argued that sales growth should be damped by prohibiting promotion. TURN's argument was a formula for utility downsizing. According to the TURN scenario, as allowable profit rates were curbed, utility stock prices would fall, and large electricity users would be forced to invest in efficiency on their own account, rather than giving the utilities a DSM cut off the top of those investments. While utilities downsized as a proportion of total economic activity, they would cause a stimulative effect on the California economy akin to a tax cut—if utility rates were cut.[44] "The attempt to offset the regulatory and financial biases toward sales increases with a DSM incentive for sales decreases, can be characterized as driving with one foot on the gas and the other on the brake.... It wastes gas, [and] requires frequent brake jobs ... [to get] the driver to the same place at the same time as another driver using less gas.... there are clearly better and more efficient ways to drive." In other words, if the true intent of the PUC is to reduce demand for electricity, the commission should eliminate utility profits for DSM expenditures (take its foot off the brake) and reduce advertising and allowable profit margins (tread more lightly on the accelerator).[45]

The Natural Resources Defense Council disagreed vigorously with this prescription. If there was an intrinsic conflict of interest in paying the utilities to unsell their most important product, NRDC didn't acknowledge it, and once again the organization argued for subsidies of utility demand-side activity because "scarce management and staff resources flow to profit centers."[46]

With its electric rates set by high-cost Diablo Canyon nuclear power, with California in a deep recession, and with increasing competition from industrial cogeneration, Pacific Gas & Electric has attempted to maintain market share by cutting the rates of its large industrial customers by as much as 10 percent, by stepping up its energy marketing to industry, and by handing out millions of dollars in industrial rebates to prevent large power users from dropping off the utility's grid.[47] By agreeing that PG&E could supply large industrial customers with low-cost power, the California PUC was in effect providing a rate subsidy for industries that use electricity wastefully. Meanwhile, residential customers and small businesses, which lack the market leverage to cut special deals with the utility, get stuck paying for the high fixed costs of wires and substations, rather than sharing those costs with several large users in their region.

Model Homes for Those Who Can Already Afford Them

PACIFIC GAS & ELECTRIC's ActSquare was one of the most tantalizing projects to support the principles of the Collaborative Process.[48] The idea for the project came from Amory Lovins, recalled Merwin Brown, ActSquare's first director, and its board of directors included Lovins; Arthur Rosenfeld, a professor of physics at the University of California at Berkeley and head of an energy-efficient building program at Lawrence Berkeley Laboratory; and Ralph Cavanagh of NRDC. For Brown, the involvement of Lovins and Rosenfeld was a matter of "credibility. . . . We knew that if we carried out the project and didn't get the results they said we could get, they would say we did it wrong, and we didn't want them sniping from the outside. . . . So far," observed Brown in 1992, "we have found that Lovins and Rosenfeld are right on a technical basis; we *can* [Brown's emphasis] save 75 percent of electricity usage. The main barriers are technical and legal." [49]

Conceived in 1989, ActSquare was launched in 1990 with two employees, and in its heyday had seven people associated with it. Ultimately, four houses and three commercial buildings were built under ActSquare's auspices, along with one agricultural pumping station, and the buildings were scheduled to be carefully monitored until PG&E terminated the project in the fall of 1996. According to Lance Elberling, who succeeded Merwin Brown as ActSquare project manager, actual documented energy savings in the houses were just over 51 percent compared with conventionally built houses. [50]

But there is little in the ActSquare program that would help renters and the poor. According to the Washington Center for Metropolitan Lifestyle and Energy Surveys, poor people spend 11.1 percent of their income on electricity and natural gas alone, while wealthy people spend 1.9 percent of their income on those same energy needs.[51] Also troubling is the disproportionate impact of home energy-efficiency programs upon the well-to-do. According to a 1990 Federal Energy Administration study, only 1 percent of households with incomes under $5,000 had benefitted from residential utility conservation programs, compared to 8 percent of households with incomes of $50,000 or more.[52] Poor people are usually renters who live in inefficient old buildings that landlords have no incentive to upgrade, and it is the tenants who pay the utility bills. In general, the work of ActSquare continues this practice of ignoring the energy problems of those with the greatest needs.

PG&E's marketing department, in a project not connected with ActSquare, helped design and market a super-efficient 2,900-square-foot suburban house in conjunction with a building-industry association[53] and the California Energy Commission— a free-standing structure that would cost a minimum of $400,000 to build even if it hadn't been designed to incorporate special energy-efficiency measures. The "PG&E Energy-Wise Showcase Home," designed by Benjamin Tuxhorn Homes in exurban Sonoma County and lauded as a model of energy efficiency for public inspection, prominently advertised its dependence on "economical natural gas" and high-tech windows and lights.[54] The

"Energy-Efficient Environment" (E3) house, also known as PG&E's "smart house," would cut energy bills from $800 per year to $250, according to a newspaper account. However, these savings of a dollar or two per day are small change for a family with the kind of income needed to make house payments of $3,000 per month. ActSquare associate Grant Brohard remarked that the controls for the "smart house" are made by a small firm and are so expensive that the only people who have bought them are Hollywood types such as Michael Jackson.[55]

Neither solar water heating nor photovoltaics are part of the ActSquare vision: ActSquare explicitly excludes buildings with on-site electricity generation from its studies, even at its model house in Davis, a city known for its trendsetting Village Homes solar community.[56] One can only surmise that PG&E is not eager to encourage technologies that would undermine its role as the monopoly supplier of those services that electricity and natural gas provide. Apparently, no one representing PG&E or its ActSquare project was present at the First International Solar Electric Buildings Conference in Boston in March, 1996. By contrast, Donald Osborn, Senior Project Manager of the solar program of the neighboring Sacramento Municipal Utility District, made a well-attended presentation, and Ed Smeloff, a very active member of SMUD's Board of Directors, attended the conference.[57]

A "Bridging Strategy" to Nowhere

Pacific Gas & Electric's $8 million Pacific Energy Center in San Francisco, which was opened in the early 1990s, illustrated the realities of the very limited PG&E energy reform agenda. The downstairs walk-in exhibit, open to the public, featured energy-saving devices with ingenious hands-on displays of low-flow shower heads; low-watt, screw-in fluorescent light fixtures; and a device that indicated seasonal differences in the sun's angle and intensity. The implied message was that if you use too much energy it's your own fault alone, because there are bulbs and tools and gadgets you can buy to reduce energy use, including fiberglass water-heater blankets and energy-efficient refrigerators (sold at that time with PG&E rebates of up to $200).

Already-viable renewable energy technologies including photovoltaics, solar water heating, and wind turbines were not even mentioned at the Pacific Energy Center. When a group of architects visiting the Center asked about this solar omission, a PG&E technical coordinator told them that solar hot water heating was "production-side" and thus "our side of the meter," presumably meaning that energy-generating technologies had no place in a demonstration of "demand-side" options.[58] Likewise, a PG&E program in conjunction with the Bank of America and Fannie Mae (a federal mortgage guarantee agency), to make loans of $2,500 to $10,000 to low-income homeowners for energy efficiency, specifically excludes solar water heating.[59]

Pacific Gas & Electric publicly espouses a "bridging strategy" to encourage customer efficiency for the next twenty years while clean alternatives are developed.[60] Yet, with gas and electricity prices holding steady, the talk of a "bridging strategy" to renewable energy has seemed more like diversionary rhetoric than policy. Despite continued declines in the cost of photovoltaics, that "transition" is repeatedly deferred to the ambiguous future.[61] With a predicted annual growth rate of 1.4 percent in the usage of fossil-fuel–generated electricity, by 2040 there would be twice as much fossil-generated electricity to replace.[62] California governor Deukmejian's termination of solar tax credits and the crash in natural gas prices has drastically curtailed new wind and solar installations in the state.[63] Meanwhile, four fifths of California's natural gas comes from out of state—a third of it from Texas. (And Texas utilities are now importing coal from out of state to burn in their power plants because of their concern about the price and "deliverability" of gas over the long term.[64]) How PG&E's "conservation revolution" would lead to a solar

Energy Colony?

California's long-term energy picture is cloudy. The state's internal production of oil and natural gas is declining, making us more dependent on Alaskan and foreign oil supplies. After the huge oil spill off the Alaska coast in 1988 the reliability of the flow of oil from Alaska was called into question.... What has saved the day for California is that other states in the West and our Canadian neighbors have large natural gas and electricity surpluses and are happy to sell those surpluses to the state.

California Almanac, 1991

transition—in the absence of an independent solar movement to pose the tough questions—has been left to the imagination.

Progressive Voices and Public Accountability

THERE ARE COMPELLING reasons why the energy utilities, in contrast to the automobile/petroleum complex, can potentially be made more responsive to democratic control. Because utility monopolies are legally chartered by state legislatures, the principle of public supervision and accountability has already been established with the establishment of public utilities commissions, although in practice that supervision may be lax.[65] No comparable independent public body exists to exercise legal oversight over the policies of automobile or oil companies. In addition, the financial health of the utilities, with their electric wires and gas pipelines, is tied to the economic health of particular regions. Once the infrastructure of transmission wires and pipelines is built, the main remaining task is to read the meters, keep the system maintained, collect the money, and try to keep customers on-line. For example, after the 1992 Los Angeles riots, Southern California Edison pledged $35 million over five years to help rebuild the devastated areas.[66] According to RLA (formerly Rebuild L.A.), SCE has already spent $11.5 million for a revolving small-business loan fund, a jobs skills development center in Compton, and a TV series for the Sesame Street age-group called "Puzzlework," which is designed to encourage interethnic understanding.[67]

Progressive elements within utility R&D departments have conceded the theoretical possibility that fossil fuels will "run out" at some time in the distant future and argue publicly for more investment in renewables. *Bringing Solar Electricity to Earth,* a video produced for the Electric Power Research Institute, featured Amory Lovins and Carl Weinberg, PG&E's research director at the time, along with images of photovoltaics and energy conservation in action. In the video, Weinberg makes an impassioned declaration about the dangers of relying on fossil fuels and the hope of a solar future:

> When you use fossil fuels, you are essentially depleting a reservoir, and there's nothing sustainable about that. It's eventually going to

run out, maybe 100 years . . . maybe 300. . . . So really, the only power source that the world has . . . that is sustainable in any sense is the sun.

In the same video, Dr. Martin Green of the University of New South Wales in Australia argues that "photovoltaics have the potential to supply most of the world's energy requirements. It's going to take a long while before you can get to the stage where that could occur, but I think the technology has that inherent capability."[68]

One could argue that Carl Weinberg's public focus on a fossil-fuel shortage decades or centuries in the future was a way to nudge his employer in the right direction without provoking an open battle. When PG&E began to cut 10 percent of its work force in the spring of 1993, Weinberg opted for early retirement, apparently preferring to let his successors dismantle those programs that he had helped build in photovoltaics and energy efficiency than be required to dismantle them himself. No one who has seen *Bringing Solar Electricity to Earth* or who has had personal contact with Weinberg would doubt the sincerity of his personal belief that our present means of producing and using energy must undergo a profound transformation.[69] When push comes to shove, however, PG&E's research staffers for renewables are evidently expendable, and innovative uses for "customer-sited" photovoltaics, which Weinberg helped pioneer—even those which are apparently cost-effective for the utility—will be shelved.[70]

PG&E's PR Problem

PG&E IS A MONOPOLY with big public-relations problems, as its PR director, Larry Simi, admitted in 1992 at a candid moment:

> In order to supply the electricity needs of California, we burn fossil fuels, we discharge hot water into San Francisco and Monterey bays and the Pacific Ocean, we cover open hillsides with noisy, ugly windmills, we dam wild and scenic rivers, we drill geothermal wells that release arsenic, our power lines emit electromagnetic fields— and we run a nuclear power plant. And you think your clients have image problems? . . .[71]

Without public pressure, the chances that genuine demand-side *reduction* will occur are minimal, and the potentially important contributions of utility projects such as ActSquare will be forgotten. In fact, ActSquare itself may represent a cynical attempt by PG&E's top management to bolster the utility's image and buy off the reformers, or at least observe them up close and keep them out of the anti-utility camp for as long as possible. When all is said and done, perhaps Carl Weinberg and Merwin Brown are merely technicians. The critical decisions will be made by others, elsewhere. ActSquare's Brown himself notes that "the purpose of R&D is to buy future options; if and when global warming comes, [PG&E] will be in a position to deal with it."[72] If nothing else, the success of ActSquare in proving that the consumption of gas and electricity can be cut by 70 or 80 percent makes it impossible for PG&E or any other utility to deny the possibility of such energy savings.

Ultimately, the best way to buttress the crucial work of progressive utility solar engineers and efficiency experts such as Carl Weinberg and Merwin Brown is to create a broad-based citizens' movement that can challenge utility control of solar technology. Utility economist Eugene Coyle, who visited California's leading utilities in the early 1980s, was impressed with the "competence and genuine interest in PV technology and energy research in general" and the "well-conceived" evaluations of new energy technologies. But "therein lies the rub," he continued, in a commissioned report. Utility PV research is paid for mostly by ratepayers and taxpayers, yet the utility "makes the decisions about what to test ... [and] retains the information ... [for] long-range planning."

Compared to the California Energy Commission or the California PUC, the utilities have "the only hard data" about cost and efficiency of the new technologies. Without charging misbehavior on the part of the utilities, Coyle reminds the readers of his report that "information is power" and that information developed by the utilities is "affected by their own perspective, which doesn't necessarily coincide with the public interest."[73] A new solar citizens' movement would need to demand that the technical and financial results of ActSquare and other environmentally sound energy projects be made fully available to the public in an entirely understandable form, and must agitate for independent assessments and public discussion of the actual potential for solar from the customer's rather than the utility's point of view. Whether solar activists should push for direct election of PUC commissioners, municipalization, or some other strategy depends on local circumstances, but in some way the control of programs to develop energy efficiency and solar must be reclaimed from the utilities. An active, independent solar movement would also strengthen and validate those doing innovative work inside the utilities.

The Collaborative Process, on the other hand, encourages secrecy and backroom dealing. Policy debates and negotiations that formerly took place in public forums are now resolved behind closed doors, and the results are presented to the public as a *fait accompli*. And as electric power systems become integrated on a national and international basis—as power is produced in one state or country and sold in another—the very possibility of local public accountability will be destroyed. Large quantities of power are now being produced by independent power producers, who are free to cut their own deals with large corporate users. Big utilities such as PG&E are competing in self-defense by offering special cut-rate deals to keep the big customers from building their own cogeneration plants or leaving the state, whereas residential and small business customers have no such choice. PG&E has been exploring ways of loading onto the residential ratebase the costs of the Diablo Canyon nuclear plant and the expensive electrical grid. A 1993 PG&E internal memo regarding the "1982 Residential Rate Revolt" disclosed the strategy:

> As part of an effort to examine various incentive regulation scenarios, including retail wheeling, and the possible need to provide as much of a rate break as possible to industrial electric customers for competitive reasons, a task force is trying to assess what levels of rate increase the residential class will bear before a rate revolt is triggered. . . . We already have data on the . . . rate increases at that time, which consisted of about three or four consecutive years of very large increases, especially electric. But if anyone has any old memos or articles, or a good memory, Collette [of Market Planning and Research] would like to identify all possible contributing factors, such as weather, events with Diablo Canyon at that time, general media treatment of PG&E, gas versus electric levels of

problem causation, data on the impact of the inverted-tier residential rate structure at that time, political, regulatory, or any other factors, such information would be helpful.[74]

When TURN, the ratepayers' advocate group, held a press conference to denounce the memo, PG&E's News Department quickly denied that a rate increase was in the offing and asserted that it was "working diligently to identify . . . cost savings . . . and to develop the best method for passing those cost savings . . . directly to customers."[75]

Was there more to California's Collaborative Process than met the eye? Was PG&E trying to divert public attention away from its plans to expand out-of-state once the Public Utilities Holding Company Act of 1935 was overturned, or were company officials nervous about charges of incompetence or corruption in overpaying Canadian natural-gas suppliers?[76] Were NRDC and the Collaborative Process, in effect, being used to grease the PR skids so the next PG&E rate increase would slide through the PUC more smoothly? Did the Collaborative Process in effect allow PG&E to greenwash itself at the public's expense, for example, with its "Conversations with the Earth" television spots? Was the NRDC eager to reassure Wall Street that the media frenzy the organization had stirred up with Meryl Streep and *60 Minutes* over Alar— which had cost conventional growers an estimated $140 million in apple sales, after Alar was denounced as a childhood carcinogen—was only an aberration?[77] Was Sylvia Siegel, the plain-talking retired founder of TURN, right to call the whole Collaborative Process "a crock?" "My good friends in the environmental movement don't give a damn about the cost," she said, and as far as she could tell, the agreement would allow PG&E to "run cute little educational spots on TV at the ratepayers' expense under the theme 'Smarter Energy for a Better World,' which really constitutes a form of institutional advertising."[78]

In the summer of 1992, two years after the completion of the Collaborative Process, PG&E came back to the California Public Utilities Commission with a request to raise its return on equity from 12.65 percent to 13 percent, an increase that would oblige ratepayers to ante up another $63 million per year to PG&E stockholders.[79]

Rather than attacking the utilities head-on during a period of capitalist ascendancy in material and intellectual life, well-funded environmental and legal or lobbying organizations such as NRDC had decided to charm the utilities with offers of cooperation. Wasn't it a form of ecological malpractice to allow PG&E to use NRDC's name to celebrate an award from George Bush, one of the most consistently anti-environmental presidents in recent history? Perhaps most important, wouldn't the creation of secret "collaborative process" negotiating forums between utilities and certain environmentalists undermine public utilities commissions and make yet another area of decision-making invisible to the public? Would NRDC, a privately financed group legally responsible only to its own board of directors, a group without an internal democratic process, begin to supplant elected PUCs with collaborative processes around the country? And is it likely that utilities will continue to "collaborate" when total electricity consumption begins to fall and their stock prices and earnings take a beating? Conspicuously absent from the full-page PG&E ad celebrating its 1991 presidential award were arguments in favor of solar and other renewable sources of power. As utility executives in other states would be sure to notice, the Collaborative Process was exceedingly useful to PG&E and other utilities in disarming the California Public Utilities Commission, greenwashing their corporate image, and delaying implementation of solar hot water heating and photovoltaics until the indefinite future.

Ultimately, the motives of NRDC and the Energy Foundation appear to be politically conservative and elitist; their funding base is well-off, and their preferred arenas are corporate offices, federal courts, and congressional anterooms. When faced with a true grassroots movement, such as referendums sponsored by Don't Waste Oregon and the Do It Yourself Committee to shut down the leaky Trojan nuclear power plant near Portland, Oregon, NRDC's reaction was to wheel and deal with the utility. NRDC's Ralph Cavanagh helped sponsor a study that would have allowed the utility to keep Trojan open until 1996, rather than shutting it down immediately, infuriating grassroots campaigners. "Utilities won't do the right thing if you inflict the fiscal equivalent of capital punishment on them," said Cavanagh, in discussing the NRDC's opposition to the

November 1992 ballot propositions in Oregon, which would have required utility shareholders to pay for the costs of decommissioning Trojan. Since its commissioning in 1976, Trojan had operated at only 54 percent of capacity, and had been plagued with repeated leaks of radioactive steam and questions about earthquake safety.[80] Though ballot measures 5 and 6 failed to pass by a small margin, Trojan was shut down a week after the election because of a release of what Portland Gas and Electric authorities said were "trace" amounts of radioactive gases from cracked steam tubes.[81] After documents given to the Union of Concerned Scientists revealed how serious the safety problems really were, Portland Gas and Electric closed Trojan for good in early January 1993.[82]

When it comes to utilities, NRDC and the Energy Foundation function as go-betweens and fixers who presume that all controversies have a negotiable solution. In deciding what to do at Trojan, NRDC did not consult its local chapter, because NRDC doesn't have local chapters. The organization has no internal democratic process involving the membership and no intention of developing a base of mass participation and action. As noted earlier, NRDC's board of trustees was founded and is dominated by Wall Street lawyers. In energy affairs, they try to convince the private utility monopolies that they would make *more* money if they followed NRDC's advice and invested in conservation rather than in new power plants.

For NRDC, this alliance with the utilities generated a heady new sensation. Instead of the typical environmentalist predicament of interminable years of nagging and lawsuits pursued from the outside, NRDC was able to feel like an influential player at the main table. With an unassailable sense of self-righteousness that was fortified by generous annual funding from the Energy Foundation, NRDC had created a lucrative sideline that attracted new funding and revenues, rather than costing the organization the support of big corporations and foundations. As early as 1991, the Energy Foundation had pledged $1.15 million to NRDC, its biggest grant recipient, allocated over a period of years.[83] And for California utilities, DSM and the Collaborative Process had bought them some credible buffers against the "continuous salvos of anti-nuclear activists," and possibly a few years of political quiescence and risk-free profits while they pursued their expansion plans elsewhere.[84]

The definitive history of these alliances between mainstream environmental groups and the energy industry has yet to be written. But the closed-door deliberations of the collaborators, who were pledged to silence about the substance of their discussions and enjoined "to speak with one voice" to the citizens at large and "not to use the process as a lever . . . to undercut other participants' interests," represent an attempt to suppress debate.[85] By providing political cover for the utilities, NRDC and the Energy Foundation are undercutting the public accountability of these private utilities. Although utilities are chartered by the state, which has the right to withdraw that charter, utility ties to a given region or city are even further diluted as utilities build more unregulated facilities outside their regulated territories.[86] Carl Weinberg believes that state governments should have the ultimate responsibility to plan for society's energy security. Weinberg would "hate to see a whole series of coal plants built now," even though coal is the most abundant fuel, and he believes that it is folly to become too dependent on natural gas in California. "You must look at energy options like a financial portfolio: some expensive Treasury bonds with a guaranteed yield, some blue chips, and one or two high-tech high flyers. Wind power is money in the bank, at 5.0 to 5.5 cents per kilowatt-hour," he argues, "even if it's a little more expensive than gas turbines, because you know that the fuel is free."[87]

A public truce seems to have been struck: the major environmental groups now encourage "energy-efficiency" programs under the control of utilities, who will pocket generous investment "incentives" (extra profits) in the bargain. Apparently by consensus, solar energy has become unmentionable until the utilities choose to contemplate solar technologies in their own good time, most likely as a result of generous public subsidies. Meanwhile, the utilities continue to promote the perception among benighted millions of electricity consumers that solar power is "not yet cost-effective."

Since the Republican take-over of Congress in 1995, the public themes at NRDC have shifted from "partnership," "collaboration," and "dialogue,"—especially "dialogue with industry," in 1994—to alerting members of the urgency of "holding the line against a congressional majority determined to dismantle twenty-five years of bipartisan legislation." Yet even if NRDC were tempted to reject its origins, the organization's work in energy would always have to "obey the unspoken dictum: Do nothing to jeopardize the value of your benefactor's endowment." Backed by millions of dollars in foundation money, the sensibility of collaboration was dominant at NRDC through the mid-1990s.

The mainstream environmental groups, with ongoing transfusions of money from the Energy Foundation, will undoubtedly continue to prosyletize before state PUCs for more collaborative processes and utility-controlled demand-side management programs in the major power markets.[88] As part of the arrangement, NRDC will refuse to confront the utilities over existing nuclear power plants and look the other way when subsidiaries of the large utilities build unrelated fossil-fuel plants in the United States and overseas. It all seems so simple: progress without conflict, change without transformations in power relationships and profits. As the public fiscal crisis starves citizens' groups and government at all levels, solar advocates are expected either to go along with the programs of the major energy interests and the Energy Foundation, or get out of the business.

Moving the Debate: From Behind Closed Doors to the Public Forum

EXAMINED CLOSELY, the alliance between PG&E and NRDC looks less like an effort to restructure the energy debate and solarize the utilities than an attempt to restrict debate to terms that would allow as little change as possible and paint the utilities with a new green facade. In this case, two very wealthy private interests—the utilities and NRDC—get to define the terms of public debate and cut the other interests—independent installers of solar water heaters and photovoltaic arrays, unions, utility critics, and the public at large—out of the discussion. In *Who Will Tell the People*, William Greider invented the concept of "deep lobbying" to show, in another context, how the big-time opinion game is played by well-heeled allies:

> The larger point is that an informal alliance was being formed by two important players . . . to massage a subject several years before it would become a visible political debate. There was nothing ille-

gitimate about this. . . . But the process that defines the scope of the public problem is often where the terms of the solution are predetermined. That is the purpose of deep lobbying—to draw boundaries around the public debate.[89]

The message of PG&E and NRDC to citizens is "leave the driving to us" with respect to the direction of the entire energy system. In other words, energy production is the exclusive domain of the utility, while purchasing energy-conservation and efficiency measures, such as compact fluorescent light bulbs, energy-saving refrigerators, thermal windows, and insulation, are suitable activities for the green homeowner. The giant utility's avuncular TV spots and two-page *Newsweek* cartoon spreads celebrating "smarter energy for a better world" clearly imply that PG&E is the one with the smarts—and that neither solar water heating, nor photovoltaics, nor wind power are adequately "bright ideas" in PG&E's prescription for the future.[90]

Yet sometimes the public-relations mask slips off in unexpected venues: PG&E's ad for the 1992 Solar Energy Expo and Rally (SEER) Fair, which the company co-sponsors, spent a full page touting the virtues of the natural-gas vehicles and fueling stations it was "working with major oil companies" to create.[91] A cynic might argue that PG&E's main goal by this point was to hook the state of California on natural gas. Certainly, the utility giant had come a long way since the solar heyday of the early 1980s when its promotions had advised customers: "Water heating is one of the best uses of solar energy. It works especially well where fuel costs are high and sunny days are numerous. Like California."[92]

Putting the private utilities in charge of efficiency and solar programs, as we have seen, is simply setting the fox to guard the chicken coop. Aside from the problem of conflict of interest, would smart venture-capitalists entrust notoriously timid and top-heavy organizations such as utilities with their money in the fast-moving field of energy efficiency? Why not take the demand-side money, contribute it to a Solar Bank with a publicly elected board, and have the bank make loans for conservation and solar programs? Why not let this Solar Bank give grants directly to groups such as the Lovinses' Rocky Mountain Institute and have RMI supervise independent projects in the loci of innovation: Silicon Valley in

California; Austin, Texas; Route 128 outside of Boston; or, better yet, California's North Coast, where the solar vision was kept alight during the nuclear winter of the 1980s? Why not mandate that a fixed percentage of energy producers' gross—whether 2 percent or 10 percent—go into a revolving fund, which would be obliged to loan money for renewable-energy projects such as solar water heating, grid-connected photovoltaics, and traffic-calming reforms to make the streets bicycle- and pedestrian-friendly? Surely our descendants will be dismayed by the delays and obfuscation that we permitted in allowing utilities and oil companies to own and control the introduction of solar technologies.

The Solar Lobby argued seventeen years earlier for the creation of a publicly controlled Solar Energy Development Bank.[93] As a matter of fact, it has been politically impossible to create a Solar Bank independent of the energy monopolies. "Sunny Mac," California's sad $240,000 excuse for a Solar Bank, had a short, troubled life and an untimely death in the early 1980s.[94] An Iowa public-power advocate observed that municipal utilities with elected boards were already engaged in demand-side management, since they were directly accountable to the voters, and added that paying privately owned utilities such as Iowa Power to conserve energy was "like paying the rat to eat the cheese."[95] But for the time being, the question "Who owns the sun?" has been lost in the hubbub of collective self-congratulation surrounding the alliance between the wealthiest power brokers in electricity and environmentalism.[96]

CHAPTER 6

Labor, Solar, and the
Energy Economy

THE SOLAR ENERGY MOVEMENT grew out of
the anti-nuclear movement. Yet the anti-nuclear
movement in the United States took years of
organizing to evolve from its original demand for "safe nukes" to a position that staunchly advocated "no nukes." Once this major shift in
emphasis had been made, solar energy appeared as the most natural and
logical alternative to nuclear power. By the mid-1970s, groups such as
Environmentalists for Full Employment (EFFE) had begun to reinter-
pret the energy issue as a social struggle against centralized, capital-
intensive energy strategies that "favor the few at the expense of the
many."[1] These activists advocated small-scale solar and renewable-energy technologies, which would create more jobs and distribute wealth
more equally. "What is being contested," wrote one activist, "is the ownership and control of energy."[2] Gradually, the anti-nuclear movement
began to realize that technical solutions utilizing conservation and solar
could become a reality only through greater citizen control over "financing, production, distribution, [and] employment" in energy.[3]

Naturally, the major energy companies opposed attempts to
replace nuclear and fossil-fuel power. But it turned out to be easier to say

135

"no" to nukes than "yes" to solar, and the few attempts to take over private utility monopolies failed. Workers and unions involved with operating plants took the side of management and fought bitterly to keep open existing projects in the fossil and nuclear energy sectors. In fact, nuclear and fossil-fuel power were providing high-paying jobs—both union and non-union—to hundreds of thousands of engineers, government regulators, and blue-collar workers who had few alternatives to maintain their standard of living. For these workers, visions of planetary ecology would have to wait because the mortgage had to be paid today.

EFFE activists Richard Grossman and Gail Daneker wrote *Energy, Jobs, and the Economy* in the late 1970s to make the connections between fossil-fuel energy and job scarcity explicit.[4] "Short-run survival in this society means having a job," wrote Harvey Wasserman in his introduction to the book. "Where income is not guaranteed, unemployment means families don't eat.... For corporate executives, that translates into political leverage. As long as working people are insecure, those doing the hiring can manipulate them. That simple reality has been used to divide the labor and environmental movements for years."

"The multinationals get us coming and going," wrote Wasserman. "They use the short-term jobs on power project construction sites as bait to get us to build machines designed to put us out of work. Then they fire up the machines with electricity for which they pay preferential rates, often below cost, at the expense of the very people whose jobs are most threatened."[5] Grossman and Daneker pointed to a number of examples to prove their point about the relationship between energy and employment:

- The aluminum industry in the Pacific Northwest consumed 25 percent of the region's electricity but provided only *0.5 percent* of the jobs in the region;
- In 1970 the number of jobs in agriculture was one third of the total in 1920, yet energy use in agriculture had quadrupled;
- The total number of jobs in the U.S. increased by 41 percent between 1950 and 1971, while the total number of jobs in the energy-producing industries increased by only 5.5 percent, despite a doubling of energy production. (Most of the increase in

energy jobs through 1971 was among filling-station attendants. Soon afterwards, filling stations began shifting to self-service pumps.)[6]

The goal of Environmentalists for Full Employment was to create a social movement to break down what they saw as the artificial barriers between labor and environmentalists on the energy question. Grossman and his EFFE colleagues argued that it was possible to replace the relatively few high-paying fossil/nuclear jobs with a great many more jobs in the conservation and solar sectors, jobs that would pay almost as well.[7] The most famous study of the job-multiplying effects of the "soft energy path" was *Employment Impact of the Solar Transition* (1979), commissioned by the Joint Economic Committee of the U.S. Congress. The study's authors argued that an investment of $65.6 billion per year in energy efficiency and solar technologies would cut U.S. annual consumption of fossil-fuel and nuclear energy by 18 percent in twelve years and save $118.8 billion in avoided expenditures on fossil and nuclear fuels by the year 1990. Study director Leonard Rodberg claimed that the 1.1 million jobs eliminated by the cutbacks in fuel-producing and electricity-producing industries would be more than offset by 1.8 million jobs created in the new solar and energy-efficiency sectors. An additional 1.9 million jobs were projected to be created by the capital freed up by the investment in the new forms of energy and conservation—for a net of over 2.5 million new jobs nationwide. The jobs to be created would have the added advantage, compared to traditional energy jobs, of being "dispersed widely across the country," generating little pollution, slowing the greenhouse effect, and contributing to the economic development of local communities.[8] Innumerable studies made since this study was issued have agreed that investments in solar and energy-efficiency solutions create two or three times as many jobs as the fossil-fuel and nuclear alternatives they replace.[9]

Eventually, according to EFFE and its allies, these new, community-controlled jobs in energy would become the foundation for a high-wage, high-skill, full-employment economy, an economy that would reduce net energy imports to zero. The problem was in building the political support to make such a technological and social revolution possible. Not surprisingly, the fossil-fuel barons and their counterparts in banking and the heavy construction industry have tried to co-opt and destroy the low-impact renewable-energy industries. Also, in the absence of readily available alternative ways to make a living, most energy and construction workers have decided that "a job in the hand is worth two in the bush," and have stood, sometimes reluctantly, beside their employers in the energy industry in attempts to maintain the status quo. Union members, beseeched by environmentalists to "think globally and act locally," saw a low-wage "you flip my hamburgers/I'll cut your hair" niche for themselves in the green economy posited by the solar activists.

The fault lines in the labor movement over energy policy nearly split open at the First National Labor Conference for Safe Energy and Full Employment in 1980. Unions from both the industrial and service sectors participated, but the building trades and the Oil, Chemical and Atomic Workers were conspicuous by their absence. Among the construction trades, only the sheet-metal workers' union—the likely installers of solar water heaters—showed unadulterated enthusiasm for the new solar paradigm.[10] Coal miners at the conference were perfectly willing to denounce nuclear power, so long as the new battle cry was "No Nukes! More Coal!" Conference organizer Richard Grossman also brought in energy economist Charles Komanoff to talk about the technologies available to clean up high-sulfur union coal mined in the eastern coalfields. At one point, when members of the International Brotherhood of Electrical Workers (IBEW), who worked in nuclear plants, burst in to disrupt the proceedings with placards such as NO NUKES ARE KOOKS, NUCLEAR POWER IS OUR BREAD AND BUTTER, and THE NATION NEEDS NUCLEAR POWER, a knock-down, drag-out slugfest seemed imminent. Despite assurances that the conference supported guaranteed employment at union wages for all displaced nuclear workers, the disrupters from IBEW Local 5 didn't seem convinced that the participants had any power to make good on their promises. The IBEW disrupters refused to leave until the hall burst into a lusty rendition of the labor anthem "Solidarity Forever."

For the mining, oil, and construction unions, there were major problems with the EFFE formulation of the energy problem:

- Union workers from the construction trades constituted the highly paid workforce that would build nuclear and other big energy projects;
- Union electrical workers, often from the IBEW, expected to staff the relatively few but highly paid operating jobs in the new nuclear power plants;
- In the 1970s, new jobs in insulation, energy efficiency, wind power, and solar had been largely low-wage non-union jobs often hiring undocumented immigrants;
- Widespread use of solar energy seemed certain to accelerate the loss of jobs in coal mining and other kinds of fossil energy; and
- Unions were skeptical of the promise that solar thermal power plants—compared to nuclear power plants—would employ five times as many workers per kilowatt-hour of electricity generated.[11]

In the 1970s, most unions in construction and energy—Operating Engineers, Carpenters, Laborers, and Electrical Workers—believed that to support the soft-energy program of the anti-nuclear and solar advocates would be to sign their own death warrants. The construction trades in California enthusiastically allied themselves with Pacific Gas & Electric, Bechtel, and other construction and utility giants to form the California Coalition for Environmental and Economic Balance (CCEEB)—headed up by Mike Peevy, the former research director of the California State Federation of Labor. CCEEB's goal was to fight the environmentalists and the grassroots anti-nuclear activists and to streamline the approval process for giant construction projects. Most unions—both construction and industrial—left the question of production technology up to the corporations, so long as their choices didn't harm workers financially. In Northern California, for example, the Laborers' contract states that "No rules, customs or practices shall be permitted that limit production or increase the times required to do any work. There shall be no limitation or restriction of the use of machinery, tools or other labor-saving devices."[12] In return, Laborers get "preference" in the training and operation of any "new or different machine," "mechanized process," "material," "method," or "technology."[13] Unionization and the resulting rises in wages and skills are both a cause and effect of the drive to mechanize. Laborers, considered the least skilled of the heavy construction trades, started at $17 to $18 per hour in 1989, over five times the legal minimum wage, with full health and pension benefits. As long as construction boomed, the job-killing effects of the ongoing revolution in mechanization and productivity could be partially masked, and wages could continue to rise for those already working.

Construction unions—like all unions—find themselves in an ambiguous relationship with both technology and the public: the higher the hourly wages, the more people want to get into the union, and the harder management tries to mechanize and reduce the number employed. Ordinary people, as taxpayers or homeowners who occasionally employ painters and carpenters, resent paying union scale, yet as workers themselves, they want to get into those same unions and earn union wages and benefits. As wages and benefits rise in comparison to non-union wages, and as jobs (especially union-scale jobs) become scarcer, the union becomes perceived as a private club that keeps others—notably women and ethnic minorities—out of the good jobs, rather than serving as a beacon of hope for the unorganized.

Periodically, legislation is introduced to end the legal requirement that contractors bidding on government work pay the "prevailing wage" for state and county construction work. The unions argue that the higher pay is justified by the relative efficiency of highly trained union labor, and by the difficulty and danger of the work. Since construction work is sporadic, most union members rarely put in full work years of 1,700 to 2,000 hours at "prevailing wage" rates. Small non-union construction firms, sometimes owned by ethnic minorities or women, claim that the major purpose of the "prevailing wage" requirement is to keep them from bidding on lucrative government contracts usually won by large multinational corporations.[14]

For many years, the long-term political strategy of the construction unions has been to push for more highways, bridges, large real-estate developments, and power plants—all to be built by union contractors. Construction unions make unreliable allies for environmentalists who are trying to stop new highways or power plants, because often these unions will oppose a project on "environmental" grounds only until the contractor agrees to hire exclusively union labor. A regional office of the plumbers and pipe fitters union in Northern California maintains an attorney whose main task is to file environmental challenges against non-union projects. As in any immature industry, many of the newly created solar jobs would be non-union, no matter what their skill level. Environmentalism, in the view of many trade union leaders since the 1970s, has become one more threat to the building-trades unions, which are constantly under pressure from a shrinking construction market, non-union employers, a hostile legal climate, automation, and from non-white ethnics and women who have been trying to beat down the doors of predominantly white, male industries through affirmative action programs.

Today, it is more difficult than ever to pass construction skills and "good" union jobs as a birthright down from father to son. The notion of the union as a "brotherhood" (as in the International Brotherhood of Electrical Workers or the International Brotherhood of Teamsters) has became harder and harder to maintain as union membership contracts and unions are legally forced to broaden their recruitment pool to include racial minorities and women, who quite literally are not "brothers."

Automobile manufacturing is another industry that has lost its capacity to create well-paying new jobs for the average person. Just before Christmas in 1991, General Motors announced that it would lay off 74,000 employees, one fifth of its North American workforce, to stanch a multibillion-dollar cash hemorrhage.[15] And *Business Week* reported in 1993 that Detroit's Big Three would continue to shift "North American" auto production to Mexico, with its cheaper ($10 to $20 per day), younger, and more malleable workforce. By the year 2000, with the North American Free Trade Agreement in place, GM, Ford, and Chrysler expect to produce three million cars per year in Mexico.[16] Even a boom in automobile sales does little for employment in the industry: while American auto production soared by 18 percent between February 1983 and February 1994, the number of auto workers employed rose only 4 percent.[17] Instead of hiring workers, management has piled on the overtime, forcing 11,000 autoworkers to strike at the giant Buick plant in

Flint, Michigan.[18] The best that the United Auto Workers union can do is try and hold on to what it has. In the four years ending in September 1994, General Motors had cut its hourly workforce from 330,000 to 247,000 people.[19]

SCEcorp, Mexico, and Carbon II

MEANWHILE, THE RUSH to coal continues in the U.S. and abroad. Texas—formerly the oil capital—now uses more coal than any other state, much of it imported, and since 1970, the proportion of Texas electricity that is coal-fired has shot up from zero to 46 percent. Since 1970, coal's share of the U.S. electric power generation market has risen from 55 percent to 62 percent, because of coal's perceived low cost at the point of sale (environmental damage doesn't enter into the equation), and an "assured" three-hundred-year supply at current rates of use. Nationwide, notes the *Wall Street Journal*, the fifty-four largest utility plants on the drawing board are coal plants, and 70 percent of the 15,382 megawatts of planned capacity additions through 2001 will be coal-fired.[20] But if coal burning continues to double every thirty years or so, easily accessible reserves will be used up in less than a century—in addition to the effects upon the global climate and the rubble left behind when the mining operations finally cease.[21]

The 1992 Energy Policy Act freed the electric and gas utilities from federal restrictions on their subsidiaries' initiatives to build new power plants. As the rate of growth in energy consumption slowed in the United States, energy corporations prospected abroad for new coal- and gas-fired power projects, the greenhouse effect notwithstanding.[22] The U.S. Generating Company, owned 50–50 by Pacific Gas & Electric and construction-giant Bechtel (both headquartered in San Francisco), has become the second-largest "non-utility" builder of coal-fired plants in the United States, with seven new units totaling 1376 megawatts on the drawing board or under construction. Mission Energy, the unregulated subsidiary of Edison International, until recently led the pack with 3,630 megawatts abroad and 443 megawatts supposed to come on-line in the United States.[23] At a time when unrestrained burning of fossil fuels threatens to cause more global warming and a higher American trade deficit, private electric utilities are bursting the bonds of state and federal regulations and becoming—like the oil companies—transnational conglomerates responsible to no one but their banks and stockholders, and free to roam the earth in search of higher profits.[24] This overseas expansion has been facilitated by the structural adjustment policies of the International Monetary Fund and the World Bank, which since the 1970s have battered down barriers to multinational investment all over the world.[25]

Edison International (until early 1996 known as SCEcorp), the Irvine-based holding company, typifies the new greenhouse-effect-be-damned rush to coal. To build new coal-fired plants out-of-state and overseas, Southern California Edison (SCE) created SCEcorp as a holding company, which owns SCE itself and its unregulated subsidiary, Mission Energy Company. The leadership of Edison International is hardly naive about either environmental or political matters. As explained in chapter 5, chairman and CEO John Bryson was a cofounder of the Natural Resources Defense Council at Yale Law School, and Democratic governor Jerry Brown chose him to head the California Public Utilities Commission when he was in his early thirties. (Warren Christopher also served on SCEcorp's board of directors until President Clinton appointed him U.S. secretary of state in 1993.) Chairman Bryson boasted at the prestigious California Institute of Technology about Edison's "unique" program to reduce greenhouse-gas emissions within Southern California.[26] In its 1992 annual report, SCEcorp proudly "pledged to reduce carbon dioxide emissions by 20 percent in the next two decades to address global warming concerns," by sponsoring efficiency programs and by burning natural gas in place of oil and coal at its power plants inside California.

Omitted from the annual report is acknowledgment of the fact that at least 11 percent of Southern California Edison's electricity is generated at gigantic coal-fired power plants east of California in tiny towns like Shiprock, New Mexico and Laughlin, Nevada, combining to produce a shroud that Edward Abbey called "the sole human artifact visible from the moon."[27] Edison's Mojave Power Plant in Laughlin, Nevada, "has no pollution control equipment," reports the *Washington Post,* and "emits 40,000 tons of sulfur dioxide annually—the largest single uncontrolled source [of haze-generating pollution] upwind of the Grand Canyon." "After intense lobbying from Edison," notes the *Post*, the ad hoc Grand Canyon Visibility Transport Commission, composed of state governors and representatives of Indian tribes and environmental agencies in the southwestern states, "decided against recommending pollution controls at the [Laughlin] plant and instead advised [Edison] and the EPA to work out a solution over the next two years."[28]

Coal for Edison's Laughlin plant arrives in a 250-mile *unpermitted* slurry pipeline from Peabody Coal's gigantic Black Mesa open-pit mine in the middle of the Navaho reservation. The pipeline drains a billion gallons per year from the Navaho Aquifer, the primary source of water for the Hopi nation in an arid region, and dumps the water when the pipeline reaches Laughlin. Peabody Coal, our country's largest coal mining company, is a subsidiary of Hanson PLC, a British holding company. In denying Peabody a permit to extend its Black Mesa operations, Administrative Law Judge Ramon Child noted that the coal company had failed to obtain written permission to mine within 300 feet of residents' homes, had mined through eleven Navaho graves in 1993 alone, had damaged houses with its blasting, and had caused sheep to die from drinking polluted water. Both Peabody Coal and the Department of the Interior argued that written permission wasn't required because the people affected were Indians on a reservation.[29] The words of Native American scholar-activists Winona LaDuke and Ward Churchill in regard to "radioactive colonialism" apply perfectly to the Black Mesa debacle: "Those who control the land are those who control the resources . . . whether the resource at hand is oil, natural gas, uranium, water. . . . Social control and all the other aggregate components of power are fundamentally interrelated."[30]

In the same annual report, SCEcorp touts Mission Energy's participation in "The New Frontier" of giant coal-fired electric plants in Australia, Mexico, Indonesia, and the United States.[31] (If SCEcorp's "New Frontier" includes a solar strategy, it was not important enough to even mention.) Greenpeace has estimated that the 27 million tons per year in increased emissions (from the 4,062 megawatts in new coal-fired capacity that Mission Energy was helping to build overseas) would be 4.5 times greater than the 5.9 million tons per year in reduced carbon-dioxide emissions for Southern California, about which Edison was boasting.[32]

Until October 1993, Mission Energy owned 49 percent of Carbon II, a 1,400-megawatt, coal-fired electric power project in Rio Escondido, Mexico, about fifty miles southeast of Big Bend National Park. The project was started by the Comisión Federal de Electricidad (CFE), the government-owned power company, then turned over to Energan, a private binational consortium, when CFE's money ran out.[33] Fifty-one percent of Energan's ownership is split equally between Mexico's Grupo Acerero del Norte and the Organización Autrey, both of which have extensive coal and steel interests that expanded dramatically during the "privatization" wave of President Carlos Salinas de Gortari.[34] A controversial special decree by President Salinas made the privatization of the Carbon II project possible, even though by Mexican law the CFE has the exclusive right to generate, transmit, and supply electricity in Mexico.[35]

Environmental Integrity at Mission Energy

Respect for the environment has been a vital concern to the company since its inception. Its power production facilities use advanced pollution-control technologies. In addition to protection of air quality, project facilities are designed and sited to protect sensitive plant and animal species. Mission Energy has contributed $500,000 to establish 500 acres of dedicated habitat for use by the kit fox and other endangered animals, including the giant kangaroo rat, the blunt-nosed leopard lizard, and the antelope squirrel.

SCEcorp, *Annual Report,* 1992

Scott Royder of the Sierra Club's Texas chapter has called Carbon II the "perfect example of how . . . [NAFTA] could . . . end up making things worse in the environmental arena," and two Worldwatch staffers wrote to SCEcorp's chairman John Bryson to inform him that his company's coal-fired expansion plans overseas were "clearly *not* providing a sustainable vision for electric utilities at the global level."[36] If run as designed, by SCEcorp or by some other owner, Carbon II is certain to reduce visibility and add to air pollution in Texas as well as Mexico, since the prevailing winds are southeast or southwest, and since the sulfate aerosols which would be released are the perfect-sized particles to create haze. Two decades of monitoring at Big Bend National Park in Texas show that visibility and air quality in that region have been declining for twenty years, and that most of the pollution comes from Mexico.[37] In former days, visibility of 150 miles had been common. Mission Energy had not installed sulfur-dioxide scrubbers on the new plant, alleging that scrubbers were not necessary because the facility would burn "low-sulfur coal," which contains only 0.8 percent sulfur by weight. An EPA staffer, nonetheless, noted "a typical new power plant in the western U.S. emits in the range of 0.1 to 0.2 pounds per million Btus—an order of magnitude cleaner than the Carboelectrica facility."[38] If it had been erected in the U.S., Carbon II would have been required to install scrubbers. By the time the first 350-megawatt unit began test operations in April 1993, the EPA—despite three months of inquiries—had yet to receive any technical information about Carbon II's emission characteristics from SEDESOL, EPA's Mexican counterpart, though such information was supposed to be furnished under the 1983 La Paz agreement between Mexico and the United States.

The English-language weekly of *El Financiero*, Mexico's major financial newspaper, reported that the owners of Carbon II "readily admit that the plant will not meet U.S. 'New Source Performance Standards,'" mostly because the plant's design "does not include scrubbers" that would remove the excess sulfur dioxide from the air. Such scrubbers would be routine on any new coal-fired generating plant in the United States, but retrofitting the already completed sections of the plant would have cost hundreds of millions of dollars, according to Mission Energy's Special Projects Manager Tom Reed. SCEcorp Chairman Bryson told the *Wall Street Journal* that Carbon II was "a modern, relatively clean facility" by world standards, one that would not violate U.S. visibility regulations, and he argued that it was the Mexican government's responsibility to decide what kind of pollution-control devices should be required.[39]

At first, the World Bank's International Finance Corporation (IFC, a branch of the bank that lends to private enterprises) and five other investment banks promised to lend money for Carbon II, but these lenders were concerned that the plant might be forced to shut down if obliged to install sulfur-dioxide scrubbers at a cost of hundreds of millions of dollars to meet U.S. air-pollution standards.[40] By mid-July of

1993, the financial closing for the Carbon II project had been postponed from September 1993 to February 1994, and the IFC had pulled out of the deal—perhaps in response to negative publicity, congressional criticism, and environmental concerns expressed by Greenpeace and the Sierra Club.[41] The NAFTA side accords on the environment and labor—heavily touted by President Clinton—would have no power over Carbon II, making it a prime example for the *Texas Observer* of why NAFTA constitutes a "border polluters' pact."[42]

In September of 1993, Paul Volcker, former head of the Federal Reserve Board, was hired by SCEcorp "to enhance its credibility," but the controversy over Carbon II continued to sour investors on Wall Street.[43] On October 6, Natwest Securities downgraded SCEcorp from "hold" to "underperform." "While the stock is presently yielding somewhat in excess of the industry average," argued Natwest, "we believe that further price erosion is likely if Mission Energy, SCE's IPP [independent power producer] subsidiary and the source of projected future earnings growth, is unable to reach a satisfactory agreement with the Mexican government regarding the construction of the Carbon II generating station." Other brokerage firms and investors had their own doubts too, and SCEcorp's stock price continued to sink.[44]

On October 11, 1993, without explanation, SCEcorp pulled out of Carbon II and accepted an $18-million loss, after sinking $358 million into the project. Faced with a "battered" Southern California economy, which had resulted in slowed electricity consumption at SCEcorp subsidiary Southern California Edison, Mission Energy's overseas projects had been SCEcorp's "best bet for earnings increases," according to a *Los Angeles Times* reporter. Following the pullout, Morgan Stanley and a flock of other brokerage firms recommended that investors sell their SCEcorp stock.[45] In less than a month, SCEcorp's stock price had lost over 15 percent of its value.[46]

One thing was certain: Chairman John Bryson's reputation as a "green" utility executive was destroyed, and his leadership in the Carbon II episode seemed to call into question the very idea of a "green" utility in the private sector. If a co-founder of one of the nation's leading environmental groups couldn't be trusted to run an electric company in an environmentally sound manner, could anyone be trusted? Ultimately,

Bryson's defense of his actions was almost puerile in its simplicity: "Mexico is now on its way to achieving the economic prosperty that will allow it to afford greater environmental efforts," he wrote, "but it can achieve that prosperity only with expanded electrification." In other words, countries had to get dirty in order to get rich, and once they got rich they could afford to get clean.[47] Environmentalist opposition in the United States during the NAFTA debate played a part in Mission Energy's pullout from Carbon II, and apparently helped turn Mexico's federal electricity monopoly, the Comisión Federal de Electricidad, against the deal. "We found we didn't have a very good reading on Mexican politics," Bryson later told *Forbes* magazine.[48] Sadly, neither the solar energy nor energy-efficiency options ever figured in the debate, although these alternatives would have promised huge benefits to Mexico, given the potential economic boon from recovering wasted energy through conservation (as was done in the U.S. in the 1980s).

However "green" his personal inclinations might be, Bryson's performance at the helm of SCEcorp would be judged by his ability to "grow" the company; otherwise the capital market would abandon his company and the stock price would fall. "SCEcorp is a well-capitalized company with free cash flow," commented a Salomon Brothers analyst. "The real issue for SCE and the [utility] industry is learning to use this capital to grow earnings and dividends."[49] Chairman Bryson's organizational response to the Carbon II crisis was to replace the head of Mission Energy with someone who had more international experience and was better able to exploit the expected doubling of electricity consumption every six to fifteen years in Asia and Latin America. "The best possible business to be in is where your market is enormous and growing fast," said Edward Muller, the new president of Mission Energy.[50]

With "green" utility executives hell-bent on cutting jobs in the United States and on building new fossil-fuel power plants around the world, clean power and ecological jobs for workers in the U.S. and abroad would have to come from somewhere outside the traditional energy sector. It was only coincidence, but the same week that Mission Energy pulled out of Carbon II, Pacific Gas & Electric proudly told investors that its three-year plan to cut three thousand employees from the payroll would be completed two years ahead of schedule.[51] Since the

Carbon II fiasco, John Bryson has continued to wax rhapsodic about SCEcorp's chance to "hit paydirt" in Mexico and to participate in the coal-fired electricity expansion in China, predicted to equal "*17,000* megawatts *each year* ... almost the total capacity of Southern California," through the year 2000 (Bryson's emphasis).[52] In these burgeoning foreign markets, Mission Energy could take advantage of cheap coal still mined by pick-and-shovel miners who earn as little as $55 per month. Hong Kong's *Sunday Morning Post* reported that it is not uncommon for Chinese coal to be loaded by women with babies strapped to their backs, despite an official ban on female miners.[53]

But the moves overseas have not yet paid off for SCEcorp. In one year, the company's stock price fell by a half to 12¾ and in June 1994, SCEcorp reduced its annual common-stock dividend from $1.42 to $1.00 to reflect new economic realities.[54] By mid-1996, the corporation's stock price had risen, closing at 17⅞ on September 13, partly in response to the California legislature's nuclear bailout.

With Mission Energy out of the picture and NAFTA signed, the Mexican government's Comisión Federal de Electricidad decided to finish Carbon II on its own, using the same 1970s air-pollution control technology that had frightened away Mission Energy. By the summer of 1995, with 1,000 megawatts of Carbon I and 700 megawatts of Carbon II on-line, complaints about a white haze over Big Bend National Park were coming in constantly. One tourist told a reporter for the *Fort Worth Star-Telegram,* "I think this is going to be my last visit, because I want to remember Big Bend the way it was."[55] José Cisneros, the superintendent of Big Bend National Park, told of a trip up the Rio Grande valley in late November: "As we approached Eagle Pass, we could see Mexico across the way. There was this incredible layer of pollution hanging over Piedras Negras, just sitting on top of it, that appeared to come from the [electrical power] plants. It was like L.A. or something. The haze stayed with us as we moved west."[56] Andy Sandom, executive director of the Texas Parks and Wildlife Department, noted that air quality at Big Bend has gone from two hundred or three hundred clear days per year to ten or fifteen.[57]

The position of the Mexican authorities is that the air pollution at Big Bend is not caused by the coal-fired plants in Rio Escondido, and that in any case it would cost $500 million to bring both Carbon I and Carbon II into compliance with U.S. air-quality standards. "We are asking for new understanding," said Leonora Rueda, director of border affairs for the Mexican foreign ministry. "We have much less money for the birds than for the people. If we can do it together—the people and the birds—great, but we have to think first about solving the social problem." For his part, Texas governor George W. Bush Jr., remarked that he had heard "numerous comments about the deteriorating air quality of one of Texas's true treasures." Governor Bush recommended that both coal-fired power plants be converted to natural gas.[58] With the EPA in negotiation with its Mexican counterpart over the possibility of funding a joint study of the problem, it seemed unlikely that a solution to the problems of Carbon I and Carbon II would be reached at any time in the near future, particularly since Carbon II had begun full operation (1,300 megawatts) by the spring of 1996.[59]

Coal Mining Without Coal Miners

IN THE BATTLE to preserve jobs, coal miners have faced even worse threats than construction workers: though total coal production more than doubled between 1970 and 1990, mechanization has rapidly reduced the number of miners, and retired United Mine Workers members now outnumber the 60,000 active miners by two to one.[60] By 1960, employer-led "automation and consolidation . . . [had] . . . *permanently* idled an army of once-valuable coal miners," wrote Harry Caudill thirty years ago, and, despite a brief employment renaissance in the late 1970s, the process has only accelerated.[61]

How times had changed since the days when "coal was king"! In June 1902, when 145,000 coal miners went on strike in the Pennsylvania anthracite pits for a 10-percent raise and union recognition, coal provided 73 percent of the nation's energy needs. As a grim winter of empty coal bins and freezing cities loomed, President Theodore Roosevelt backed off from military action against the strikers, and instead called for arbitration of the dispute. After seven months, the miners got their raise. It was the first time a U.S. president had recognized the legitimacy of a strike, though he did not oblige the operators to grant blanket union recognition.[62]

Mining employment peaked in 1923, when 862,500 people were needed to hew 657.9 million tons of coal out of the earth and prepare it for sale—one worker out of 50 in the entire country—compared to one in 850 today.[63] Although by 1923 oil and automobiles were playing a substantial role in the transportation market, coal provided a full three quarters of the energy needed to run the country that year.[64] Yet for decades, the miners union would battle for recognition in "Bloody Harlan" and dozens of other coal counties.[65]

During the labor wars of the 1930s, the United Mine Workers of America (UMWA) finally unionized the coal fields, under the charismatic leadership of President John L. Lewis and the benevolent neutrality of Franklin Delano Roosevelt's White House. Under the brilliant, sonorous direction of "John L.," as he was known, the United Mine Workers founded and funded the Congress of Industrial Organizations (CIO) to organize the mass production workers in the steel and automobile industries. This new organization was expelled from the more conservative American Federation of Labor, but the CIO eventually prevailed.[66]

As World War II ended, coal was still a primary and highly visible energy source. Filthy, hissing, romantic steam locomotives outnumbered diesels, and anyone then living in the East or Midwest can recall the acrid smell of coal smoke and the spectacle of glistening lumps of anthracite rushing down shiny chutes into the coal bins under every apartment building on every block. Flecks of soot and fly ash would pepper the snow as soon as it stopped falling, but the black smoke pouring from brick chimneys meant that steam was hissing into hot radiators all over town—and that some family was living in a "free" basement apartment next to the coal bin, keeping the furnace stoked.

As recently as 1978, there were a quarter of a million coal miners in the country, most of them unionized. Since then the number of coal miners has fallen by half, while coal production has climbed to nearly a billion tons per year. "Only half the coal miners are unionized now," says Mike Buckner, research director of the UMWA and himself a former miner, "and the non-union mines don't pay retiree pensions or health benefits. But we have to fight for them; they're our fathers and grand-fathers. The non-union mines don't bear that burden."[67] The union casts a smaller shadow today, but it hasn't lost its spirit. When Pittston Coal cut off retiree health benefits, the union launched a strike and help mobilize thousands of supporters through mass sit-ins and plant occupations. After nine months of court injunctions, multimillion-dollar fines, and thousands of arrests, the mineworkers' union and its allies forced Pittston to back down in late December 1989. Typically, perhaps, the fight was over retiree rights rather than the rights of active miners.[68]

Coal as a fuel source has become almost invisible of late, even as production and profits soar. Much coal is now burned at remote mine-mouth generating plants operated by utilities. The electricity they generate is transmitted at extremely high voltage over hundreds of miles. There have been innumerable protests over the health dangers from high-voltage electromagnetic fields (EMFs), a dispute that provoked Minnesota farmers to sabotage power pylons and transmission towers in the summer and fall of 1978.[69]

Today, the biggest short-term threat to the United Mine Workers union is competition from strip-mined western coal, which contains less sulfur than the eastern underground coal. Wyoming has become the world's coal capital. According to the National Mining Association, Campbell County in Wyoming produced 205.6 million tons of coal in 1994 from its seventeen surface mines, well over three times the 63.5 million tons produced in all of Germany that year.[70] Arco Coal's Black Thunder Mine in Wright, Wyoming, is the most productive coal mine in the entire world. At Black Thunder, 150 feet of overburden must be removed to get at the 70-foot-thick coal seam. It takes only 450 workers (including supervisors) to mine 1.25 tons of coal per second . . . or 40 million tons per year,[71] compared to the 500 workers it took to produce 33 million tons of coal from the same mine three years earlier.[72] At western strip mines, the minority of coal workers who are unionized belong to construction unions, whereas the main tools of the mining operations are mechanical: monstrous draglines, giant bulldozers, $4.5-million trucks with ten-foot tires, and power shovels with buckets big enough to hold a housetrailer. The labor cost of western coal is "under $1 per ton, compared to about $10 per ton labor costs for eastern coal," notes Mike

Buckner. Nationally, surface miners dig 6 tons of coal per hour, compared to 2½ tons per hour for underground miners—a tripling of productivity in the West since the early 1970s. Today, the United Mine Workers can no longer dictate the pace and terms of mechanization for the benefit of its members. Only two fifths of coal now comes from eastern underground mines, where the UMWA is strongest, and only half of those mines are unionized.

The United Mine Workers union advocates for high-sulfur eastern coal by lobbying—together with coal companies and utilities—to ensure Department of Energy support for a multiyear, $4.6-billion Clean Coal Technology program for removing sulfur, particulates, heavy metals, and other noxious coal-burning wastes in order to create what is supposed to be a $50-billion–per–year clean-burning coal market by the year 2000. (Imagine the impact upon the solar and wind industries if nearly five billion dollars in government R&D subsidies was actively championed and allocated.) Hazel O'Leary, the current secretary of energy, is avidly pushing "clean-burning technologies" that are expected to be a major export item to countries such as China (the world's largest coal consumer), Russia, and Australia.[73] These three countries and the United States together possess three fifths of the world's coal reserves.[74]

As coal production now vaults to historic highs and takes over an ever-increasing share of the electric utility market, the miners themselves are disappearing. Between 1980 and 1993, the number of active coal miners in the UMWA plummeted from 160,000 to 60,000. The union has tried to forge an "employee involvement program" as an alternative to formerly antagonistic management-labor relations, and to share the workers' ideas on how to operate more efficiently. But often the union loses ground anyway. Union "cooperation" at the Maple Creek Mine in western Pennsylvania "increased productivity 200 percent and cut the employers' cost [per ton] by over 50 percent," said United Mine Workers president Richard L. Trumka, and the workers' "reward" was "quicker, faster, more sure unemployment."[75]

Between 1978 and 1993, the UMWA's overall share of U.S. coal production declined from 65 to 30 percent.[76] The UMWA relies mainly upon negotiating strategies, along with occasional, last-ditch strikes when the coal companies threaten the union's fundamental right to exist.

On May 10, 1993, two thousand mine workers struck Peabody Coal, the country's largest producer. The dispute was over "jobs—the main and only issue," and was based on the fact that no miners spend their entire working lives in one mine. Peabody, still 95-percent unionized, was trying to deny union members first crack at jobs in new mines as the old mines play out—a de-unionization strategy known as "double-breasting." But to the media and the public at large, the mineworkers' struggle was almost inscrutable and almost invisible. When coal labor unions are discussed at all, it is during the context of a strike, and the resulting two-inch story usually appears twice on some inside page: once when the strike erupts and once when it is settled or broken—unless someone is killed on the picket line.[77]

By mid-July of 1993, seventeen thousand U.S. miners were on strike. The strike lasted until December 1993, when union members voted to settle for a guarantee of three fifths of the new jobs at non-union mines owned by members of the Bituminous Coal Operators Association. In return, union members agreed to allow operators to run their mines seven days a week with up to four 10-hour shifts required of workers, and without obligating the companies to pay overtime on Saturdays and Sundays. As before, overtime pay of time-and-a-half would begin after forty hours on any given week. Nevertheless, with a three-fifths foothold in new eastern mines, the United Mine Workers believed they could unionize more mines and slow the decline in membership.

Some UMWA leaders felt that their best long-term hope would be to seek new members among factory and government workers whose parents and grandparents once toiled in the mines and fought for their rights under the United Mine Workers banner. As to rumors that the windy mountains of West Virginia would be an ideal site for wind farms, and therefore a source of jobs for UMWA members, Mike Buckner asked, rhetorically, "Would our members be willing to give up a $25-an-hour job to build windmills at $15 an hour?"

Job cutbacks in coal are not peculiar to the United States. In Germany sixty thousand union miners and support personnel of Ruhrkohle AG conducted a warning strike and marched in the streets on September 22, 1993 to protest a proposal by Ruhrkohle's supervisory board to cut eighteen thousand of the present eighty thousand employees by mid-

1994.[78] Germany's government coal subsidies total over $6 billion per year—or about $43,000 per year for each job in coal.[79] With the subsidies now threatened by drastic budget cuts, German energy companies are beginning to invest heavily in low-cost U.S. coal.

But the greatest long-term obstacle to the use of coal may be the menace of global warming. While coal advocates claim that they will eventually be able to remove most of the pollutants from smokestack gases, none is so bold as to claim that they can burn coal and still keep carbon dioxide out of the atmosphere.[80] No matter how "clean" the combustion process becomes, the basic equation remains the same: carbon (C) plus oxygen (O_2) yields carbon dioxide (CO_2), the principal greenhouse gas. Some coal producers dismiss the theory that burning carbon causes global warming, and call for more research; other coal advocates view the prospect of ecological catastrophe with calm equanimity. "Global warming will make wastelands of some populated areas but make some other places livable," L. G. Bracken told the *Wall Street Journal*. "Maybe you'll want to move some people from Zimbabwe to Siberia, for example." (Mr. Bracken is a vice president of Houston Lighting and Power Company, a subsidiary of Houston Industries, Inc.)[81] Environmentalists disagree. "Clean coal is not clean and never will be," argues Sierra Club energy lobbyist Daniel Becker. "I don't want to give the coal people heart attacks, but we have to phase out the use of coal."

The Union of Concerned Scientists (UCS) has made the battle against global warming one of its central themes, and the UCS-sponsored "World Scientists' Warning to Humanity," signed by a majority of living Nobel laureates in the sciences, highlighted global warming as one of the principal reasons that "human beings and the natural world are on a collision course."[82] In their report *America's Energy Choices*, a coalition of UCS and other mainstream environmental groups argue that there is no unresolvable contradiction between economic growth and reduced carbon-dioxide emissions, a point of view very similar to that of the World Commission on Environment and Development.[83] *America's Energy Choices* seeks to convince "the nation's energy policymakers, at all levels of government and in the private sector, to seize the opportunities outlined in this study so that our nation can realize a 'win-win' energy future . . . that combines prosperity and environmental integrity."[84] The report proposes vast changes in production technologies and in how people live and work. New electric cars would be developed, and automobile vehicle-miles traveled would be reduced by zoning changes that would increase settlement densities, especially along improved mass-transit corridors. The UCS report's list of recommendations is long and intelligent and inclusive. President Clinton borrowed from it in his unsuccessful advocacy of a "Btu" or fuel tax. The proposed technical solutions to the problem of global warming make sense, but the political and economic obstacles remain immense. Presumably, once tax incentives were changed and the price of fossil fuel raised, the oil companies would stop searching for oil, the electric utilities would stop building coal-fired power plants, and the auto companies would turn from the manufacture and marketing of gasoline cars to electric vehicles, bicycles, and trolley-buses. So far, it must be said, the energy giants have done everything in their power to make sure the scenario proposed in *America's Energy Choices* dies an early death. And the unions, whose members would benefit along with all U.S. citizens from an emergency national program to develop nonpolluting, renewable energy sources, continue to be caught in a cycle of struggling for vastly expanded use of coal—a strategy that will yield only short-term gains in employment, if that, and possibly irreversible damage to the planet's atmosphere.

Can Labor and Environmentalists Work Together?

To SPEARHEAD the transformation to a renewable-energy economy, the environmental group Greenpeace proposed in 1992 a "5-percent solution," heralding solar and wind energy as the keys to job creation and reduced use of fossil fuels, and proposing that $15 billion per year (or 5 percent) be shifted from the nation's military budget "to put the U.S. firmly on the clean energy path." According to Greenpeace, $15 billion per year should be sufficient to generate 1.5 million new jobs in the first year, and 750,000 new jobs each year thereafter. The Sierra Club, in turn, recommended "an immediate emphasis on natural gas for electricity generation and a longer-term shift to renewables: solar, wind, geothermal, and fuel cells."[85] (Regrettably, the Greenpeace Atmosphere and

Energy Project was shut down in June 1995 as Greenpeace suffered budget cuts).[86]

Anthony Mazzocchi, a high official in the Oil, Chemical and Atomic Workers Union (OCAW), was primarily concerned about preserving the living conditions of the energy and chemical workers who were being laid off because of new environmental regulations. At a joint press conference with Greenpeace held in late 1992, Mazzocchi called for a "Superfund for Workers" that would guarantee four years of education—at full tuition, with no loss of pay or benefits—to people who lost their jobs as a result of environmental shutdowns.[87] Mazzocchi, a World War II combat veteran, adopted the idea from the $40-per-month allowance the government paid to returning veterans like himself who needed financial help to ease their readjustment to civilian life. He recognized that massive shutdowns of toxic factories were coming, but argued that environmentalists should defend full-income support and retraining for veteran workers in dangerous industries to soften the impact of restructuring. The workers, after all, had had no voice in deciding what was to be produced, so Mazzocchi argued that they should not be the chief victims when factories "where the product was the poison" were shut down. "Jobs or income is our agenda," says Mazzocchi, who is scornful of government "retraining" programs, which prepare displaced workers for jobs that typically come with a 50-percent-pay cut.[88]

Unfortunately, the "labor-environmental alliance" is usually little more than a hope and a prayer in the minds of a few dedicated local activists. Where they are not outright hostile toward one another, environmentalists and labor unions manage at best a wary truce. Environmentalists pay pro forma obeisance to the notion of a Superfund for Workers and agree halfheartedly with the notion that existing plants should be operated more safely. A 1993 eight-page issue of *Chlorine Free*, published by Greenpeace, advocates the elimination of almost all chlorine-containing products, without bothering to mention the Superfund for Workers or concerning itself with what would happen to those workers who lost their jobs because of the chlorine ban.[89] Automobile, oil, and construction unions, for their part, pay lip service to environmental considerations while opposing small cars and Corporate Average Fuel Efficiency (CAFE) gasoline mileage standards. Construction unions, typically, lobby alongside of corporations such as Bechtel for more roads, higher soundwalls, longer bridges, deeper shipping channels, and more power plants.[90]

Yet on rare occasions the game changes, and unions seek out antitoxics campaigners and other environmental allies when companies try to eliminate the union and establish absolute authority over the production process. An oft-cited case of such labor-environmental cooperation is the alliance of Local 4-620 of the Oil, Chemical and Atomic Workers International Union (OCAW) in Geismar, Louisiana, with the National Toxics Campaign and the local chapter of the Sierra Club during and after a five-year battle against a lockout by the BASF Corporation.

The BASF Corporation is a wholly owned subsidiary of the German chemical giant BASF AG, one of the world's largest chemical companies. In the 1980s, the Geismar plant produced a whole series of chemicals, including Basagran, an herbicide widely used on soybeans, rice, and golf courses. Most of the BASF products used poisonous and explosive chemicals and by-products in their manufacture, which the corporation often disposed of in dangerous and illegal ways. When the BASF-OCAW contract at the Geismar plant came up for renewal in 1984, BASF management demanded the right to replace union members with outside contract workers any time, on any job, for any reason. To submit to this demand would have allowed BASF to bust the union. When the contract came due on June 15, 1984, and the union refused to capitulate, BASF ordered union members to vacate the plant, backed by the authority of company guards. The workers complied, in a state of shock, and the lockout had begun.

The 400-member union local soon regrouped, and together with the national leadership of the union it launched a no-holds-barred battle against BASF around the world. One of the strategies the union used, after a careful inventory had been conducted of all plant operations by the locked-out personnel, was to expose the company's anti-environmental practices, and ally itself with local citizens' groups and environmentalists who were fighting pollution in the chemical corridor (later dubbed "Cancer Alley" by activists) along the Mississippi between Baton Rouge and New Orleans. The national union hired legendary organizer Richard Miller to coordinate the battle against the lockout, and the union even bought space on half-a-dozen billboards near the plant that displayed the slogan, BASF: BHOPAL ON THE BAYOU, to suggest that the lockout of experienced, full-time chemical workers was endangering the entire community with a toxic catastrophe.

The lockout at Geismar, said Dick Leonard, special campaigns director of OCAW, was a "unique" struggle catalyzed by BASF's "particularly brutal and straightforward attempts to smash the workers' organization and ensure a docile work force."[91] What eventually defeated the lockout, and checkmated BASF's attempts to get rid of the union, was the massive solidarity from OCAW locals around the country, who adopted the families of locked-out workers, and the financial and organizational support of the national union, which spent over $2 million to fight BASF throughout the United States and Germany. Another factor was the success of the union in countering BASF's U.S. expansion plans. When BASF proposed building a new paint plant employing two hundred workers in Terre Haute, Indiana, OCAW cautioned the local town to beware of BASF's "bait-and-switch" strategy of siting its "Midwest Ecology Project" there, a waste incinerator and landfill that would accept and reprocess hazardous waste from all BASF operations in the United States. The revelations, backed by leaked proof of BASF's true intentions, soon killed the "Ecology Project" and the proposed paint plant.

The OCAW also produced and distributed thousands of copies in both English and German of its *Lockout* video, which called attention to BASF's past as an employer of slave labor during World War II, and published a newsletter that contained exposés of BASF's internal operations, which OCAW mailed to selected executives and customers. The union also organized shareholder protests at stockholders' meetings in Ludwigshafen, the world headquarters of BASF AG and the site of its 50,000-employee home complex, where activists spoke against the lockout and advocated settlement with the union. To help redress the power balance in Geismar, OCAW helped organize the Louisiana Coalition for Tax Justice, which called attention, parish by parish, to $2.5 billion in state tax breaks to corporations that have impoverished the public institutions of Louisiana.[92] The union's long-term strategic alliance with local citizens' and environmental groups helped tilt events in its favor in Louisiana, helped by the sympathetic neutrality of Governor Buddy Roemer, a

Democrat turned Republican, whose Department of Environmental Quality was willing to pursue and prosecute environmental problems uncovered by the union.

OCAW staffer Richard Miller believes that the lockout finally became too costly for BASF when the union's efforts convinced the state Department of Environmental Quality to withhold an operating permit for a newly built $12-million plant to produce glyoxal, a chemical used in permanent-press fabrics. The plant had been built on the contaminated site of the old Basagran plant in Geismar, and the plume of pollution seeping from the old plant threatened to contaminate the aquifer that supplied the surrounding region with clean water. Eventually, BASF managed to operate its brand-new glyoxal plant under a conditional use permit, but it was obliged to drill expensive horizontal wells to suck contaminated water from underneath the plant site.[93]

The campaign against BASF was a thousand mosquito bites and pinpricks that made life difficult for the corporation. BASF also came under pressure from other chemical manufacturers, who were getting tired of the bad name BASF was giving the chemical industry in Louisiana. Finally the damage to the company became too great, and BASF decided to settle, grudgingly, with the union.

But members of OCAW Local 4–620 have been very slow to forget and forgive BASF for the lockout, particularly since the company, suffering lower profits worldwide, has continued its pressure on the local union since the lockout ended. To this day, each of the four hundred members of OCAW Local 4–620 pays a $4 monthly assessment to the Louisiana Labor/Neighbor Project for environmental and tax justice. Organizer Dan Nicolai of Labor/Neighbor now works closely with the Ascension Parish Residents Against Toxic Pollution to oblige local chemical firms to pay for chemical-waste treatment and to open up more chemical plant jobs to local residents, often African-American descendents of sugar plantation sharecroppers and laborers. Nicolai works with attorney Yolanda Moton, an environmental justice organizer with the Jobs and Environment Campaign, who has been focusing most of her recent efforts on emergency response for Donaldsonville, an African-American community surrounded by three ammonia plants.

But the conditions and attitudes that produced the Louisiana Labor/Neighbor Project constitute a rare exception. The union is not worried about job blackmail in Louisiana because it knows that BASF is intent on expanding production at the Geismar site. Local leader Duke King says that the union gets "ten times as much good publicity and local support" through its environmental and tax advocacy than it ever got by participating in traditional charities and benefits.[94] For the time being, the Sierra Club and local anti-toxics groups accept the existence of the chemical industry as a given, rather than calling for its shutdown, and the environmental groups work together with OCAW and local grass-roots anti-toxics groups—often led by African-Americans—to take aim at lax enforcement of tax and environmental laws. So far, the rare strategic alliance between the Oil, Chemical and Atomic Workers and concerned environmentalists seems to be holding.

Compared to the Louisiana Labor/Neighbor Project, most coalition efforts around the country are "slow and frustrating."[95] The members are still "getting to know each other" and "agreeing to disagree" over key issues such as nuclear power—as if the idea of labor-environmental cooperation had been invented only yesterday.[96] Michael Merrill, a labor educator who consults for OCAW, is scornfully frank about most environmentalists' lack of sympathy for workers' concern with well-paying jobs. The labor movement's tried and true program for income protection is economic growth, says Merrill. "What place," he asks, "will working people have in the brave new green world of tomorrow? The environmental movement's failure to provide a convincing answer to this question is the principal obstacle to a meaningful labor-environmental coalition." Merrill argues that the Superfund for Workers concept should go way beyond job retraining; it should become a "green parachute" for workers who lose their income "because of environmental regulation or any other reason. . . . The only way to stop people from getting paid for destroying the environment is to pay them for not destroying it. If you want chemical workers to stop making hazardous chemicals and nuclear workers to stop making plutonium, then pay them not to make hazardous chemicals, even if they are paid to do nothing . . . or paid to do what bosses do—which is to sit around all day and figure out how to do things differently. It is precisely these ... unthinkable things that the original ... Superfund for Workers imagined...."[97]

Like the United Mine Workers, the Oil, Chemical and Atomic Workers Union has been badly damaged by the increasingly hostile anti-union climate in the U.S., and by the productivity and mechanization revolutions in its primary industries of oil refining and chemicals.[98] The raw statistics are brutal. Between 1960 and 1988, the number of production workers in "petroleum and energy products" fell from 138,000 to 108,000 while production doubled. During the same period, OCAW's membership dropped by half, from about 180,000 to 90,000—a tendency magnified by the drive of oil refiners to replace full-time refinery workers with temporary, ill-trained construction workers who don't comprehend the dangers of the petrochemical industry. (This practice of contracting out dangerous jobs was illuminated in ghastly relief by the criminal explosion that leveled the Phillips 66 chemical plant in Pasadena, Texas, killing twenty-three workers on October 23, 1989.)[99] The OCAW expects that 65 of the country's 183 oil refineries will close by the year 2000, as refiners carry out a $20-billion retooling job to produce cleaner-burning gasoline.[100] Oil refining has become so automated, with such a high proportion of management personnel and outside contract employees, that an OCAW strike can no longer shut down production. The oil companies have fought their way free of a uniform contract-expiration date for refinery workers, and currently only one half of the nation's refinery workers are union members, down from around two thirds. "To be honest," joked one staffer, oil companies are so powerful that refinery wage negotiations are more a case of "collective begging" than "collective bargaining." To admit that, he continued, is far from saying that oil workers would be better off without unions: the union can still protect against unjust firings, make sure that promotion is based on seniority rather than favoritism, and fight for safer working conditions.[101] In recent years, the union secured oil-company agreement for one-time payments of $250,000 in death benefits above and beyond workers' compensation if someone is killed on the job.

The problems that the OCAW and its members face are too far-reaching to be addressed through the collective-bargaining mechanism. Ultimately, the OCAW believes that it must build a new social movement and a new political party, one that represents the interests of workers. The OCAW, like many labor unions, has lost faith in the Democratic Party, and while not supporting an electoral alternative at this stage, is urging its members to support Labor Party Advocates (LPA) as an alternative political arena for labor. In its eight-point Working People's Agenda, the union contends that "continued unrestrained growth . . . will kill the planet we live on," and the OCAW convention called for a "Worker's Bill of Rights modeled after the G.I. Bill to educate workers and maintain their income during the difficult transition to a permanent peacetime and environmentally friendly economy."[102] The convention also instructed the union's leadership to help build Labor Party Advocates throughout the labor movement as a first step in creating an independent political voice for labor. By April 1996, LPA had been endorsed by unions representing a million workers.[103] In San Francisco, both the Labor Council and the Building and Construction Trade Council have endorsed Labor Party Advocates, and the San Francisco Labor Council has formed a permanent committee to push for a Labor Party.[104]

Successful examples of worker attempts to guide the direction of technology development are rare, usually occurring when a shutdown threatens, and these attempts have always been resisted by capitalists. Occasionally, attempts by labor to steer the technological process have been effective when labor was experiencing particular market strength—and when the efforts of labor activists have maintained the profitability of the industry in which they were organized. One example was the 1960 "mechanization and modernization" agreement between the International Longshoremen's and Warehousemen's Union (ILWU) on the West Coast. Fearing that the shipping firms would soon move to mechanize with or without the union, Harry Bridges, president of the ILWU, began to discuss the need to use labor-saving equipment throughout the whole range of the members' work situations. After three years of heated discussions between shippers and the union and within the union itself, the union and the Pacific Maritime Association (PMA) signed an agreement that withdrew all restrictions on the introduction of new technology. On the West Coast, shippers agreed to continued pay guarantees for longshore workers, shorter work hours, and earlier retirement in return for longshore workers' assent to containerization and other new technologies, which would increase productivity by twenty to thirty times for cargo formerly unloaded by hand.[105] And in return for ILWU assent to the "mechanization and modernization" agreement, the PMA recognized the principle that displaced workers had a right to share in the profits of technological transformation. Though the number of West Coast longshore workers has declined by almost 90 percent since the World War II peak of 100,000, the ILWU has maintained its integrity as a union, its control over longshore work on the West Coast, retention of the union-controlled hiring hall, and the near-sacred right of members—based on the seniority list—to sign up for work as often or as seldom as they please. Though in financial terms the productivity gains to shippers "far outstripped the gains to the Union and its members," the ILWU "has . . . survived this revolution in technology as well as any union." Though its members still find technological change "pretty scary," the Longshoremen's Union believes it must go along with management plans in order to preserve the jobs of currently employed workers.[106]

The ILWU has come under environmentalist criticism for supporting the continued dredging of San Francisco Bay, which stirs up toxic bottom sediments left over from shipbuilding, pesticide runoff, petrochemical production, and storm runoff from the streets. Though some environmentalists find the union attitude questionable, the *raison d'être* of the labor movement is to increase pay, improve conditions of work, and protect jobs. "Much of trade union organization is primarily defensive . . . the employer acts and the union reacts. In the main, unions are geared to remedy past grievances and to take care of present problems. They rarely prepare to meet the future—let alone anticipate it. There is always the temptation to drift, hoping for the best and meeting problems as they arise."[107]

Yet precedents exist for productive and resilient cooperative efforts involving labor, environmentalists, and industry. In Massachusetts, John O'Connor, formerly of Massachusetts Fair Share (and a coauthor of this book), founded the National Toxics Campaign, which pioneered the Good Neighbor Agreement as a tool for fighting toxics. The basic idea of the agreement is that the local community has a *right to know,* a *right to* inspect, and a *right to negotiate* with local businesses to encourage and even force them to reduce their discharges of toxics and to clean up toxic waste dumps. A similar process could be established for community oversight of energy policies and practices, including mediation of conflicts where environmental issues affect employment security and job conditions for local workers. In Berlin, New Jersey, the Dynasil Corporation, a high-purity glass manufacturer that makes components for the laser optics and aerospace industries, agreed to allow Jane Nogaki of the local Coalition Against Toxics—a National Toxics Campaign affiliate—to inspect its plant, accompanied by Richard Youngstrom, a consulting industrial hygienist. Eventually Dynasil signed a five-part agreement with the Coalition Against Toxics in which the plant agreed to all five cleanup recommendations proposed by the local committee after the inspection.[108] Labor organizer Rand Wilson and John O'Connor also helped organize a nationwide Campaign for Responsible Technology (CRT), which has promoted an Electronics Industry Good Neighbor Campaign. The CRT helped convince Congress to earmark $10 million toward health and safety research from a $100 million research grant to Sematech, which is the national research consortium created to improve the chip industry's competitiveness. Though CRT doesn't control the funds earmarked for health and safety research, the organization is represented on the committee which decides how the $10 million will be spent.

Fortune's Fortunate 500

"After losing money in 1992, America's biggest industrial companies earned $62.6 billion in 1993 . . . Fortune *magazine reports. . . .* Fortune *said the earnings improvement was especially significant because it was achieved despite stagnant growth in sales. The improvement didn't benefit the U.S. jobs picture, however. Total employment among the 500 fell for the ninth straight year, from 11.8 million to 11.5 million. . . ." In the last decade, according to David Birch, President of Cognetics, a Cambridge, Massachusetts, consulting firm, America's largest companies have eliminated 4.7 million jobs, or one quarter of their workforce.*

San Francisco Chronicle,
March 1994, October 1994

Hampered by the rapid rise and fall of electronics companies and production facilities, increasing unemployment in California's Silicon

Valley, and an inability to organize either the production workers or technicians in the industry, electronics activists in the Silicon Valley Toxics Campaign have sought their political base among residents who live near the plants and who are concerned with toxic emissions into the groundwater and the air.[109] The Electronics Industry Good Neighbor Campaign has participating representatives in Austin, Albuquerque, San Jose, Phoenix, and around Route 128 near Boston—all areas with significant concentrations of electronics manufacturing. The basic goal of the Campaign for Responsible Technology is to make Sematech accountable to community, labor, and environmental groups through guaranteed representation on the Sematech board. To accomplish this, CRT encourages individual plants to sign "good neighbor agreements," which give community representatives the right to know and help control the toxic problems connected with the particular production processes. *The Good Neighbor Handbook: A Community-Based Strategy for Sustainable Industry* recounts some of the successes of this community economic development strategy.[110]

NAFTA, Labor, and the Environmental Movement

THE BATTLE OVER the North American Free Trade Agreement (NAFTA) has provided the most intricate example of a strategic alliance between labor and some environmental groups. Along with the majority of organized labor, the Sierra Club, Greenpeace, the Rainforest Action Network, and Public Citizen all vehemently opposed the treaty. For these environmental groups, the pro-NAFTA forces were sounding the trumpet for a frontal assault on hard-won environmental-protection provisions at both the state and federal levels, provisions that could be denounced under NAFTA as "non-tariff barriers to trade."[111] The environmental groups were also shocked and dismayed by the extent of water and air pollution in Mexico, exemplified by the destruction of a rare Mexican sea turtle and the attempt by SCEcorp to build the Carbon II power plant in Rio Escondido with sulfur-dioxide scrubbers that were twenty years out of date. Lori Wallach, director of trade programs for Public Citizen, declared, "It's perplexing to me how any environmental group could support this treaty." At the AFL-CIO convention in San Francisco, Sierra Club executive director Carl Pope gave a rousing speech against NAFTA, vowing that "we're all gonna be lying on the track in front of the train ... to keep it from leaving the station."[112]

For both the labor and environmental opponents of NAFTA, the issue is clear: the passage of the treaty has eroded the ability of both camps to democratically influence the direction and conditions of capitalist investment and development. Over the long run, NAFTA opponents argued, U.S. corporations would invest increasingly in Mexico to take advantage of that country's lower wages and environmental standards. Washington representative Nolan Hancock of the Oil, Chemical and Atomic Workers Union considers it "almost inevitable" that NAFTA will shift oil refining from the United States to Mexico, and he has called for a "national security import fee" to keep oil refineries and jobs in the United States.[113]

Six major environmental groups supported NAFTA: the Natural Resources Defense Council, the National Wildlife Federation, Conservation International, the World Wildlife Fund, the National Audubon Society, and the Environmental Defense Fund. John H. Adams, the executive director of NRDC, told the *New York Times,* "I recognize that this is a complex and difficult issue on which reasonable people with the same values and objectives can differ." Adams said that his staff had spent thousands of hours with Clinton administration officials to help build a "solid institutional framework" that would protect the Mexican environment without weakening U.S.

Principles of CRT's "Electronics Industry Good Neighbor"

- *Conduct a "product life cycle" analysis for semiconductors;*
- *Establish a program for the substitution of all glycol ethers by 1994;*
- *Research alternative production processes to eliminate other toxic hazards;*
- *Establish a safe substitutes policy;*
- *Achieve zero emissions by the year 2000;*
- *Create a model recycling program for semiconductor chips;*
- *Ensure community, labor, and environmental representation on Sematech's board;*
- *Create a National Citizen's Advisory Board.*

The Bargaining Chip, April 1993

---

health and environmental statutes and regulations.[114] President Clinton, backed by a $30-million lobbying push from Mexico and an equivalent effort from U.S. financial and business interests, successfully shepherded NAFTA through Congress in 1993.[115]

The issue of renewable energy offers an especially promising cause around which labor activists and environmentalists can extend their anti-NAFTA coalition-building. The appeal of solar energy to people on the community level and also to some labor groups is twofold: first, solar and wind technologies generate power near the point of use and are therefore more responsive to community control; and second, renewables create more jobs per unit of energy produced than either the nuclear or fossil-fuel industries. Rebuilding the nation's railroads and mass-transit systems; guaranteeing more "energy-efficient mortgages"; and pushing for more wind turbines, photovoltaics, and solar water heating programs in every region of the country are only a few of the vital, forward-looking projects that would benefit immeasurably from the skills and perspectives of union construction workers—if only the unions could set their sights on making sure that union contractors get these jobs.[116]

But even the most ardent solar enthusiasts could not realistically claim that renewable-energy systems will supply enough well-paid jobs to productively occupy all of the country's workers without somehow redistributing those jobs to benefit more people. At some point, the labor movement and its political allies will have to raise again the questions of a shorter workweek, a shorter working life, and increased leisure time. Shorter work hours do not mean that a country cannot compete in the international arena. In 1992, for example, the average skilled industrial worker in the United States spent 1,912 hours on the job, compared to 2,080 for Japan and only 1,667 hours for Germany. Both Germany and Japan are among the world's top creditor nations, even though the Japanese work 25 percent more hours than Germans every year.[117] As unemployment rose in a newly reunified Germany, the metalworkers union and Volkswagen agreed to cut the workweek from 36 hours to 28.8 hours in order to avoid layoffs. Though average take-home pay was cut by 10 percent for the affected employees, total employee costs to Volkswagen fell by almost 20 percent. Jobs were saved, but the metalworkers union had to jettison its historic principle of fewer hours with no cut in pay.[118]

The monster issue in both the 1992 and 1996 presidential campaigns has been jobs and the economy.[119] An environmental-solar energy movement that doesn't address those concerns would be "courting irrelevence."[120] Mainstream environmental groups, including the Natural Resources Defense Council and the Sierra Club, have rarely supported union struggles for higher wages, better working conditions, and a repeal of labor laws such as the 1947 Taft-Hartley Act, which hamstrings union organizing efforts. It is folly for environmental groups that have been hostile or indifferent to labor's concerns to expect labor to support a "green" economic transformation that would accelerate the phaseout of well-paying, unionized jobs, unless those environmental groups agree to campaign for workers' rights and labor representation in the new industries that are being created. Until that happens, a labor-environmental alliance will remain a pious fantasy. But labor unions in the energy sector must also realize that the present economic model based on ever-increasing inputs of fossil fuel is also a sure path to oblivion. To break out of its isolation, labor must support community causes that transcend labor's traditional and narrow preoccupation with wages and union representation.

In this context, we should recall a letter of warning written by cybernetics pioneer Norbert Wiener of the Massachusetts Institute of Technology to United Auto Workers president Walter Reuther in 1949 about the new developments in "servomechanisms and programmable machines," now known as computers:

> This apparatus is extremely flexible and susceptible to mass production and will certainly lead to the factory without workers. In the hands of the present industrial set-up the unemployment produced by such plants can only be disastrous. I would give a guess that a critical situation is bound to arise under any conditions in some ten to twenty years. I do not wish to contribute in any way to selling labor down the river, and I am quite aware that any labor which is in competition with slave labor, whether the slaves are

human or mechanical, must accept the conditions of work of slave labor. For merely to remain aloof is to make sure that the development of these ideas will go into other hands which will probably be much less friendly to organized labor.

Professor Wiener promised full cooperation with Reuther in formulating a labor response to the question of automation, but got no response.[121]

CHAPTER 7

Solar Homesteaders or
Solar Sharecroppers?

*"Impossibility" is a word found only
in the dictionary of fools.*[1]

—NAPOLEON

USTOMER-LOCATED SYSTEMS—utility jargon for home-scale solar water heaters and
rooftop photovoltaic (PV) modules—tend to
frighten utility executives. If people actually produced electricity nearly
maintenance-free from sunlight falling on their own rooftops, the utilities
would experience even greater popular resistance to rate increases,
because either the utilities would be irrelevent entirely, or else their role (in
grid-connected arrangements) would be reduced to mere coordination
and storage of electricity produced by the customers themselves.[2] With
unobtrusive photovoltaic roof shingles, cladding, and even windows now
becoming available, the Swiss have begun to sheath roofs and walls in elec-
tricity-generating photovoltaic cells that keep out the weather as well.
Design work is going on to incorporate solar water heaters into electricity-
generating photovoltaic arrays. A new generation of PVs will incorporate
an "inverter," so users will no longer need a separate electronic unit to convert the solar-generated direct current (DC) to typical household alternating current (AC).[3] The National Association of Home Builders'
Research Center is designing a flexible photovoltaic roof shingle that
could be installed on existing roofs, even if the roofs are uneven.[4] Yet the

superb video produced by the Electric Power Research Institute (EPRI), *Bringing Solar Electricity to Earth*—which shows successful stand-alone PV applications in the Dominican Republic, French Polynesia, and on isolated ranches in the United States—has not been publicized to audiences outside the utility industry.[5] And so far, EPRI has not made a video about the ticklish questions, for utilities, of customer-located, grid-connected PV systems, although a new video made by author and teacher Rob Roy provides an excellent basic introduction to homeowners intrigued by the prospect of harvesting their own energy from sunlight and wind.[6]

Since 1972, PV module costs have declined one-hundredfold, to around $5 per peak watt of power. With the rest of the "balance-of-system" costs (inverter, charge controller, batteries, etc.), photovoltaic systems can be installed today for as little as $8 to $12 per peak watt, or about 40 cents per kilowatt-hour.[7] The limiting factor is not simply the cost of producing the photovoltaic cells. Traditional crystalline silicon cells are the most efficient type for capturing the sun's rays, and the most durable, but they are also the most expensive to produce. As a result, the DOE's National Renewable Energy Laboratory and individual manufacturers have undertaken major efforts to develop thin-film "amorphous" silicon cells, which are not as efficient as crystalline cells, but which can be manufactured much more cheaply.[8] And yet, the costs of balance-of-system components are equivalent to those of the PVs themselves, which means that the price of solar cells could be cut dramatically while reducing total costs only slightly.

The largest solar electric generating plant in the world is the 355-megawatt LUZ International "solar thermal" plant, located between Los Angeles and Las Vegas, which delivers its power to Southern California Edison. Not a photovoltaic plant, LUZ is a 100-acre field of parabolic trough collectors in the Mojave Desert; these mirrors track the sun throughout the day, are repositioned by computer, and concentrate the reflected rays on a boiler to drive a conventional turbine-generator unit. Between 1984 and 1991, LUZ brought down the price of electricity its installations produced from 24 cents to 8 cents per kilowatt-hour. But LUZ went bankrupt, a victim of the cuts in tax incentives for renewables, a cost overrun as the organization hurried to meet a tax-credit deadline, a declining utility buyback rate for the electricity it produced, and an unexpected hike in state property taxes.[9] These uncertainties, combined with a 78-percent decline in the price of natural gas for competing sources of electricity, spooked major investors into pulling out of the project. Bereft of institutional support, and culpable for over-hyping its own commercial viability, LUZ filed for bankruptcy in 1991, though its component installations are still producing electricity.[10]

So-called technical decisions on the installation of and billing for solar electricity mask highly political questions about the potential for local democratic control of energy. How much will a utility pay for overflow rooftop power, "net excess production" in utility parlance? Will the utility own the PV panels and inverters like the phone company used to own the phones? How much generating capacity should be installed on each house? How much overflow electricity will the modules predictably and reliably generate for sale back to the utility? Is there any real danger when the electricity flows in a reverse direction, through the inverter into the power grid? How will electricity use be billed? Will the existing electric meter simply be run backwards when a household generates an overflow—the "net-metering" system—or will the utility insist on installing expensive racheted meters that record separately the flows to and from the customer, and then bills and credits according to time of use? Will utilities—as the Salt River Project in Phoenix or Southern California Edison have done—charge three times as much for the electricity it sells as for the electricity it buys from people

Technological "Breakthroughs" and the Price of Photovoltaic Cells

To get the installed price down will require more than just the development of a new cell technology using metallurgical-grade silicon. It is common among people who have a stake in a particicular technology to portray it as a breakthrough that will dramatically cut the cost of [solar cells], but the fact is that many improvements in many different areas of technology will be needed. The cost of PV is no longer dominated by any one process or material, but is made up of many unrelated components: silicon, glass, metallization, interconnections, insulators, lamination, inverters, installation, etc. Even a dramatic breakthrough in one of these areas is going to result in only an incremental reduction in the total [installed cost of PV systems].

CHRIS KEAVNY, photovoltaic cell designer

with PV arrays . . . or will the utility—as in Europe and Japan—pay a premium over retail rates for PV-generated electricity to encourage household installations? Finally, can a revived citizens' solar movement catalyze massive, increasing photovoltaic buys to create a true mass market and bring down unit costs?

Net metering is a method of electrical billing that allows customers with grid-connected electrical generators (including wind and microhydro turbines, as well as photovoltaics) to register instant credit for the excess electricity they are producing, since at such times the household wind or solar system is functioning exactly like a miniature power plant, providing energy to the interconnected grid. Where net metering is permitted by state regulation and utility practice, people keep their existing meter and hook up a line from the inverter on the customer side of the meter, so that the meter runs backwards if more electricity is being produced at a given time by the solar array than is being used in the house. Any kilowatt-hours of electricity produced are thereby credited against consumption. From four states in 1989—Massachusetts, Minnesota, Texas, and Maine—there are now thirteen states that have some form of net metering provisions.[11]

Minnesota and Wisconsin have the most customer-friendly net-metering regulations. Both states require the utilities to buy back "net excess production" from small renewable-source customer-producers at the retail rate per kilowatt-hour. The municipal-power city of Palo Alto, California, home of Stanford University, has a near-identical regulation. Imagine a situation with monthly billing in which a family closed up and left their house on June 1 with a 3-kilowatt panel on the roof, yet left the electricity turned on. In Palo Alto, they could expect to come home July 1 with a credit of $30 to $40 worth of electricity added to their account.

The Minnesota regulation was passed by the Minnesota Public Utilities Commission in 1983 at the behest of Jacobs Wind Electric, which has produced small wind turbines, or wind chargers, for decades. The frank intent of the Minnesota regulation was "to give the maximum possible encouragement to cogeneration and small power production consistent with protection of the ratepayers and the public." The arrangement has encouraged wind power by side-stepping a lengthy process of determining the "avoided cost" for each individual wind generator, and by providing a fair, easily understood price to farmers who were expected to benefit most from the policy.

The exact language of the regulation on "cogeneration and small power production" bears repeating, because the issue of utility buyback rates will become urgent as soon as rooftop and backyard renewable-energy installations become more common. The regulation states: "a qualifying facility [such as a wind generator or rooftop PV array] having less than 40-kilowatt capacity may elect that the compensation for net input by the qualifying facility into the utility system shall be at the *average retail utility energy rate*" [authors' emphasis].[12] Therefore in Minnesota, small PV and wind producers connected to the grid are not forced to become sharecroppers of the power they generate. Utilities, both public and private, are required by law to buy back surplus power at the average retail rate (in 1994) of about 7 cents per kilowatt-hour. In 1991, thirty-two qualifying wind-generating systems were in place under these regulations.

Most users have bought their wind turbines with a seven-year loan at about 7.5 percent, just like any other piece of farm equipment. Since the turbine's actual lifetime production is 35,000 to 60,000 kilowatt-hours of billable electricity, the real payback time for a $50,000 machine is ten to fifteen years, when accelerated depreciation and a 1.5 cent per kilowatt-hour federal tax credit is figured in. Once the device is paid off, owners can expect to net $2,000 to $4,000 per year in profits from reduced electric bills and electricity sales to the power company. Like most states, Minnesota imports most of its fossil fuel, so the state government has made it official state policy to take advantage of local renewable resources, such as the high average winds near the southwestern town of Holland, to keep energy dollars circulating in the state.[13]

Sharecropping the Sun:
Southern California Edison and Photovoltaics

IN CALIFORNIA, the Public Utilities Commission historically required privately owned utilities to pay 6 cents per kilowatt-hour for wind-generated energy, but allowed them to pay only the "avoided-cost" rate of 3 to 4 cents per kilowatt-hour for excess household photovoltaic energy. Solar businesspeople, such as Steve Coonen of Photocomm, complained that it was discriminatory for the rate structure to compare environmentally clean PV power to the 3 to 4 cents per kilowatt-hour of avoided cost for gas-fired fossil fuel. Coonen's ideal scenario would have been for excess rooftop PV power to be bought back by utilities for at least 10 cents per kilowatt-hour since PV-generated electricity doesn't add to air pollution or the burden of greenhouse gases. But, he lamented in 1992, "there is no way we solar people can match the political strength of the utilities in Sacramento on this issue."[14] And yet, as we shall see below, the solar times are changing in California.

In 1991, Southern California Edison (SCE) began publicly to discuss plans to deploy rooftop photovoltaic panels—designed and manufacturered as a joint venture with Texas Instruments (TI)—on new houses. Because the cells would be made from cheap metallurgical-grade silicon at $1 per pound, compared to the $35-per-pound cost of contemporary semiconductor-grade solar cells, SCE claimed it would be able to install a 100-square-foot grid-connected solar array for no more than $1,500 to $3,000, as opposed to the $8,000-to-$12,000 installed cost of comparable crystalline silicon modules, the typical technology. If projects such as SCE's can really bring the price of PV electricity down to $1.50 per peak installed watt, such electricity would be as cheap as that generated from coal-fired plants. There would be no pollution and no greenhouse gases generated once the units were in use.[15] According to SCE, each module, installed on a south-facing roof in Southern California, could generate the equivalent of 2,000 kilowatt-hours per year of electricity, enough to supply one third of the electrical power needs of the average family. And, if the residence were designed and equipped as efficiently as most off-the-grid homes are, those 2,000 kilowatt-hours per year would be enough to supply *all* of the home's annual net electricity needs. Despite all the ballyhoo in the early 1990s, however, SCE's estimated commencement date was repeatedly pushed back until the venture was cancelled, raising the question of whether the whole proposal was merely a PR stunt.[16]

John Bigger, a photovoltaics expert for many years with the industry-funded Electric Power Research Institute (EPRI), was less sanguine than Southern California Edison about future price reductions for photovoltaic technology. In 1992, Bigger said he believes that with large-scale, continuous, and orderly increases in PV purchases, the installed price of photovoltaics can come down to $5 to $7 per peak watt by the year 2000 and to $3 to $4 per peak watt by 2010.[17]

According to Nick Patapoff, the engineer in charge of photovoltaics for SCE, the proposed TI/SCE solar modules could be built into new homes, "just like a microwave," and would feed power directly into the SCE grid after conversion from DC to AC. Patapoff—like many other utility planners—obviously views the TI/SCE arrays as a supplement to the existing grid, rather than as individual or neighborhood systems designed to stand on their own, and he says that the utility is not interested in contracting out for retrofits of old houses because of "reliability" problems analogous to those encountered with household solar hot water heaters installed in the 1970s and 1980s. After conducting a computer-assisted aerial survey, the utility concluded that the best place to begin to install PV modules would be on flat industrial and warehouse roofs. Restricting the residential PV systems to new houses alone would seem to be a perfect formula for sandbagging or delaying the widespread installation of photovoltaics, since the vast majority of customers live in existing buildings. However, since 1978 SCE has demonstrated at least token interest in stand-alone PV technology, operating a 2-kilowatt photovoltaic system at its general office in Rosemead, outside of Los Angeles, and experimenting with the installation of PV systems for remote, off-the-grid customers who had previously supplied themselves with diesel power. [18]

Ken Koch, the former president of the Electric Vehicle Association of California, has sought support from Southern California Edison for electric transportation and solar electricity. An electric-car pioneer, Koch commuted thirty-one miles per day, or 5,000 miles per year, from 1986 until 1991, in an electric vehicle of his own design that can exceed 60 miles-per-hour on the freeway. Until recently, says Koch, SCE had done nothing to encourage small-time electric-car users. In 1994, however, SCE proposed that the California PUC allow the utility to offer a special off-peak rate of 4.5 cents per kilowatt-hour from 9 PM until noon the next day, a rate available only to those charging electric cars. To be funded by ratepayers at large, the program would cost $150 million between 1995 and 2000, and would include, among other provisions, $37 million to subsidize the cost of electric-car batteries. SCE planners like the program because it would encourage sales of off-peak power.[19]

A battery-operated electric car driven 10,000 miles per year might use between 1,440 and 3,500 kilowatt-hours of electricity per year, according to sources in the electric-car business—equal to the production of one or two 10-foot by 10-foot PV arrays of the type SCE has been planning. By comparison, an electric-assisted bicycle travelled those distances could be expected to use only 33 kilowatt-hours. Everyone in the whole family could ride electric-assisted bicycles for a year with $15 worth of electricity.[20] And the new wave in solar architecture is to make the entire roof, the south-facing walls, and even the windows out of photovoltaic materials, turning the house itself into a solar electric machine.[21] But when SCE engineer Patapoff was asked, "Why not plan for four or five 10-foot by 10-foot arrays, rather than just one, so that the modules would generate *all* of a family's present electric needs plus the juice to run an electric car?" he replied that "aesthetics" would preclude the installation of so many solar modules.[22] The real reason, as John Bigger, the PV expert then of EPRI conceded, is that there is "no incentive to size your PV system to sell electricity back to the system" if the buyback rates are only a third of the price per kilowatt-hour that the utility charges for its power.[23]

John Bryson, the chairman of SCE, says that photovoltaics "can fill a niche that is an ideal fit for us. The sun shines in the hottest part of our service territory when air quality is worst."[24] SCE's attitude is typical: rather than designing the entire electrical grid around renewables such as solar and wind power, the renewables would be deployed to fill the gaps in baseload fossil-fuel power. The reason is simple: despite a plethora of information about the use of hydrogen or compressed air for electricity storage, those are never considered when utilities debate plans to promote renewables. The basis for our entire energy infrastructure is a system of fossil-fuel and nuclear plants that are paid for or being depreciated, and that are factored into the rate base and therefore continually producing profits.[25] The last thing the utility giants want to encourage is the idea that whole communities can become self-sufficient in terms of electricity.

In 1993, SCE announced a new "Partnership with the Sun" solar strategy, which is premised upon the construction of huge parking lots roofed-over with its new solar cells, where commuters' electric cars could be recharged during the working day without adding to the peak afternoon power demand.[26] More recently, the utility has stopped talking about residential generation of solar electricity, and is concentrating its promotional efforts on utility-owned electric-vehicle recharging stations, on a transportable off-the-grid photovoltaic station for the desert research center of California State University at Fullerton, and on Solar One, a $39-million, 10-megawatt solar thermal generator—half-funded by DOE—in the desert near Barstow, California.

Southern California Edison is all for solar so long as the ratepayers, via "green pricing," or the taxpayers, via the government, pay the way. Chairman Bryson has made an urgent plea for "broad-based public support and financing" to make the new solar projects a reality. According to *California Energy Markets*, SCE would pay for $35 million of the $120-million "Partnership with the Sun," and the other $85 million would be contributed by the South Coast Air Quality Management District, the Department of Energy, and a number of other partners. "With the help of public/private partners such as the Air Quality Management District and the California Public Utilities Commission," said Bryson, "we can establish in California a new solar energy industry that can create jobs, increase our energy independence, and clean our air."[27] Bryson made no mention of relinquishing utility control. Increasing the buyback rates for electricity from grid-connected, customer-owned PV systems is clearly not part of the plan.

If the SCE planners have their way, three conditions will have to be met for household solar electricity to become commonplace in the utilitiy's service area:

1. The government and ratepayers must pay for renewable-energy R&D;
2. SCE must meter the current, and charge substantially lower rates for the "net excess power" it buys from PV-producers than customers pay for the power the utility sells to them; and

3. Photovoltaic and wind-generated power must be used primarily to supplement, rather than reduce or replace, peak demand.

Those are very narrow goals on the agenda of the Photovoltaics for Utilities "working group," which has begun to hold meetings around the country to promote utility use of photovoltaics as the price of solar technology drops.[28]

In essence, Southern California Edison's aim is to turn photovoltaic power producers into sharecroppers of their own sunlight—whether the PV systems are grid-connected or freestanding. By late 1994, a battle had commenced within PV4U, the California branch of the federal Department of Energy's TEAM-UP program, which subsidizes utility expenditures in photovoltaics, over the issue of the ownership of customer-located PV systems. Utility representatives insisted that utilities must own the systems, no matter where they are located, while consumer groups and a small trade association of off-the-grid photovoltaic installers adamantly opposed utility ownership.[29]

SCE had initially tried to use its financial power to muscle into off-the-grid photovoltaic markets it had never served in the past. According to a proposal presented to the California Public Utilities Commission in late 1993, customers would "lease" a $20,000, 1-kilowatt system over a fifteen-year period with monthly payments of $320—regardless of how much power is used—and would then own the system at the end of the period. SCE's payback terms represent a 25-percent annual return on capital outlay. (Don Loweburg of Offline, a company that installs off-the-grid photovoltaic systems, points out that expected battery life is approximately fifteen years, so half of the value of the total installed system would be exhausted by the time it reverts to customer ownership.)[30] By contrast, if homeowners could finance a comparable $20,000 system through a home mortgage at 8 percent, their monthly payments would drop to $180.[31] But, to date, it has been almost impossible to convince the federal home-loan/mortgage-guarantee programs to finance off-the-grid houses, which constitute 98 percent of solar electric business, because photovoltaics are not considered to be "electricity" for financing purposes. Another way to foster widespread citizen ownership of solar electric systems is to make sure that the federal government guarantees home mortgages for those systems.[32] To leave the financing up to the utilities is to turn people into photovoltaic serfs.

The immediate reaction of solar contractors was to protest SCE's off-the-grid initiative, so the utility modified its proposal by promising to contract out all work through existing solar installers. About half of the affected contractors went along with the proposal, believing that they might gain some business and that it would be good for the industry if the public were to see actual PV systems being installed and used. Another faction decided to form the Independent Power Providers (IPP) in order to fight SCE's initiative before the California Public Utilities Commission. Don Loweburg, who acts as the spokesperson for IPP, called SCE's experimental proposal "anti-competitive" in a request for denial filed with the California PUC. "Southern California Edison's program," wrote Loweburg, "is in fact designed to move into an already existing market and gain control. Edison's market goal of one megawatt represents the current yearly total off-the-grid market volume in California. The three-year ... experimental ... duration of this program will certainly be sufficient to put many small companies out of business."[33]

A Question of Ownership

Photovoltaics are the first widely applicable electric power source [whose fuel] is not a commodity. Solar energy is the first power source that can break the energy companies' monopoly on power ... The big question is who owns these PV modules, and more specifically who owns the electric power these modules make? Will the energy companies hold on to their power monopoly, or will electric power become something we commonly own?

By analogy, consider the following: you live in a cold climate where vegetables, shipped from a distant place, are only available at supermarkets. Someone invents a greenhouse that allows local vegetable production. Imagine inviting the supermarket to come to your homestead, erect a greenhouse, grow a garden, then sell you the vegetables.

RICHARD PEREZ, *Home Power*

A one-year report to the Commission Advisory and Compliance Division (CACD) of the California Public Utilities Commission noted that SCE's first-year revenues were only $3,000, compared with expenses of $271,000, mostly in regulatory costs. In the first year, SCE could show only one signed contract for installation of an off-the-grid PV system. At first, the CACD recommended that SCE's pilot program be discontinued, but this idea was overruled by PUC members who decided to let the program run two more years to prove itself. Ray Paz, the spokesperson in charge of photovoltaics in SCE's Dispersed Energy Applications program, indicated in August 1996 that the utility intended to maintain the complete program,[34] despite rumors that it might spin off the PV installation business to an unregulated subsidiary. So far, in any case, the independents have been winning, by offering superior products and services at lower prices, and SCE is starting to let bids out to the independents to build and install the PV systems that the utility itself sells.

Meanwhile, the California Public Utilities Commission has guaranteed that the ratepayers will pay for SCE's nuclear mistakes. Through the year 2003, says Rich Ferguson, the energy chair of the Sierra Club, it will cost the ratepayers $1.4 billion more to operate the two San Onofre nuclear plants than it would have cost to shut them down in 1996.[35] It makes sense to assume that the utility's long-term goal in entering the off-the-grid market is to establish its credibility as a player in household, grid-connected photovoltaics, since the price of photovoltaic electricity continues to fall. According to 1992 estimates from the American Solar Energy Association, the delivered price of photovoltaic electricity may dip below 10 cents per kilowatt-hour by the year 2000, less than the present cost of San Onofre power, and SCE wants to avoid future political challenges to its control of part of the market for solar electricity generated in remote sites on people's roofs.[36] But the off-the-grid market is miniscule compared to the long-term potential for grid-connected, household photovoltaics.

In Idaho and Nevada, private utilities are likewise trying to buy their way into the off-the-grid photovoltaic market.[37] Idaho Power will set up a photovoltaic system in areas not served by the grid, but most potential customers aren't buying their equipment from the utility, since the catch is that Idaho Power will own the system even after it has been paid off. According to Gary Beckwith, a PV technician for Real Goods Trading Corporation, most Idahoans interested in off-the-grid photovoltaics, domestic solar water heating, and other renewable technologies, prefer to take advantage of generous state tax deductions of up to $5,000 per year over four years to pay for installing their own systems or hiring independent local contractors. In addition, the Idaho Department of Water Resources offers loans of up to $10,000 for residential systems and up to $100,000 for commercial and industrial renewable-energy systems, repayable at 4-percent interest over five years.[38]

Oregon is one of the most generous states when it comes to solar tax incentives: Oregon allows its citizens to divert up to $1,500 of money that they would normally pay for state income taxes for a whole range of household renewable-energy projects, including photovoltaics, solar water heating, and wind generators. In some areas of Oregon, the utilities offer rebates of $400 to $800 for the installation of domestic solar water heaters.[39] It must be noted that low-income people who don't own property—those also hit hardest by high utility bills—have no way to participate in most of the state's programs.

What's Yours Is Mine and What's Mine Is Mine: PVs in Phoenix

IN THE NEAR FUTURE, the biggest barriers to household photovoltaic electricity will be the utilities themselves, who have a natural aversion to a technology which makes their centralized power plants and dams irrelevent. The biggest battles will be over money and property rights: who owns the sunlight, who owns the PV panels, and how and at what rate should the electricity be metered. All of those questions have arisen at the Solar One project in Phoenix.[40] Developer John F. Long—a contractor who has built an astounding forty thousand homes since 1947—decided to develop and supply an entire neighborhood of twenty houses from renewable-energy sources. Taking advantage of federal tax credits before they ran out in 1985, John F. Long Properties spent a whopping $1.6 million on the installations. Solar One's photovoltaic array is tied into the electrical grid owned by the Salt River Project, a federally owned irrigation and power network governed by an elected board of large agricultural growers. Long's firm promised "free electricity" to Solar One home buyers, a promise, however, that was never put in writing. The solar hot water heaters in each house have "worked out great" so long as the owners use a water softener to prevent the systems from plugging up. There have been few technical problems with the

PV modules and with the inverters that convert the raw DC to AC to conform with the existing electric grid (and with most appliances). The biggest design mistake in the project was to locate all of the PV modules on the equivalent of two building lots at the entrance to the development. Not only were two building lots wasted, and expenses incurred building a wall around the lot, but the modules have been repeatedly vandalized by rock-throwers who have smashed equipment worth tens of thousands of dollars.

The photovoltaic panels themselves generate 280,000 kilowatt-hours per year. The excess power is sold to the Salt River Project. Magnus Jolayemi, an agricultural engineer and head of the Solar One Homeowners Association (SOHA), has said that he and others were originally attracted to the project by the developer's verbal promise of 500 to 1,200 kilowatt-hours per month of "free electricity," a promise that was largely kept until late in 1990, when Long sold the PV array to the homeowners association for a nominal fee of $100.[41] Without the wealthy builder's powerful advocacy, the relationship with the Salt River Project began to sour. According to Rosemarie Williams, who volunteered to read meters for the homeowners association: "This thing works, but I think Salt River is trying to shut it down." As the volunteer monthly meter-reader, Williams noted great discrepancies between the amount of electrical power generated by the PV array and the amount Salt River Project paid for electricity bought back from the system.[42]

But the most expensive problem has been in the Salt River Project electric bills themselves: Rosemarie Williams' bill jumped from -$24.12 (the utility paid *her* money!) in April 1990 to +$1.71 in April 1991, and from -$18.18 in July 1988 to $57.76 in July 1990, even though Solar One, now owned collectively by SOHA, produced as much electricity each year as it consumed.[43]

The basic reason for the gyrating bills was the Salt River utility's strict observance of that hoary business principle: "Buy low and sell high." Rather than simply allowing Solar One homeowners to run their meters backwards, the entire solar array and each house were equipped with a very expensive system that separately monitored both incoming and outgoing current according to time-of-use (TOU). Then the rates were rigged in favor of the Salt River Project (see table 7–1), so that dur-

TABLE 7-1: Time-of-Use Rate Structure Set by Salt River Project for Solar One Homeowners Association (in cents per kWh), Summer 1991*

	A photovoltaic electricity bought from homeowners	B utility electricity sold to homeowners	A-B less to homeowners	A/B
Off peak 11 PM–11 AM	1.6 cents	3.8 cents	-2.2 cents	0.42
Shoulder peak 11 AM–2 PM	2.2 cents	9.5 cents	-7.3 cents	0.23
On peak 2 PM–7 PM	2.4 cents	9.5 cents	-7.1 cents	0.25
Shoulder peak 7 PM–11 PM	2.2 cents	9.5 cents	-7.3 cents	0.23

*Summer Rate, applicable May 15–Oct 14, 1991
Source: Rosemarie Williams, Solar One Homeowners Association.

ing sweltering Phoenix summers, when the Salt River Project should have been grateful for the extra peak-levelling power, the owners of the neighborhood photovoltaic facility were paying a whopping 7-cent spread: they were buying power from the utility for 9.5 cents and getting paid only 2.4 cents for their power!

There were further disputes over exactly how much electricity was actually being produced by the Solar One PV array: in August 1991, the array reportedly produced 25,560 kilowatt-hours but was credited with producing only 6,300 kilowatt-hours (at an average of 2.7 cents per kilowatt-hour) by the utility. In the same month, the twenty homeowners bought 24,660 kilowatt-hours from the utility at an average price of 7.7 cents per kilowatt-hour. No wonder the Solar One homeowners were furious. Although their solar array had produced 900 kilowatt-hours more than they had consumed during hot afternoon peaks in August 1991, they still ended up paying $1,908.89 to the utility.

The problems of the Solar One Homeowners Association show the lengths to which a utility will go to sabotage a project it doesn't understand or wants to destroy. Originally part of a visionary scheme by John F. Long Associates to create an entire solar town in the sunny desert outside Phoenix, Solar One was thrown together rather quickly to take advantage of federal tax credits that were about to lapse. The decision to build a centralized photovoltaic array on two building lots was a marketing decision. Long Associates believed, mistakenly, that people would not appreciate the aesthetics of on-site photovoltaics, whereas most homeowners are proud of their PV systems and love to show them off. The larger Phoenix project was scratched, not because of technical considerations, but because the housing market collapsed and federal tax incentives for clean energy ended in 1985. Once again, as in the case of solar hot water heating in California, the technology itself was the least of the problems. "The panels work fine," says Rosemarie Williams, "the problem is getting the Salt River Project to stop screwing us."

Despite its reputation as the nation's sunniest state, and its history as the site of the first functioning solar water pump, Arizona corporate leaders have been hostile or indifferent to solar. "In Concert with the Environment," an expensive teaching module made available gratis in 1991 to high schools by the Arizona Public Service Company, the state's largest private utility, included a video with a homily by APSC president Mark DeMichaele about how "each of us can impact our environment" by "learning wise energy-use habits now."[44] The state government has allowed over $20 million of Exxon gasoline overcharge funds to be squandered in series of real-estate adventures and utility subsidies that had nothing to do with the supposed goal of subsidizing energy for the poor and promoting energy self-sufficiency.[45] Magnus Jolayemi, president of the Solar One Homeowners Association, is bitter about this experience with a supposedly "publicly owned" utility. "We are guinea pigs here. We are experimental test subjects for solar energy in Arizona. But we have gotten ripped off by SRP [the Salt River Project]."[46]

As we have stressed, knowledgeable observers no longer question the technological feasibility of photovoltaics. The most interesting recent breakthroughs have related to the financial assumptions about integrating photovoltaics, wind power, and other intermittent renewable-energy technologies into the utility grid. Professor Shimon Awerbuch has argued that the traditional system of utility accounting, which focuses on dollars per kilowatt-hour of newly installed capacity, underestimates the value of relatively simple, decentralized technologies such as photo-

TABLE 7-2: Comparison of New Generating-Capacity Impacts

	Rooftop photovoltaics	Gas-fired combustion turbine
Fuel cost	zero	unpredictable, will rise
Fuel availability	total	unpredictable, increasingly imported
Maintenance costs	low	low but rising
Cost per unit	low	high
Emissions	none	carbon monoxide, nitrogen oxides, etc.
Greenhouse gases	none	carbon dioxide, methane
Pollution control	none	production site
Line losses	negligible	5%–15%
Fuel inventories	none	substantial
Siting costs	negligible	substantial

Source: Shimon Awerbuch, adapted from several original sources.[47]

voltaics, which have high upfront costs and low, risk-free operation costs. In comparing home-based photovoltaic electricity to a natural-gas–fired combustion turbine, a number of factors beyond simple upfront cost have to be compared as well (see table 7-2).

Awerbuch makes the striking argument that photovoltaics are cost-effective for utilities today for peak capacity, and are very close to being competitive for baseload capacity, even without counting the environmental costs, which up until now have been borne by the public at large. Gas-fired combustion turbines may be cheap to build, but future volatility of natural-gas prices makes them a bad risk over twenty or thirty years. "If you want a secure income of $150 a year for the next twenty years," Awerbuch asks rhetorically, "do you plunk down $1,000 for a junk bond or do you go out and buy a Treasury bill for $2,000?" For Awerbuch, because of the impossibility of forecasting fuel prices, installing a gas turbine is like buying a junk bond and installing photovoltaics is like buying a T-bill.[48] Another major advantage to photovoltaics is that they can reduce the need for long-distance, high-voltage transmission wires, which monopolize space and disturb the visual harmony of the landscape.

The conventional utility theory of photovoltaic diffusion holds that PVs will spread from small, stand-alone applications—such as those that have been used to power the pumps for isolated stock-watering troughs or to electrify houses located miles from utility lines—to rural villages and islands (such as French Polynesia), to installations built to provide peaking power on hot summer afternoons, and finally to bulk power stations.[49]

Engineers in the business world have already recognized that photovoltaics (usually combined with battery storage) are an efficient, low-maintenance power source where small amounts of electricity are needed at remote locations. For purposes such as water-pumping for cattle-watering or irrigation, using PV is much cheaper than maintaining a small diesel or running a supplementary power line. M. C. Russell of Ascension Technologies in Waltham, Massachusetts, points out that it is now almost impossible to drive more than five or ten miles on major highways between New York and New Haven without seeing a photovoltaic device that powers a remote telephone or neutralizes the electric current on a bridge to prevent it from rusting. John Bigger, formerly with the Electric Power Research Institute, has published a list of sixty-five different uses for photovoltaics *within* any major utility's operation, and he spends a great deal of time traveling and helping install PVs for those applications. A typical example would be using PVs for powering air-sampling equipment near a large power plant. Photovoltaics will take over many tasks, the engineers acknowledge, as the technology becomes cheaper and more flexible and moves from isolated, exotic uses to cheaper, mainline electricity supply functions.

Solar skeptics often focus on the problem of storing electricity from "sporadic" sources. And yet, according to a U.S. government study released in 1990, "when intermittent sources of power exceed about 10 percent of a utility's generating capacity, . . . energy storage can prove critical in matching supply and demand. By 2030, having adequate storage available could double the production of electricity from solar and wind resources. However, [research] shows that between now and 2010, storage will not be a prerequisite to penetration by intermittent renewables."[50] Another study conducted by Pacific Gas & Electric showed that careful management would allow up to 35 percent of electric power production to be achieved with intermittent sources without any increase in electricity storage capacity.[51] By treating home-based photovoltaics and other renewable-energy applications primarily as tools for grid management rather than as sources of power, utilities are trying to assert their right to own customer-located PV arrays.[52]

The pace at which utilities adopt the views of their top researchers and PV specialists will probably depend much more on political pressures and commitments from fuel suppliers than on technical feasibility. No one except coal producers doubts that the world's supply of fossil fuels will run out eventually, but, for the time being, fossil-fuel electricity generation will always appear cheaper than adding new photovoltaic capacity. The U.S. Department of Energy's "America's 21st-Century Program" is perfectly willing to contribute $650,000 of U.S.-produced PV panels, inverters, and batteries to light up a couple of thousand buildings in Brazil, and it has much grander PV projects in mind with World Bank sponsorship for India ($140 million) and Indonesia ($120 million). But such projects are mostly for foreign consumption.[53] Whether photovoltaics will soon be installed on a mass basis in the United States depends on the nitty-gritty details of ownership, cost, and control. All of those issues have continued to occasion bitter dispute.

Manufacturing Photovoltaics: Another Toxic Time Bomb?

SOLAR ADVOCATES often forget that photovoltaic cells are difficult and hazardous to manufacture. Solar cells are a form of semiconductor, and, like any other semiconductor, production is "dependent on a variety of toxic and dangerous materials which only recently have come into large-scale use."[54] Solar-cell fabrication consists of three basic steps: wafer cleaning, cell structure deposition, and cell processing. Solvents and acids are used for wafer cleaning; flammable liquids, solvents, and toxic gases are used for cell deposition; and solvents, acids, and bases are used for cell processing. This alchemist's brew justifiably represents a lucrative bonanza for industrial hygienists and designers: appropriate handling requires systems for solvent recovery, acid neutralization, toxic-waste isolation and collection, and waste-gas scrubbing.[55] Silane (SiH_4), a gas that catches fire or explodes when exposed to air, may be the most dangerous raw material used. Arsine gas (H_3As) is a highly poisonous as well as flammable arsenic compound stored under pressure. Though dozens of substances used in the production of different kinds of photovoltaic cells are acutely toxic, explosive, or highly flammable, photovoltaics designer Chris Keavny reminds us that individual factories produce a single kind of PV cell, so under no circumstances are all of these potent chemicals present in a particular manufacturing process. Arsine, for example, is used exclusively to make gallium arsenide cells for satellites and for research. Acids and alkalis are used to produce crystalline silicon cells. Silane and other gases are used to produce amorphous silicon cells, whose production processes require few or no acids and alkalis.[56]

The government, aware of the hazards, has created the Photovoltaic Environmental Health and Safety Assistance Center at Brookhaven National Laboratory to deal with these issues. Paul Moscowitz, the chief PV safety expert at the center, is familiar with the German-owned Siemens Solar factory in Camarillo, California, and he believes that the hazards there have been reduced to "acceptable" levels. "With due diligence," he adds, none of the health and safety problems are unmanageable. Solar advocates can only hope that what Moscowitz says is true. Still in its infancy, PV production is growing at an impressive rate—18 percent per year from 1986 to 1991—and the pace promises to quicken. So far, federal health and safety efforts have lagged; Moscowitz estimates that only 1 percent of government photovoltaic expenditures are on health and safety, compared to 10 percent for other DOE-funded energy projects.[57]

Unless caution is exercised in the early stages of setting up PV manufacturing processes, the problems will probably be reminiscent of the "toxic underbelly" of Silicon Valley's "clean" electronics industry. By the early 1980s, it had become evident that chemicals were poisoning electronics workers at three times the average rate for California industry, and that some of the same toxins were also "lurking in the kitchen tap" and might be causing birth defects and cancer. Slowly, pressed by the Silicon Valley Toxic Campaign and other community and labor organizations, government responded with the enactment of the world's first hazardous-materials storage ordinance to prevent groundwater leaks and explosions, with a pioneer right-to-know ordinance, and with legislation to control the emissions of toxic gases and of ozone-layer–threatening CFCs.[58] And slowly, too, industry has begun to change its practices. As explained in chapter 6, the example of the Silicon Valley Toxic Campaign has catalyzed a nationwide Campaign for Responsible Technology (CRT), which convinced Congress to earmark $10 million for health and safety from the $100 million authorized for Sematech, a nonprofit consortium of fourteen U.S.-owned semiconductor companies that is trying to reclaim leadership in microchip design and production from the Japanese.[59] The proportion of government photovoltaics research funds earmarked for health and safety should also be bumped up to 10 percent, and PV health and safety efforts should include active participation by community and labor groups in the governance of the Photovoltaic Health and Safety Assistance Center at Brookhaven. So far, since many of the photovoltaics and microchip production processes are similar, the tendency of the still-rather-small photovoltaics industry has been to adopt the hazard-education materials of its larger cousin.[60]

Information about toxics exposures and emissions in specific photovoltaic plants is sometimes hard to come by. The Siemens Solar plant in California, the world's biggest producer of crystalline silicon photovoltaic cells, in 1990 reported releases (mostly up the stack) of 11,960 pounds of 1,1,1-trichloromethane (TCA), a suspected carcinogen and proven stratospheric ozone-eater, and substantial amounts of methylene chloride, which is a probable carcinogen and is associated with a higher risk of heart attacks in exposed workers. Both of these chlorinated hydrocarbons have caused serious groundwater contamination problems in Silicon Valley. Other Siemens Solar releases include substantial amounts of hydrochloric and sulfuric acid and sodium hydroxide.[61] Solvent disposal declined by 70 percent between 1987 and 1990, according to the Toxic Release Inventory. These problems notwithstanding, Paul

Moscowitz considers Siemens Solar a model operation that is doing the best it can to protect its workers and the public.

Solar advocates and the PV industry have a special responsibility to insist that photovoltaics, which are proudly touted as a "clean" power source, be produced in a clean manner.[62] As the industry continues to grow, its competitors and critics can be counted on to use every weapon possible to slow its expansion and splinter its advocates in the environmental community. To demonstrate good faith, the nascent PV industry should begin by signing "good neighbor agreements" with local neighborhoods, in which companies agree to make available accurate information about the problems as well as the triumphs of photovoltaic production and allow community organizations to partipate in inspections. The PV industry should also seek inspection and advice from electronics industry unions and from organizations such as the Cambridge-based Jobs and the Environment Campaign, the Campaign for a Responsible Technology, and the Silicon Valley Toxics Coalition—all of which have widespread experience in assessing the dangers of semiconductor production. Eventually, the total production of PV semiconductors may dwarf the production of microchips. And solar activists and the PV industry must constantly reassess the photovoltaic technologies that are still at the research stage, and that promise to be much less hazardous to produce.[63] To do otherwise would be folly and a betrayal of one of the most important reasons for environmentalists to support solar electricity: the fact that it is clean, waste-free, and emission-free at the point of electricity production.

The Giants of the American Photovoltaics Industry

THROUGH THE MID-1980S, U.S. oil companies were the chief players in solar, but by 1990 *Barron's* had begun to detect a new trend in the field: the sale of the photovoltaic business from the hands of the major oil companies to electronics firms.[64] Though the parent firms sometimes change hands, domination by giants—foreign and domestic—continues. Siemens, Germany's largest electrical equipment company, bought Arco Solar and renamed it Siemens Solar. The fact that Arco Solar had recently won a $2.6 million federal R&D grant for "high-

TABLE 7-3: U.S. Photovoltaic Module Shipments, 1995 (by company, in megawatts)

Siemens Solar (formerly owned by Arco)	17.0
Solarex (formerly independent, now owned by Enron and Amoco)	9.5
Solec International (formerly independent, now owned by Sanyo Electric and Sumitomo)	2.6
Astropower	2.5
ASE Americas (formerly Mobil Solar, now owned by a German utility and electrical equipment company)	2.0
Three smaller U.S. companies	0.9
TOTAL: U.S. production	34.5
TOTAL: European production	21.1
TOTAL: Japanese production	19.2
TOTAL: World production	113.0

Sources: PV News 16, no. 2: (February 1996); information on ownership of companies from "U.S. Industry Harnesses the Sun and Exports It," *New York Times* (June 5, 1996, pp. C1, C18).

efficiency, amorphous-silicon-based submodules" probably helped raise the selling price.[65] Siemens Solar produced 17.0 megawatts of photovoltaic panels in 1995, half of the U.S. total. Solarex, a joint venture of Enron (the largest natural-gas producer in the U.S.) and Amoco ($38 billion in oil sales in 1993) was second, producing 28 percent; and formerly independent Solec International, now owned by Sanyo Electric and Sumitomo, was responsible for another 8 percent of total U.S. production of photovoltaic panels. The top two producers of photovoltaics in the United States control 78 percent of the market; and the top five control over 99 percent.[66] In July 1994, Mobil Oil had sold its Mobil Solar subsidiary to German interests, which renamed it ASE Americas,

Inc. With 75 percent of American-produced photovoltaic cells sold overseas, according to the *New York Times*, ASE Americas apparently expects to find its markets there as well, especially in Latin America.[67]

During the 1980s—the Reagan-Bush decade—the major energy companies were almost totally successful in halting the implementation of solar and other renewables as large-scale solutions to America's energy needs. A few figures tell the story. In 1984, federal energy subsidies, mostly in the form of tax breaks, totaled $42 billion: $22 billion (52 percent) to fossil fuels, $16 billion (38 percent) to nuclear, and only $4 billion (10 percent) to renewable energy. In addition, privately owned utilities were allowed to deduct the cost of fuel as an "operating expense," a federal tax break that saves them $11 billion per year, according to an unpublished study by the U.S. Department of Energy.[68]

In addition to reinforcing the fossil-fuel status quo through direct grants and the tax structure, the Reagan-Bush administration did its best to preempt the future by eviscerating the research budget for solar. The Department of Energy research and development budget in 1989 totalled $324 million for fossil energy and $345 million for nuclear, compared with a mere $35 million appropriation for photovoltaics.[69] Between 1980 (the last year of the Carter administration) and 1990, DOE outlays for R&D of all renewable-energy sources crashed from $557 million to $81 million, a near 90-percent reduction in terms of constant dollars.[70]

Thus the Reagan-Bush forces had won that round of battles, if not the war. This victory can be tallied by commonly available energy statistics: two decades after the first Earth Day, the contribution of solar energy (including wind power) to the nation's usable energy supply was still infinitesimal: less than one part in ten thousand. Of a total of 82 quads (quadrillion Btus) of energy consumed in the United States in 1988, only 0.05 quads were derived from solar thermal plants and 0.02 quads from wind generators. When federal and California tax breaks ended in 1985, the number of new solar hot water heaters and wind installations nosedived. The U.S. Export Council for Renewable Energy (US/ECRE), closely tied to the Solar Energy Industries Association, tries to put a cheerful front on the matter by lauding aggregate figures for all the "renewable" energy sources: hydropower, biomass (mostly wood waste burned by lumber and paper companies), solar (60 to 95 percent derived from LUZ International's troughs in the Mojave Desert), geothermal (not actually a renewable-energy source), and wind power. But hydroelectric dams are nothing new and are hardly ecologically benign; nor is the burning of wood waste for energy "new"; woodburning plants account for most of the increase in "renewable" energy chronicled in the US/ECRE 1990 report.[71]

Photovoltaic production has been soaring in the 1990s. Production in the United States led the way, increasing from 25.6 megawatts to 34.5 megawatts from 1994 to 1995, a 35-percent rise in one year alone. Similarly, U.S. production had increased almost 2.5 times from 1990 to 1995, according to *PV News*, the industry's statistical bible. Total world production more than doubled to 113.0 megawatts between 1990 and 1995. As table 7–3 confirms, the United States has been the global leader in photovoltaics, followed by Europe and Japan. Only 8 percent of world production occurred elsewhere. Outside of the U.S., Europe, and Japan, India was by far the largest producer, with almost 4 megawatts of PV production to its credit.

By 1995, 62 percent of "American" production actually occurred within foreign-owned (mostly German) companies, a trend that seemed to have begun with the 1990 purchase of Arco Solar, the world's biggest PV producer, by the German corporation Siemens. According to Paul Maycock, about 75 percent of U.S.-produced photovoltaics are exported. Of the 8.8 megawatts used domestically, about 2 megawatts are installed on off-the-grid buildings and only 1 megawatt on buildings connected with the utilities' electric grid.

To a casual observer in the early 1990s, the American solar energy industry, abandoned by the government and squeezed by corporate preemption of government funding, seemed to have died, at least in terms of being an alternative social force with its own clearly defined mission and heroes and enemies. The main problem with that gloomy appraisal was the fact that people were buying solar devices and low-energy household appliances like never before. As the price of photovoltaics came down, the impetus for adoption came from two separate currents: those who saw PV as a commercial-industrial tool for engineers and other experts, and those who treated it as an indispensable part of a grassroots techno-logical movement for self-reliance and ecologically sound independent living. In fact, a new solar energy movement appears to be stirring up the ashes of the old movement, and is finding live embers to rekindle.

Certain survivors of the political movement for solar energy have made a temporary peace with the utilities and oil companies and found niches in promoting low-wattage, screw-in fluorescent light bulbs for municipal energy-conservation commissions, or advising the utilities on energy conservation. Peter Barnes, who had founded and co-directed San Francisco's Solar Center, went on to cofound Working Assets, a socially responsible investment brokerage firm. Denis Hayes founded a group called Green Seal, Inc., one of several private agencies that is checking and certifying consumer products for their environmental suitability. At present, he heads the environmentalist Bullitt Foundation in Seattle, and chairs the board of the Energy Foundation.[72] Dr. Charles Komanoff, a New York economist, earned himself a national reputation in the 1970s by proving that nuclear plants are so unreliable that they remained on-line only 55 percent of the time.[73] He now supports him-self as an energy consultant, and several years ago started a group called Transportation Alternatives to make New York, his native metropolis, more friendly to walkers, cyclists, and mass-transit users.

Solar activism and grassroots technological ferment has continued steadily only in those remote and rugged regions of the country where a critical mass of counterculture rebels had made solar a matter of person-al and political principle—and where some homes are so isolated that solar has been cheaper than utility hookups that would have necessitated line extensions for the existing grid.

"I think there are two main factors," says David Katz of Alternative Energy Engineering:

[A utility hookup] takes away from your independence. If you tie in with the utility company, then you've got wires under your house; you've got a group of people in trucks who are going to come every four to six months and trim the trees, follow the wires, and you're going to have a path to your door. To get utility power, you have to put a septic system in, you have to deal with the county, have inspectors come to your house. You bring yourself into the mainstream and you become part of the culture you ran away from.

The other part is that there are a lot of people who don't like where the power comes from, the fact that they are mining uranium to make nuclear power on Native American land, or that they are burning coal and causing acid rain, causing global warming. People don't want to put that energy into their house.

The Northern California counterculture, with its widely scattered settlements and multi-talented inhabitants, has fostered the use of solar electricity more effectively than have proponents in any other region of the United States. Most people in that region have a passing familiarity with photovoltaics and know its drawbacks. Yet even a purist like David Katz is honest enough to concede that a principal use of the photovoltaic modules he sells is to keep up with David Letterman and *Seinfeld*:

One of the reasons people moved up here was to get away from the death cultures of the city, to get away from television, to get away from the government, to be free, to be in a place that really wasn't a part of the nation, that wasn't controlled by anybody. The electricity made it easier to bring that stuff in . . . you go around the hills, and you see satellite dishes everywhere, televisions everywhere. That definitely wasn't good. It also made it easier for people to move here . . . people could sell their home in Los Angeles and spend $10,000 on a power system and have all the same conveniences in their houses here, and maybe that wasn't so good. Even the people who have all these [conveniences] live closer to the land than people in the city. You live in the trees. And if people didn't have the solar panels, they'd all be running generators all day.[74]

Through the 1990s, the social composition of the off-the-grid homesteaders has been becoming more mainstream. Photovoltaic contractor Don Loweburg of North Fork, California, notes that the day of cultural rebels is gone. Most of his customers are solid middle-class people in remote areas who like the idea of energy independence and find a utility hookup too expensive. And yet so far, these people have not made the commitment to activism and advocacy that would have brought renewable-energy issues into the mainstream political discourse.

Integrating Solar Electricity Internationally

IN 1992, SWITZERLAND had about four hundred grid-connected PV systems, ranging in size from 500 watts to 560 kilowatts. The norm is a household system of 3 kilowatts with an average annual production of 3,000 kilowatt-hours of electricity. According to Bill Maag, the Swiss department of energy has allocated, from carbon-tax revenues, about 100 million francs per year (roughly $83 million) for five years to support renewables such as photovoltaics, biogas, and wind—a figure that is not much below the entire 1991 U.S. budget of $129.5 million for renewable energy.[75] As a small, mountainous country, Switzerland has no land to waste, so the Swiss federal energy office has allocated the equivalent of over $4 million to developing building-integrated PV systems. Pioneering PV-architect Steven J. Strong of Solar Design Associates in Massachusetts reports that one of the cleverest projects being funded is a "novel PV roof tile system" with "quick-connect, plug-in electrical connectors" that can be used to replace existing roof tiles, or used on new construction.[76] The goal of Switzerland's national Energy 2000 program is for fossil-fuel use to peak in 1995 and to begin declining by the year 2000.[77] To match the Swiss level of spending, given its larger size, the United States would have to spend almost $3 billion a year—about the size of the program proposed in the Solar Lobby's 1979 *Blueprint for a Solar America*. By 1992, Switzerland, with ⅟₃₇th the population of the United States, already had one half as much installed, grid-connected photovoltaic capacity as the U.S.[78] In both Germany and Switzerland, utilities promote photovoltaics by assigning retail or near-retail buyback rates for the net-excess production of electricity from the solar electric modules installed by home and apartment owners. The town of Burgdorf, Switzerland, pioneered the concept of "rate-based incentives" for photovoltaics in 1991, when voters decided to set aside 1 percent of utility revenues to create a fund that would pay people premium rates of 1 Swiss franc ($0.70) per kilowatt-hour of photovoltaic electricity generated, an amount that is recorded by a special meter that costs about $100 to buy and install. This system was soon picked up by the city of Geneva, and has been widely copied and adapted in Germany. (For more description of Germany's rate-based incentives, see chapter 8.) In most jurisdictions in the United States, as table 7-4 shows, any net-excess production is bought back by the grid at low "avoided-cost" rates that are one third or one fourth of retail, which allows the utility rather than the homeowner or resident to profit most from the photovoltaic installation.

TABLE 7-4: Estimated Retail versus Utility Buyback Rates for Renewable Electric Power from Customer-Owned, Grid-Connected Systems (in U.S. cents per kWh)

| | Retail Rates | Buyback Rates | |
		Rooftop PV	Wind
UNITED STATES:			
California[79]	12.0–14.0	4.0	6.0
Southern California: SCE[80]	11.0–14.0	4.0	—
Arizona Salt River Project, Phoenix[81]	9.5	2.4	—
Minnesota[82]	7.0	7.0	7.0
Wisconsin[83]	7.0	7.0	7.0
FOREIGN:			
Germany[84] (nationwide: 90 percent of retail electricity rate)			
Aachen[85]	18.0	124.0	22.0
Saarbrücken[86]	15.0	18.0	18.0
Japan[87] retail and buyback rates equal in Japan			
Great Britain[88]	—	—	20.0
Switzerland[89] (summer)	13.0–16.0	11.0	10.0
(winter)		15.0	—
Burgdorf	13.0–16.0	70.0	—

In Germany, the federal government is paying 70 percent of the installation costs for one thousand photovoltaic systems on one- and two-family houses.[90] Despite the budget strain of national reunification, Germany allocated about $60 million in 1993 alone for photovoltaics. An important part of the effort, as in Switzerland, has been development of building-integrated systems. According to Dr. Olav Hohmeyer, of the Fraunhofer-Institut für Solare Energiesysteme in Freiburg, Germany, "The city of Aachen is planning to pay 100 percent of the standardized costs of PV and wind energy (about US$1.24 per kilowatt-hour for PV and US$0.22 per kilowatt-hour for wind) for 1,000 kilowatts of demonstration plants. This is a decision of the city council, which is presently negotiating the plant with the local [city-owned] utility company. I think it will take about a year to finalize the plan."[91] Utility buyback rates throughout Germany for net-excess power from photovoltaic generators have been set by the German government at 90 percent of retail, and the Fraunhofer-Institut has built a highly efficient home powered by sunlight alone, which functions smoothly despite the fact that at Freiburg's latitude the sun provides only one seventh of the energy in winter as in summer. Electrolytic hydrogen from water is stored under pressure and supplies a fuel cell that stores and generates electricity. Hydrogen is also used as a cooking gas at the Fraunhofer solar house.[92]

In 1993, the government of Japan budgeted the equivalent of over $2.5 million on basic research and codes related to PV, and over $12 million on building demonstration projects. As in Switzerland and Germany, much of this effort has gone into developing attractive, cost-effective roof tiles that can be integrated into existing buildings as well as new construction. Under Japanese regulations, utilities are required to provide net metering for small PV producers, with a buyback of net excess electricity at the retail rate. Utilities are "encouraged," writes Steven Strong, "for the good of the country," to offer a 10-percent premium on renewable electricity sold back to the system. To encourage the installation of customer-located home PV systems, the Japanese Ministry of International Trade and Industry (MITI) announced a program in January 1994 to subsidize grid-integrated systems with as much as $27,000 per system for installations up to three kilowatts.[93]

In the United States, the most advanced solar designs integrate photovoltaic components directly into the building, a concept known as "building-integrated photovoltaics" (BIPV). In the Impact 2000 house, designed by Steven Strong, the photovoltaic modules function as both a roofing material and as a grid-connected power source. But this option hasn't caught on yet in the conservative building industry, especially with fuel prices having remained low since the early 1980s.[94] In 1996, only 3 percent of photovoltaic production was being installed by the U.S. domestic building sector (about 1 megawatt per year out of the 34.75 megawatts in total production). Alan Paradis and Daniel S. Shugar of Advanced Photovoltaic Systems argue that "integrating conventional construction materials into PV module design allows the use of standard installation practices and design load conditions," which they view as necessary conditions for the widespread use of photovoltaics in the building industry. The goal of those promoting integrated applications is to make photovoltaics a barely perceptible feature of the building, generating clean power through roof tiles, awnings, skylights, windows, and curtain walls, all of which mesh with standard structural, heating, and lighting requirements, instead of an awkward afterthought.[95]

Perhaps it should come as no surprise that Japan, Switzerland, and Germany—the three nations most active in promoting photovoltaics—are heavily populated, highly developed countries that import almost all of their petroleum. None of these three nations is host to transnational oil companies with extensive foreign holdings, one of the consequences of not being a victor in World War II.

U.S. Government Policies to Encourage Photovoltaics

A S NOTED ABOVE, the production of photovoltaic modules in the U.S. increased eightfold in the 1980s, despite their abandonment by most public officials, who were determined to prop up existing fossil and nuclear energy systems. For instance, until 1985 the federal government had granted income-tax rebates of up to $3,000 for the installation of domestic solar water-heating systems. The loss of such incentives had a devastating effect on customer-located solar installations, although a few state and utility programs remained in place in the wake of the federal cancellation.[96]

Because four fifths of the annual federal research support for photovoltaics was cut off during the 1980s, it should come as no surprise that the American share of world PV production fell during the same period, from 60 percent in 1983 to 33 percent in 1988, even before the sale of Arco Solar. In constant dollars, Reagan-era DOE support for research and development on solar buildings plummeted from $100 million to $1 million.[97] In 1991, however, the government's Solar Energy Research Institute (SERI) was renamed the National Renewable Energy Laboratory (NREL), fulfilling a long-time promise to raise the agency's status and visibility to the level of national labs such as Los Alamos, Oak Ridge, and Brookhaven. According to NREL insiders, credit for reviving federal interest in photovoltaics should go to J. Michael Davis, assistant secretary of energy for conservation and renewable energy under President Bush. Davis had worked at SERI during its first heyday in the late 1970s and later founded a solar equipment firm in Denver.

As promised during the 1992 presidential compaign, the Clinton-Gore administration continued to increase expenditures on photovoltaics and other renewables, but by 1995, the Republican-controlled houses of Congress were pulling the plug again. Larry E. Shirley, chair of the American Solar Energy Society, complained in the spring of 1996 that federal spending on renewables research was like a Disney World roller coaster. From $719 million in fiscal year 1980, the budget crashed to lows of $111 million in 1989 and 1992, when allocations began to climb again under President Bush, to a high of $344 million in 1995 under President Clinton. In the spring of 1996, Shirley warned, "We are now headed downward again, with a cut of 29 percent in the current budget and more cuts planned by some Congresspeople."[98] By January 1996, 160 employees had "voluntarily" resigned from the National Renewable Energy Laboratory, and 30 more had been laid off, reducing the number of full-time–equivalent employees at NREL from a high of 923 in 1994 to 650 by 1996.[99] Meanwhile, German and Japanese companies have continued to buy out U.S. photovoltaic companies, and in both countries, government and private funding of renewable energy continues to increase.[100] Although the United States was the country that originated most of the major innovations in solar technology, in the present political climate the prospects for government leadership in support of research and development and mandatory solar purchase orders seem illusory. Without pressure from an activist solar movement demanding that the federal government turn the nation's energy policy sunward, the government will presumably follow the lead of traditional fossil-fuel lobbies and allow photovoltaics, solar hot water heating, and wind power to languish.

Altogether, tax-supported institutions comprise 18 percent of the gross national product and buy $1 trillion per year worth of goods and services. One of the challenges of a new solar movement would be to demand that governments begin to undertake actual purchasing of renewable-energy technologies. The Government Purchasing Project, founded by Ralph Nader, has continued to hammer away at the idea of using concentrated public buying power to create the mass markets that would drive photovoltaics and other energy-efficient technologies into the mainstream.[101] In July 1992, the Government Purchasing Project began to publish the quarterly periodical *Energy Ideas*, the purpose of which is to inform purchasing agents at all levels of government what they can do to encourage their offices to take advantage of ultra-efficient and solar technologies. A subscription to *Energy Ideas* is free to government employees. In *Photovoltaic Cells: Converting Government Purchasing Power into Solar Power,* Holger Eisl and Barry Commoner argue that solar activists should pressure federal and state governments to buy progressively larger amounts of photovoltaic arrays to institutionalize and rationalize the demand for photovoltics. By way of analogy, they point out that the price of integrated-circuit chips fell by 95 percent in only five years when the Pentagon began to place huge orders for them, enabling the industry to take advantage of economies of scale and speed up its learning curve. Large photovoltaic purchases nearly became federal industrial policy in the late 1970s, but in the end, President Carter refused to authorize the project. Eisl and Commoner recommend that the purchases begin with photovoltaics for battery recharging, and proceed to generator sets, to remote-location pumps, to a small but fixed proportion of grid-connected federal power additions, and then to electric-vehicle recharging systems. As production volumes double and redouble, they argue, photovoltaics would become cost-effective for a significant proportion of residences in smoggy Southern California.[102] By a similar process, governments could jump-start the mass production of electric cars by requiring that the 100,000 light-duty vehicles they purchase every year be electric.[103]

The Department of Defense (DOD) has installed approximately 3,000 small- to medium-sized PV systems for the Army, Navy, and Air Force. PVs are used for remote building applications (such as ventilation, lighting and refrigeration, outdoor and runway lighting, and communications and meteorological equipment). DOD's PV systems average about 1 kilowatt in size, and range from 10 watts to 350 kilowatts. Since 1984, the Coast Guard has been leading the way in using photovoltaics in place of disposable batteries to power navigational buoys. Over 15,000 such systems are in operation. Although small in comparison to the potential market for residential photovoltaics, these government-funded purchases (like the use of PVs for remote, off-the-grid homes) have helped create interim markets and have demonstrated the technical viability and reliability of presently available but still unfamiliar solar technologies.

To shift the center of gravity in the electrical system from the utilities toward the residential customer, customers need credit and organization. The National Association of Energy Efficient Mortgage (EEM) Service Companies, backed by the Sierra Club and the Natural Resources Defense Council, has begun to pressure President Clinton and the U.S. Congress to allow mortgage limits to be increased for energy-efficiency improvements that would reduce utility bills. President Carter's Executive Order No. 12185 had directed federal mortgage guarantee programs to grant EEM mortgages in 1979, but the mortgage guarantee agencies threw up so many barriers that the new regulations were immobilized.[104] The newly proposed EEM programs, says Jim Curtis, head of the association, would fund one million energy-efficient mortgages per year; create 125,000 new jobs in construction; save $1.8 billion in energy costs; cut down smog emissions; and avoid the emission of 3.6 million tons of carbon dioxide. Such mortgages could easily be extended to include off-the-grid photovoltaics and solar water heaters, and later, grid-connected PV systems, as the price of photo-voltaics comes down. At present, says Curtis, of all the federal home mortgage guarantee programs, only the Veterans Administration will insure loans that would be paid off through reductions in utility bills. Thus, a loan of $5,000 for energy-efficiency improvements might add $34.96 to monthly payments on a 30-year mortgage but would reduce monthly utility bills from $150 to $100, for a net saving of $15.04 per month to the homeowner.[105]

The Clinton administration's briefly touted Climate Change Action Plan was supposed to reduce American emissions of the greenhouse gas carbon dioxide to 1990 levels by the year 2000. Action No. 8 of that plan promised to "promote home energy rating systems and energy-efficient mortgages," a step that could have financed photovoltaics and other customer-located solar energy systems directly.[106] In early December 1993, Jim Curtis and other energy-efficient mortgage advocates met at the White House with representatives of the White House Office of Environmental Policy to help plan a serious effort to implement energy-efficient mortgages. Four months later, wrote Davis, the people with whom he had met at the White House were no longer there, and Katy McGinty, head of the Office of Environmental Policy at the White House, "doesn't answer phone calls or letters.... Silly us! We thought there were going to be some major changes in how Washington does business. Is it still mostly 'make-believe equals survival' in D.C.—or are there changes still to come?" By July 1994, energy-efficient mortages seemed a dead issue at the White House.[107]

Monopolies in the Driver's Seat: Photovoltaics and the Utilities

THE UTILITIES HAVE shown every intention of maintaining tight control over the development of photovoltaics, a potentially disruptive new technology with respect to the status quo.[108] The Department of Energy sees the electric utilities as the prime market for photovoltaics and seems ready to hand over the decision-making power concerning DOE funds to the Utility Photovoltaic Group, providing for minimal oversight. The Utility Photovoltaic Group has written the TEAM-UP proposal (Building Technology Experience to Accelerate Markets in Utility Photovoltaics), which would commit the DOE to a $160-million program—supposedly to be matched with about $350 million from utility customers or stockholders—to subsidize the price of photovoltaic modules and allow the installation of 32 megawatts of large-scale, utility-controlled PV arrays through the year 2000.

A subsidiary part of this project seeks DOE funding to maintain a "market aggregation" office, to help the utilities negotiate wholesale PV purchases from manufacturers, and thus effectively drive independent PV installers out of business. According to the utility group's proposal, the ideal TEAM-UP program director must be an "experienced utility industry manager ... [who is] ... well accepted by the utility constituency," and utilities will be guaranteed a majority on the board of directors and working committees which advise and consent on all major decisions. Basically, representatives of the utility industry will get to channel the public's money wherever they choose, with no formal public oversight and no representation from homebuilders, independent PV installers, environmentalists, labor unions, or even the Department of Energy. Though the TEAM-UP grant proposal conceded that there was "no guarantee" that the grant would create "widespread and sustainable commercialization ... and lower prices" for photovoltaic manufacturers, TEAM-UP estimated government subsidies would double domestic PV sales and double or triple the number of utilities that normally include photovoltaics in their product mix.[109]

An energized, democratic solar movement would use government leverage to accelerate the establishment of a vigorous photovoltaics market, which would contribute to local economic development and local control over the economy. The genius of photovoltaic technology is that it fosters a genuine "free market"—more than any other form of energy production, because sunlight falls everywhere and is free. While the TEAM-UP program of the Utility Photovoltaic Group would increase short-term buys of photovoltaics, it is anathema to the goal of a democratic solar transition. The reason is simple. Turning over photovoltaics to the electric utilities perverts the decentralized potential of PV technology in order to support the needs of a highly centralized system of finance and control. Instead of becoming a tool to encourage local self-reliance, photovoltaics would thereby serve the opposite tendency.

The Geopolitics of Solar Electricity in the Tropics

PHOTOVOLTAICS ARE of particular interest to the developing world, because of the enormous amount of territory that has not yet been electrified.[110] When the United States went through its stage of rural electrification in the 1930s, photovoltaic technology was not yet available. Since extending the electric grid appeared to be the only option, this was done, with federal subsidies covering most of the cost. All

---

El Higueral, A Typical Remote Installation

In 1993, I had the privilege of installing a photovoltaic system in a remote village in El Salvador. The people of El Higueral (population 82) were building a new school . . . , and their sister city of Belmont, Massachusetts, was assisting them in buying the materials. When we brought up the possibility of electric lighting, the community leaders said that it would be especially useful in this school because they wanted to use it to teach classes for the adults as well as the children, and the adults didn't have time during the day. So, with their help, at a total materials cost of $1,400 (which could have been less if more had been bought locally), we planned and installed a system with two 48-watt solar panels and four 20-watt fluorescent lights. The installation took only a day and a half, with one engineer and several helpers, and it has functioned flawlessly now for a year. . . . In many respects, El Higueral was an ideal site. The village has two men who have considerable interest in and experience with electricity, so maintenance is possible, and the climate is nearly ideal; even during the rainy season there is usally some sunlight each day. It is also located on the top of a mountain, several kilometers from the nearest road (the solar panels had to be packed in on muleback), so there was no question that solar was the only alternative. In some less remote villages, the people have been less interested in such a system because they are hoping to persuade the government to bring the grid to them.

CHRIS KEAVNY[111]

TABLE 7-5: Typical Third World Household Electric Demand

Application	kWh per year
Lighting	58
Television (black & white)	13
Water pumping	9
Refrigeration (5 cubic feet)	400
TOTAL: Typical Third World household	480
TOTAL: Typical U.S. household	8,500

decisions about generation options that are made in the United States today take into account the fact that this grid already exists. In most of the Third World, however, large areas are not connected to an electric grid, and so expansion of conventional power generation usually requires both new generating plants and new distribution lines. Photovoltaics can render such centralized investment unnecessary.

Another reason why solar energy looks most attractive to energy planners in some parts of the world is the uneven distribution of energy reserves. The United States has been able to exploit its large reserves of coal and natural gas and a moderate supply of oil, and the country's relatively low population density has made it possible to dam a large number of rivers without giving up too much important agricultural land. In other words, the American abundance of land and other energy sources has made it difficult for solar energy to compete.[112] In many other countries, the situation is very different. Most of Central America, for example, is a geologically recent formation and consequently has no fossil-fuel resources at all. Hydropower potential exists in some areas, of course, but is not without its drawbacks: in El Salvador, the displacement of large numbers of people by the artificial lakes of hydroelectric projects was an important factor leading to the eleven-year civil war that ended in 1992. Smaller, less mountainous lands, like the Hawaiian Islands and the nations of the Caribbean, don't even have the hydroelectric option;

although tidal power and geothermal are under scrutiny in these areas, generally their power plants burn imported oil, the most expensive of all conventional sources.

A typical U.S. household uses almost twenty times as much electricity as its equivalent in the Third World. Table 7–5 gives the requirements, in kilowatt-hours per year, for the basic electrical services that would make the most difference in quality of life to people in remote villages who have had to get along until now with candles, kerosene lamps, and flashlights. The chief reason why solar power is more attractive to people in the poorer countries is not that it costs them less than it costs those in the wealthy countries, but rather that all of the alternatives are more expensive. This is true partly because many of the hidden costs of fossil-fuel generation are now more apparent, and partly because the wealthy countries have now bid up the price of oil to the point where the poorer nations cannot afford it. In developing countries, the electric utility is most often an agency of the central government. In the current political climate (under the rubric of the so-called neoliberal model, which is being promoted by the World Bank and other capitalist-oriented organizations), there have been proposals to privatize this function.[113]

A study done in Mexico by the national utility found a wind- or PV-diesel hybrid to be the most cost-effective option for electrification of a typical rural village. This

Gaviotas: A Renewable-Energy Utopia in Colombia

A few miles from Gaviotas, I see the first signs of the new civilization.... What appear to be bright aluminum sunflowers begin to dot the landscape. They are windmills, unlike any I've ever seen—light, compact units whose blade tips are contoured like airplane wings to trap soft equatorial breezes. They were designed by engineers ... lured here from Bogotá's finest universities to create the right technology for the tropics.

When we arrive at the open-air Gaviotas preschool, children are on the playground. Their seesaw is actually a pump in disguise. As they rise and descend, water gushes from a vertical pipe into an open cement tank. Over the years, Gaviotas technicians have installed these in thousands of schoolyards, using kid power to provide the village with clean water.

ALAN WEISMAN,
National Public Radio, 1994

study assumed electrical loads of 1.6 kilowatt-hours per day for each household—much less than in the U.S., but actually an abundant amount by Third World standards. The total life-cycle cost was $12,000 per household over 25 years, or $0.62 per kilowatt-hour, compared to $17,300 for the 25-year cost of an all-PV system.[114] Mexico, of course, has oil reserves, and is better off economically than many other countries. Mexico has also attracted unusual attention from electrical-industry multinationals following the passage of NAFTA. In fact, the huge potential of the Mexican photovoltaic market may well have been a major consideration behind the purchase of Mobil Solar Energy Corporation by a jointly owned subsidiary of the Germany utility consortium RWE AG (a group that includes Germany's largest electric utility) and, at least initially, Daimler-Benz AG, manufacturer of Mercedes automobiles.[115]

There has been considerable interest in setting up photovoltaic manufacturing plants in Third World countries to meet internal demand, as well as for export. India, with almost 4 megawatts of photovoltaic modules shipped by four companies in 1995, is the largest PV manufacturer in the tropics; domestic production totals over 40 percent of the 9 megawatts installed in 1995, according to Paul Maycock's figures. China produced 1.5 megawatts in 1995. Brazil's Heliodinamica company, privately owned by Bruno Topel of São Paulo, produced 0.4 megawatts in 1995, up from 0.1 the year before, but far below the country's reported PV production of 1.0 megawatt for 1991.[116] Unfortunately, the strategy of expanding domestic production in order to meet the rising domestic demand for PVs has not greatly reduced the cost of photovoltaics in the developing world.

Because the manufacture of solar modules is not a labor-intensive process, the low labor costs give manufacturers in the developing world only a small advantage, which is largely offset by their lower productivity (due to less-advanced manufacturing technology) and the need to import all of the source materials. Large, technically sophisticated tropical countries such as Brazil and India, for example, could have become the world leaders in photovoltaics if their leaders—both political and scientific—hadn't squandered their countries' talents and scarce foreign exchange on nuclear power and automobiles in imitation of the United States and Europe. After all, India installed 9 megawatts of photovoltaic modules in 1995, more than any other country, and more than twice what the country produced. In 1984, professor and writer Clóvis Brigagão charged that Brazil's multibillion-dollar nuclear power program was "characterized by a total lack of future vision," and called attention to the fact that Brazil at the time had almost no formal program to take advantage of solar energy, a lack that is now being rectified.[117] In the 1950s, Brazil went car-crazy. The capital of Brasília, established in 1960 by President Juscelino Kubitschek, was designed around the automobile, and Kubitschek founded the "Brazilian" automobile industry by inviting General Motors, Volkswagen, and Ford into the country. From that time on, Brazil became permanently dependent on oil imports to feed the thirst of its middle class for an "American" lifestyle. The attempt to replace gasoline with alcohol fuel was an expensive failure, unaccompanied by efforts to reduce car size and driving.[118]

Geopolitical considerations have also had considerable influence on smaller countries in Latin America. For example, Cuba and Nicaragua during the 1980s were supplied with oil at low cost by the Soviet bloc. While this allowed considerable improvements in the quality of life and industrial development, it also reinforced dependency. Now that this source of subsidies is gone, these nations are stuck with oil-fired generating plants, without the cash to buy oil. The result has been massive power shortages and rationing. Cuba has responded by turning to bicycles and organic agriculture, and may be climbing out of its oil dependency.[119]

The developing nations of the world could actually learn from our mistakes, instead of emulating those mistakes. With the proper kind of leadership, they can build an electrical infrastructure that will help rather than hinder their struggle to break the cycle of economic dependency and environmental degradation. As in Gaviotas, a technologically utopian village in Colombia, people can design and install technologies that use the sun, the wind, and even "kid power" to pump water and provide electricity.[120] But under present circumstances, tropical countries that opt for solar energy will end up exchanging their dependency on imported oil and coal for a dependency on imported photovoltaic cells produced by Siemens or Enron. Sunlight may be free and constant in the tropics, but not yet the tools to capture it.

Moreover, the situation that faces U.S. citizens, with their vastly more lavish lifestyles, is not really so different. Will local people own and manage the means to produce their own electricity, or instead be beholden to undemocratic and supranational energy cartels whose primary goals are growth and increased profits?

CHAPTER 8

Fighting for a New Solar Society

"I speak for the trees."[1]
—Dr. Seuss, *The Lorax*

THE SOLAR PATH IS SIMPLE: use less energy, and make sure that the energy comes from generating with renewable sources, as close to home as possible, rather than from mining and burning the world's accumulated energy capital. The production and distribution of electricity could become part of the solar path. Instead of an "alien, remote and ... uncontrollable technology run by a faraway, bureaucratized, technical elite," electricity could be "an understandable neighborhood technology run by people you know who are at your own social level."[2] The arguments of Amory Lovins's famous *Foreign Affairs* article "Energy Strategy: The Road Not Taken?" still reverberate. Lovins argued that the only viable long-term energy policy was a conversion to a "soft energy path" based on decentralized, renewable resources derived from the sun. And, he asserted, many of the world's energy problems are peculiarly suited to local, homegrown solutions. Lovins wrote, "an affluent industrial economy could advantageously operate with no central power stations at all!"[3]

By contrast, the "rapid expansion of centralized high-energy technologies" to produce electricity would create "insuperable" problems: sooner or later the world would run out of extractable coal, uranium, and oil. At some unknown point, the excess of carbon dioxide in the atmosphere might trigger an irreversible greenhouse effect. In thirty pages of carefully documented argument, Lovins emphasized that it was ridiculous to assume an obligatory link between increases in energy consumption and higher living standards or GNP. In fact, the opposite effect is more probable: to maintain the 6-percent annual increase in energy production of the post–World War II period through almost universal nuclear- and coal-fired electrification would suck up three quarters of this country's investment capital and impoverish the rest of the world. Furthermore, by incorporating excessive energy costs into its industrial products, the United States would price itself out of many international markets. Lovins pointed out that foreign competitors such as Japan and Germany used half of the energy per unit of production as did the U.S.

Lovins reminded his readers that the act of burning fossil fuel to make electricity, and then sending this current great distances, wastes two thirds of that fuel's energy in the production and transmission process. The problem is really one of fitting each use of energy to its most appropriate use. For 90 percent of energy uses, electricity is an indefensible luxury. Clearly, home heating is a problem that should be solved by "passive" solar building design; houses should be tightly insulated, and the location of windows and heat-absorbing walls should take maximum advantage of the sun's energy.[4] For Lovins, splitting the atom at a temperature of thousands of degrees to generate electricity (much of which will be lost in transmission) in order to raise the temperature of a house by 30 degrees in midwinter made about as much sense as "cutting butter with a chainsaw."[5] Worse yet, Lovins asserted that precarious and highly centralized power grids encourage "disruptions" during times of "social stress": "Even a single rifleman can probably black out a typical city instantaneously." Borrowing terminology from the anti-nuclear movement, Lovins noted that protecting the centralized power grids of the "hard" energy path provided the rationale for the social controls of a "garrison state"; and he cautioned his readers that, if nuclear power plants continued to be built around the world, more and more potentially hostile countries would acquire the expertise and the plutonium needed to build nuclear weapons of their own.[6] Lovins argued that only by use of hidden subsidies have fossil fuels and nuclear energy remained "cost-effective"; if the tax advantages, environmental costs, and military expenses actually attributable to fossil fuels and nuclear energy were accurately tallied and explained to the public, the idea that they constitute "cheap energy" would be revealed as a colossal fraud.

Although in discussions about energy and utilities emanating from conservative think tanks such as the Heritage Foundation and the Cato Institute, Lovins's name is rarely mentioned because his arguments attack the dominion of oil and nuclear-power interests, he has inspired many utility insiders with his ideas about energy efficiency and self-sufficiency and demand-side management (DSM).[7] Despite his stinging criticisms of traditional utility practices, Lovins has been able to develop a significant niche as a technical consultant to the utilities, because of his technical inventiveness and his ability to explain and inspire specialists in compelling language. His often-expressed faith in the steering power of properly designed free-market incentives legitimizes him to people whose minds are closed to scientists such as Barry Commoner, who question the very premise of private control of energy.

Lovins has generally refused to address whether the notion of "designing . . . free-market incentives" is a contradiction in terms, as if the "designed" market he proposes weren't the product of political struggle. Neither does he express any skepticism about conceding control over demand-side management to private utility monopolies, although, as we have argued in chapter 5, there is no evidence that utility-controlled DSM has ever caused electricity sales of a utility to cease growing and then decline. But long-time anti-nuclear activists fondly remember Lovins's brilliant critiques of nuclear power. Supposedly, dozens of utility engineers and PR flacks were assigned in the late 1970s the full-time task of refuting those charges. Lovins's "soft energy" paradigm is now an everyday concept among energy activists as well as utility planners worldwide, and because of him, utilities are obliged to profess at least a pro forma interest in energy efficiency and renewable energy. The income generated by Lovins's Rocky Mountain Institute and its consulting business have enabled his group to advance a credible technical critique of the traditional uses of fossil fuels in electricity generation, in transportation, and in building.[8]

Competition, Wheeling, and the Breakup of Electricity Monopolies

GIANT INDUSTRIAL corporations have supplied the money and political muscle behind the national campaign for utility deregulation and "customer choice" for over a decade. As natural-gas and other fossil-fuel prices have plummeted since the mid 1980s, the average price of electricity produced at older plants is two to three times higher than the price of electricity from modern new power plants, particularly in areas with high-priced nuclear power, such as California, Illinois, Michigan, Ohio, and New England. As a result, large industrial customers began to demand special deals from utilities. When lower rates were denied, big industrial customers threatened to bypass the utilities by building their own plants or dealing directly with less expensive, new independent power providers. To make good on their threats, the industrial customers and the independents had to break the utilities' legal monopoly over the production, transmission, distribution, and sale of electricity. Key to this rupture was the right to "wheel" massive amounts of electricity—for a fee—over existing utility transmission wires: a practice called "wholesale wheeling" when power is wheeled to a utility for resale, and "retail wheeling" when it is wheeled to a final customer. The Electricity Consumers Resource Council (ELCON)—whose thirty members include General Motors, Ford, Du Pont, and Anheuser-Busch—is the leading Washington voice in the battle for cheaper wholesale electricity.[9]

In alliance with independent power producers and some consumer groups, ELCON helped pass the Energy Policy Act of 1992 (EPACT), which opened up power lines to wholesale wheeling, an activity under the purview of the Federal Energy Regulatory Commission (FERC).[10] On the state level, many industrial interests have worked to convince state public utilities commissions to implement retail wheeling and competition in electricity.[11] In Washington, ELCON's main focus has been to fight utility attempts to gut expanded federal regulation of utilities.[12] In their battles with utilities, big industrial users have often entered into temporary alliances with utility consumer groups such as California's

Turn Toward Utility Rate Normalization (TURN), because both share the goal of lower electricity rates, as well as hostility to electric utilities.

Experience with other former monopolies, such as American Telephone & Telegraph (AT&T), shows that workers and their unions are the first casualties of deregulation, and indeed, deregulation constitutes a corporate offensive against organized workers by facilitating wage cuts and speed-ups. The scaled-down successors to the broken-up monopolies can provide goods and services for less, because as a rule they pay their workers less, regardless of the technology involved. Between 1988 and 1993, the "Baby Bells" that succeeded AT&T eliminated 60,000 jobs, and AT&T itself laid off tens of thousands of workers. Top hourly wages for a unionized operator at AT&T, represented by the Communications Workers of America, were $13.80 in 1993, compared to $10.20 at non-union Sprint, according to Jack McNally and Eric Wolfe of Local 1245 of the International Brotherhood of Electrical Workers, whose members work at Pacific Gas & Electric and other West Coast utilities. According to studies cited by McNally and Wolfe, real wages have dropped for pilots (by 10 to 20 percent), airplane mechanics (as much as 17 percent), and cabin attendants (by 25 to 40 percent) since the deregulation of the airline industry. Railroad and trucking companies have cut wages by 20 to 35 percent as deregulation proceeds, and accident rates have risen abruptly, especially in trucking, as drivers and independent truck owners drive longer and longer hours to maintain their incomes.[13]

The Price of Deregulation

In 1986, Congress first deregulated the cable TV industry. Two years later, its accounting arm found that monthly rates for the most popular cable services had jumped by 26 percent—to $14.77. Prompted by public outrage over those soaring rates, Congress nationally regulated the industry again in 1992. The Federal Communications Commission estimates customers since have saved more than $3 billion.

Sacramento Bee, June 28, 1996

"Who will pay for the nukes?" is the $264-billion question looming over the whole deregulation scenario.[14] Every interest group wants to shove the costs onto someone else. In California, for instance, PG&E's aging Diablo Canyon nuclear plant produces electricity for about 13 cents per kilowatt-hour, compared to 5 cents per kilowatt-hour for new combined-cycle natural-gas power plants, and 6 to 7 cents for new wind-turbine electricity. PG&E management argues that electric consumers should pay for Diablo Canyon, because Diablo was approved and certified for operation by the public authority of the California Public Utilities Commission. Organizations such as TURN, which represents residential consumers and small businesses, want to push the nuclear costs onto PG&E's shareholders. Some big corporate electricity consumers such as General Motors or Intel have threatened to opt out of the utility system altogether, and instead cut their own deals with independent power producers like Enron, who could supply all of either corporation's electrical needs throughout North America.[15]

If price per kilowatt-hour becomes the sole criterion for electricity sales, will long-term planning for clean air and water and renewable-energy technologies go out the window? Will renewable sources be in place to fill the gap when natural-gas prices soar, just as they plummeted in the mid-1980s? In New Hampshire, Representative Jeb E. Bradley, Republican from Wolfeboro and chair of the New Hampshire House Science, Technology, and Energy Committee, has expressed his concern that restructuring will lead to bulk purchases of cheap coal-fired electricity from Appalachia and the Midwest, and that the prevailing winds would blow more particulates and acid rain than ever into New England. "I don't think that any of us believe that energy restructuring should result in dirtier air than we've come to expect in New Hampshire," said Bradley, who added that he expected Congress and the FERC to maintain interstate environmental controls on power plants in other states. Bradley noted that generating plants in coal-mining states such as Virginia, West Virginia, Pennsylvania, Ohio, Indiana, and Kentucky were already held to lower air-quality standards than the Northeast under 1990 amendments to the Air Quality Act. According to New Hampshire state legislators, electricity from Appalachian mine-mouth power plants can be produced for half the price of electricity generated in New England.[16]

In California, the Center for Energy Efficiency and Renewable Technologies (CEERT)—a Sacramento lobbying group composed of renewable-energy firms and environmental groups, including the Sierra Club, the Environmental Defense Fund, and the Natural Resources

Defense Council—issued an "Earth Day Challenge" to California lawmakers and energy regulators "to require a side-by-side comparison of the environmental and economic contributions of both nuclear *and* non-hydro renewables," said Rich Ferguson, national energy chair for the Sierra Club. Both Ferguson and CEERT executive director John White pointed out that, under rulings of the California Public Utilities Commission, ratepayers would pay for $4 billion in excess capital costs and $500 million in excess operating costs for California's four nuclear plants, a de facto operating subsidy of 4 cents per kilowatt-hour, compared to a subsidy of 2 cents per kilowatt-hour for renewable-energy producers.[17]

Economist Eugene Coyle cites three reasons why residential and other small electricity consumers will likely be the big losers wherever PUC regulation is subverted by the proposed new style of competition:

1. Transaction costs, such as advertising, will be too great for the size of the potential cost savings that non-utility suppliers could offer a small-scale customer;
2. It will not be lucrative to compete for the poor and the frugal, or for people with a past history of payment problems, who will therefore remain captive utility customers;
3. California's experience with "direct marketing" of natural gas shows it to be a total failure because there is no beneficial competition for residential customers. (The smallest customers pursued by non-utility natural-gas marketer Enron are small businesses such as bakeries and hamburger chains.)[18] The long-term tendency will be to let electrical service deteriorate in poor neighborhoods. As in health care, a two- or three-class system of electrical service will develop, as utilities and their competitors strive to improve service to their largest and most affluent customers and cut service costs on the poor accounts.[19]

By late August 1996, it seemed that California would be the first state to open its market in electricity (which is worth $20 billion, annually) to competition and retail wheeling by 1998, and TURN had been mollified into neutrality by promises that residential consumers and small businesspeople would get a rate reduction of 20 percent (see page

307 for an update on California's legislation).[20] When electricity is deregulated, the poor and isolated will suffer most, if the telephone business is any indication. Before the breakup of AT&T, local telephone service was partially subsidized by long-distance revenues. "Deregulation," McNally and Wolfe have written, "reversed that arrangement; competition forced long-distance rates down, but local telephone rates rose 40 percent between 1984 and 1989." Phone installation fees also climbed steeply, causing additional hardship for poor and rural customers in particular. As a result, claims New York's Public Utility Law Project, one quarter of all low-income families in that state lack telephone service.[21]

To counterbalance the political power of California's privately owned utilities and large industrial users to foist upon small consumers the "stranded costs" of nuclear power, TURN has developed a "community access" proposal, which would allow a town, city, or county government—or other public entity such as a water district, or even the San Francisco Public Utilities Commission—to represent the interests of all residential customers and small businesses in their service area in dealing with utilities or other electricity producers. Under TURN's community access proposal, however, the town or city would not be allowed to buy the transmission and distribution system. An analogous proposal in Massachusetts would allow towns and cities to represent consumers and small businesses in their service area, with the option of setting up municipally owned systems.[22] Meanwhile, the American Association of Retired People (AARP) proposes to form electricity buyers clubs for its members—a kind of customer aggregation that would be beneficial to some, but likely to attract the most savvy residential customers while abandoning the unorganized and the poor to fend for themselves in the cold new world of retail wheeling.

Green Pricing, Photovoltaics, and "Competition"

A S A HOSTILE CONGRESS and an indifferent president cut federal R&D expenditures for solar in 1995, solar enthusiasts began to search for other sources of money. The problem was not lack of citizen interest; all polls continued to show massive support for solar energy, as they had for over two decades.[23] But that public sentiment was not getting translated into action by the private and governmental institutions that dominate energy policy in the United States. For solar advocates left penniless by the vagaries of energy policy, "green pricing" became the new holy grail. The concept of green pricing was simple: solar energy would be paid for by consumers who voluntarily agreed to pay a premium of a few dollars each month for "green" electricity in some form. Green pricing had been pioneered at the publicly owned Sacramento Municipal Utility District, where over a hundred new homeowners per year pay $6 per month extra for the privilege of having SMUD install a 2-kilowatt photovoltaic system on their roofs.[24] In the context of SMUD's overall commitment to the sustained development of solar energy, green pricing constitutes a laudable initiative.

But the sincerity of public

Who Pays for the Nukes?

The reality of wildly expensive nuclear plants can't be denied. Somebody has to pay. Neither regulatory reform, industry restructuring, nor retail wheeling will make the costs go away. That's what this fight over the future is about.

DR. EUGENE P. COYLE, April 1996

---

enthusiasm for solar energy has allowed some utilities to carry green pricing programs to surprising extremes. Detroit Edison reported that it had already signed up 280 customers at a premium of $6.59 a month for its SolarCurrents program, which claimed to supply its participants with solar electricity from a centralized plant with a generating capacity equivalent to only fourteen standard 2-kilowatt household arrays. The printed account of the project does not clarify what guarantees have been given to participants that they are actually using solar electricity in their homes, in addition to the ordinary Detroit Edison mix of fossil- and nuclear-generated electricity. The most impressive part of the Solar-Currents project is the depth of belief in solar shown by volunteers motivated to hand over their hard-earned dollars to the utility company for benefits that are largely symbolic.[25]

One might ask whether those 280 participating households could have made a greater impact on the energy politics of the Detroit area by pooling their $1,845 per month of solar surcharges, and instead founding a solar activists club to watchdog the environmental performance of Detroit Edison. A good place to start would be for one of the participants to request the names, addresses, and telephone numbers of other Solar-Currents participants in order to discover whether Detroit Edison is willing to help create an active solar lobby in its service area.[26] To have a real impact on the democratization of solar electricity, SolarCurrents participants could begin to demand that the Department of Energy create a revolving fund for low-interest loans to encourage PV installations, rather than channeling its subsidies through Detroit Edison. Or they might start a campaign to institute a system of net-metering for renewable electricity in Michigan, something that would create real incentives for residential ownership and control. Minnesota's net-metering law, for example, mandates that qualifying facilities of up to 40 kilowatts with grid-connected capacity must be allowed to charge the utility for excess solar-generated electricity by running the home meters backwards; the law also requires utilities to reimburse surplus electricity fed back into the grid at the retail rate.[27] Or they could follow the example set by Aachen and dozens of other cities in Germany, and lobby for a 1-percent tax on utility revenues, which would pay residential producers of solar power $1.24 per kilowatt-hour for the solar electricity they have generated. Creation and control of such a system of rate-based incentives (see below) will require tremendous political dedication. By contrast, most privately owned utilities offer photovoltaics (if they offer them at all) as an optional expense, with no effort at education and promotion, in order to satisfy the popular demand for solar without relinquishing any control over the technology.

Keeping It Cold: Efficient Refrigerators

THE RECENT HISTORY of refrigerators illustrates a representative victory of the Lovins-style approach to efficiency and good design. For homes without air conditioning, refrigerators use 20 to 50 percent of the electric power consumed. For many years, refrigerators were designed to waste progressively more energy. According to an MIT study, refrigerators marketed in 1972 consumed 14 watts per cubic foot, compared to 6 watts for the average refrigerator sold between 1925 and 1950, independent of new features such as defrosters and icemakers. Over time, refrigerators got bigger, further multiplying electricity consumption. The authors of *Power Struggle* remind us that for companies

TABLE 8-1: Refrigerator Electricity Demand

	Watts/Cubic Foot
1925–1950 (average)	6.0
1950–1965 (average)	10.0
1972 (one model)	14.0
1978 (*Consumer Reports* average)	11.0
1985 (*Consumer Reports* average)	8.0
1994 (*Consumer Reports* average)	4.0
1994 Gram (Denmark)	1.5

Source: 1925–1972, Ridley and Rudolph, *Power Struggle;* other data from *Consumer Reports* (February 1994).

such as General Electric and Westinghouse, "boosting demand on the [consumers'] side of the electrical outlet was good business for equipment manufacturing on the other. Consumers were unknowingly caught in the middle of this squeeze."

Federal energy-efficiency standards have forced refrigerator manufacturers to build a more efficient product and have halved the electricity consumption of the average refrigerator from 1,663 kilowatt-hours per year in 1978 to 819 kilowatt-hours per year in 1994, even though the most recent models have more shelf space (all the American refrigerators have freezers). *Consumer Reports* claims that the stingiest American-made refrigerator is the 22-cubic-foot Amana TZ21R2, which uses only 723 kilowatt-hours of electricity per year, or 3.75 watts per cubic foot of space. Some Europeans have moved even faster. Denmark's Gram is over twice as efficient as the best American-built refrigerators, and a Vestfrost fridge, commended with an award from Greenpeace in Germany, uses propane and butane as coolants rather than the standard ozone-destroying chlorofluorocarbons (CFCs).

The predicted life of a new refrigerator is fifteen years. A household that replaced a 17-cubic-foot refrigerator built in 1978 with a 22-cubic-foot Amana could expect to recoup the new refrigerator's sales price of $723 in seven years on electricity savings alone. And yet, the electricity consumption per cubic foot of the new and bigger energy sippers has taken about fifty years to surpass the performance of the average refrigerator sold between 1925 and 1950.[28]

Recently, Whirlpool won a $30-million prize from a consortium of utilities for building and marketing a "super-efficient" 22-cubic-foot refrigerator. Its chief innovation was the use of vacuum insulation panels. The competition specified that the winning refrigerator must use 25 percent less electricity than the 970 kilowatt-hours required by federal standards in 1993, and that it must avoid the use of ozone-depleting CFCs as refrigerants. The prize was the idea of David Goldstein of the Natural Resources Defense Council, as a way to encourage greater refrigerator efficiency, and unfortunately, contest rules precluded participation by the California manufacturers of the Sun Frost refrigerator, although their unit is six times more efficient than an efficient conventional model, because they did not make enough refrigerators in the qualifying year. The exact electricity consumption of the winning refrigerator was a trade secret, but it was rumored to be in the 500 to 600kilowatt-hours–per–year range.[29] The $30 million would be paid out incrementally to Whirlpool as the refrigerators were sold. However, there was no requirement that customers turn in their old, energy-gobbling refrigerators to qualify for a rebate on the new energy-savers.[30] Once the rebates of $100 per refrigerator were used up, Whirlpool withdrew its line of EnergyWise models. Vince Anderson, director of environmental and regulatory programs for Whirlpool, said that the company was "able to sell some units but sales were not outstanding … and not sufficient for us to continue the [EnergyWise] model line." According to Anderson, the sales of SERP EnergyWise refrigerators with $100 rebates continued until the $30 million allocated for rebates was expended in the territories of the twenty-four participating utilities. Anderson claimed that focus groups of likely refrigerator customers would opt for the more expensive but energy-efficient refrigerators only if they would make their money back in three years or less. Since there was nothing to oblige the company to keep selling its award-winning refrigerator once the prize money was gone, perhaps the cause of energy efficiency would have been better served by spending $30 million on a campaign for tougher refrigerator standards that match or surpass those in Denmark, for instance.

Energy Conservation, Public Power, and Economic Development in Osage, Iowa

THE ARGUMENT FOR democratic local control of energy is as timely as when Amory Lovins advanced it twenty-five years ago, though Lovins, today a major consultant to private utilities, rarely mentions this aspect now. Although only a distant dream two decades ago, the prospect of local control is beginning to capture people's imagination in new ways. The increasing efficiency of wind and micro-hydroelectric turbines, solar thermal, and photovoltaic technologies is at last providing a technically feasible base for local democratic management of energy. The municipal utility that serves the town of Osage, Iowa (population 3,800), exemplifies the potential for community responsibility and control when the primary purpose of the administrative organization is to serve local energy needs instead of the financial obsessions of anonymous investors. In Osage, the three utility commissioners are local people, chosen by the mayor for six-year terms, and authorized to run the utility efficiently, without political interference. So far, no privately owned utility has shown the friendliness and initiative of Osage Municipal Utilities in controlling utility rates and reducing electricity and gas consumption.

Energy Efficiency: The Osage Experience

Our program was started in 1974 when OPEC raised oil prices. As a utility, we could do little or nothing about the rates, however we could help our customers hold down their bills by showing them how to reduce usage. From this early start we began to see that our program was keeping a significant amount of money within the community, which was being used for health care, home improvements, and other visible benefits. Essentially, we were engaged in economic development. Then, as concern for the environment accelerated, we realized we also had launched what could be called a stewardship program. We were (1) reducing customers' bills by reducing usage, (2) keeping money in Osage (economic development), and (3) helping the environment by reducing pollution—thus we were making better use of human, economic, and natural resources.

WESTON E. BIRDSALL, 1991

According to Wes Birdsall, former general manager, and architect of the utility's energy-saving approach, "Our main goal is to keep our money at home and help our customers, who are also our stockholders."[31] Birdsall has been pushing for lower energy usage for fifteen years, and he has cut the electric rates in town five times since 1980. Between the early 1970s and 1990, average natural-gas consumption per residential customer fell almost 40 percent. Birdsall, now retired as general manager, still heads the conservation program, and he noted with pride that monthly electricity consumption in Osage averages 560 kilowatt-hours per household, rather than the 700 kilowatt-hours in comparable Iowa towns.

Osage Municipal Utilities buys its electricity from coal-fired plants in the region and passes along the cost of the power it buys. The basic residential rate was 5.2 cents per kilowatt-hour in 1996, a rate that is unchanged since 1992. While maintaining its conservation and efficiency programs, Osage has been rebuilding its distribution system using savings accumulated over twenty years, and has so far avoided debt.[32]

The company sponsors energy-saving contests among its ratepayers, and donates free water-heater jackets and shade trees for people who want them. Birdsall estimates that the utility's conservation programs save Osage's community $1.1 million per year, or $260 per person. "The overall beneficiary is the community," says one small businessman. "When you consider that our dollar for natural gas goes to Texas and our dollar for electricity goes to Wisconsin, then the more we can conserve in town, the better off everybody is. Energy conservation has given this town a tremendous shot in the arm." And most residents concede that the progress Osage has made in utility reform would have been impossible had the utility not been publicly owned.

Why Wind Power Has Worked Better in Denmark Than in the United States

DENMARK'S EXPERIENCE with windpower demonstrates the superiority of local ownership and control as a strategy for promoting the development of wind power for electricity generation. Denmark began to turn back toward the wind in the 1970s, as a response to the tripling of the price of oil during the Arab embargo in 1973. At that time, imported petroleum supplied 95 percent of Denmark's energy needs. In 1975, the Danish Academy of Technical Sciences concluded that wind energy could supply 10 percent of the nation's electricity, and since then "Denmark has never looked back," writes Paul Gipe in his indispensable book *Wind Energy Comes of Age*,[33] despite the fact that North Sea discoveries had made Denmark self-sufficient in petroleum by the early 1990s.

Denmark is serious about wind power at both the national and the grassroots levels and has propelled that committment by strong financial incentives for the owners of wind turbines rather than for the electric utilities. Wind power has thus become a true grassroots technological movement. Denmark's first national energy plan—agreed upon in 1981—included a goal of 1,000 megawatts of wind capacity by the year 2000. According to Denmark's Energy 2000 plan, decided upon in 1991, the country intends to produce 10 percent of its electricity through wind power by 2005 in order to comply with an international agreement to reduce greenhouse-gas emissions by 20 percent.[34]

Danes have a very strong tradition of being able to act individually while working together as a team. Gipe points out that while almost all Denmark's farms are family-owned and operated, most Danish farmers are active in farm cooperatives. Their capacity to combine individual initiative with community action and self-reliance has served Denmark well in wind energy. The strongest momentum for wind power development has come from agricultural cooperatives and individual farmers, who in 1993 owned 80 percent of the country's wind-generation capacity, sometimes singly on individual farms and sometimes collectively in clusters of up to a half-dozen turbines. The biggest Danish wind farm has only 100 turbines, compared with more than 4,900 at Tehachapi in California.[35] The Danish windmill owners association, founded in 1978 in response to problems in early designs, has been adamant in demanding minimum design standards from manufacturers and routinely publishes reports from its members on the output and reliability of their machines.

In Denmark, the most successful wind turbines have been designed and built by farm equipment manufacturers. Johannes Juul, the father of contemporary Danish wind-turbine designs, had little formal education and was first exposed to windmill design during World War II, at a school for windmill electricians founded by Poul la Cour, a famous inventor known as the Danish Edison. The emphasis in Danish designs has been on reliability, sturdy construction, low maintenance, and quiet operation. Since Denmark is a small country, manufacturers find it easy to stay in touch with the final consumers, a connection that has proved invaluable in developing durable, efficient machines. Though the $53-million (in U.S. dollars, equivalent) Danish government investment in wind-turbine R&D was one ninth of the U. S. government's $486 million dollars for the period from 1974 to 1992, Danish wind turbines were acknowledged to be the best in the business, with a market share of 53 percent worldwide.[36] The big winners in the U.S. wind-energy research game have been giant corporations such as General Electric, Westinghouse, Boeing, and Hamilton Standard, whose aerospace divisions are masters at extracting R&D contracts from the government, but who are not necessarily so skilled at producing and selling products for a mass market.[37] They have tended to develop lightweight, high-performance machines that extract the maximum possible energy from a given wind velocity—and also break down constantly. "With the exception of one, U.S. Windpower's Model 56-100," notes Gipe, "none of the U.S.-designed machines in California can be called a success." Danish farm equipment manufacturers have evidently been better attuned to the fact that wind turbines operate about 6,000 of the 8,760 hours in a year—twice the operating hours of the average automobile over a 10-year lifespan. Indeed, by 1992, 70 percent of California's generating capacity came from Danish and Japanese wind turbines.[38] The Danish success would seem to prove the superiority of consumer incentives and local ownership and control over centralized Department of Energy–sponsored R&D, what Carl Weinberg, former head of research at Pacific Gas & Electric, called the superiority of the *market pull* over the *technology push* strategy.[39] If the real intention is to move wind power (and by analogy, solar water heating, photovoltaics, and energy-efficient building techniques) into general use and to increase local self-reliance, a better strategy would be to create a strong market through easy credit and favorable electricity buyback rates for the ultimate owners and consumers of those products, and to make these sensible new technologies mandatory in appliance, machinery, and building codes. Such an alternative strategy, however, would break the hegemony that electric utilities and DOE now hold over the style and pace of wind-energy development.

In Denmark, incentives work in favor of the wind-turbine owner rather than the electric utility. Connected to the grids, windmill owners offset their own electricity consumption through a net-metering billing system that allows them to sell electicity they don't consume back to the grid at about 10 cents per kilowatt-hour, whereas in most of the United States the payback rate is somewhere between 2 and 4 cents. Both individual and cooperative windmill owners have an incentive to install machines as big as the law allows, because windmill income is exempt from income taxes, a particularly important consideration in Denmark, where high income-tax rates make 1.00 krone of wind-turbine income worth 1.46 krones of taxable income.[40] Since wind turbines are reliable, well-understood, and reasonably profitable in Denmark, banks routinely finance 60 to 80 percent of their cost at reasonable rates of interest. According to Gipe, the amount of wind-generated electricity generated increased steadily every year between 1980 and 1994; by the latter year, Denmark was second only to the United States in total wind-generation capacity. Together, the Danish windmill owners association and the association of Danish windmill manufacturers have combined to become an "influential voice" in Danish politics, so effective that the German wind- energy association hopes to "replicate" the Danish organizational strategy in Germany.[41]

By contrast, wind power in the United States has been a boom-and-bust affair that never created a grassroots constituency, outside of a restricted circle of wind-energy enthusiasts and financiers. Environmentalists were ambivalent, troubled by the turbines' noise and visual intrusiveness, and by their tendency to kill hawks and other prized birds of prey.[42] More than 15,000 wind turbines had been installed in California between 1980 and the end of 1994, which by then produced 90 percent of the wind-generated electricity in the United States.[43] In fact, as recently as the late 1980s, California wind turbines produced more than 90 percent of the world's wind-generated electricity.[44] Five laws and rulings created the fabulous wind-rush in California at Tehachapi and at Altamount Pass:

1. Sections 201 and 210 of the Public Utilities Regulatory Policies Act of 1978 (PURPA), which broke the utility monopoly on electricity production by requiring utilities to buy power from small, usually renewable sources;
2. The Energy Tax Act of 1978 (ETA) and its successors,[45] which created tax credits of 15 percent for investments in renewable energy from 1980 through 1985;
3. A 10-percent general-investment tax credit, applicable to wind power as well, which added up to a 25-percent federal income-tax credit for investments in wind power through 1985. These tax credits were allowed to expire at the end of 1985;
4. A California income-tax credit of 25 percent, which (unlike residential solar credits) was *added* to the federal credit, creating a total tax credit of 50 percent for several years;[46]
5. Contracts to sell electricity at about 7 cents per kilowatt-hour to California's biggest private utilities. Most of the wind power producers contracted with the utilities to buy the electricity at fixed prices for ten years, followed by twenty years at floating prices. The ten-year fixed prices specified an average "avoided cost" rate of about 7 cents, negotiated in the early 1980s, when it was expected that the price of oil and natural gas would continue to rise. Instead, those fossil fuels fell in price, and the utilities tried to wriggle out of the contracts.

For some investors, wind turbines became a fashionable investment scam, irrespective of whether or not they produced any electricity. Oak Creek Energy Systems, Inc., advised potential speculators that "on the surface, it may not matter which windturbine developer you choose," because the tax savings "nearly total the entire cost of your wind turbine." *Forbes* magazine wrote that the tax credits had "helped create the fad of the year: the wind-park tax shelter." With tax credits and five-year write-offs for depreciation, the potential return for an initial investment of $100,000 was $137,765 over fifteen years. The boom was on. In 1984 and 1985, the peak years, close to 800 megawatts of new wind-generation capacity was installed in California. Wind energy began to go bust in 1986, broadsided by the end of tax credits, by losing battles over buy-back rates, and by a precipitous fall in the price of competing natural gas—just as developers were beginning to site reliable long-lasting machines.[47] By 1988, the annual total of new installations had fallen to 50 megawatts. Despite a slight recovery in 1990 and 1991, installations of net new capacity have been negligible since 1992 in California. Nevertheless, about 1,500 megawatts of wind-power capacity remained in place, producing about 3.2 billion kilowatt-hours worth of electricity in 1995.[48] As this book goes to press, the future of the wind industry in the United States is as questionable as the future of other solar technologies in this country. The forces arrayed against wind energy's expansion have simply been too overpowering.

Why Rate-Based Incentives in Germany Work Better Than Federal Handouts to Utilities in the United States

GERMANY, DESPITE THE STRAIN of reunification on the national budget, allocated about $60 million for photovoltaics in 1993 alone. An important part of the German effort, as in Switzerland, has been the development of building-integrated systems. The extra revenue generated by such systems is significant enough to allow a homeowner to pay for the PV system in perhaps twenty years. The Fraunhofer-Institut in Freiburg has built a highly efficient home powered by sunlight alone, a house that functions smoothly despite the fact that at Freiburg's latitude, as far north as Toronto, the sun provides only one seventh the energy in winter that it does in summer. At the Fraunhofer solar house, hydrogen electrolyzed from water with photovoltaic electricity is stored under pressure and supplies a fuel cell that stores and generates electricity for the winter months. Hydrogen is also used at the solar house as a cooking gas.[49]

Utility buyback rates throughout Germany for net-excess power from photovoltaic generators have been set by the German government at 90 percent of retail—about 15 cents per kilowatt hour. As we shall see, there is a vital grassroots movement throughout Germany that advocates setting aside a percentage of utility revenues to fund a system of "rate-based incentives" that reimburses residential owners of grid-connected PV systems for the electricity they produce at three to five times the retail electric rate.[50]

The "Solar Energy Promotion Society" (Solarenergie Förder-verein),[51] led by Wolf von Faback, decided in 1993 to push for the rate-based incentives in Aachen, a city of 250,000 bordering the Netherlands and Belgium, where the local utility is city-owned. Under the Aachen plan, which was based on the arrangement used in Burgdorf, Switzerland, people who sign up for the program and install photovoltaic arrays sign a contract through which PV users can expect to recover the cost of system installation, given the city government's commitment to buy photovoltaic power at 2 deutsche marks (about $1.24) per kilowatt-hour over twenty years. Household photovoltaic technologies in Germany and Switzerland generally cost about $25,000 for a 2.5-kilowatt system, about 33 percent more than such a system would cost in the United States. According to Tom Jensen, a market analyst at Strategies Unlimited, a Mountain View, California, market-research firm,[52] the Burgdorf/Aachen plan spread very quickly in Germany: by February 1996, nineteen cities with a total population of over 5 million had instituted rate-based incentive programs for photovoltaics, including Bonn; Hamburg, with 1.8 million inhabitants, Germany's second largest city; and Munich, with 1.3 million inhabitants. According to *Solar Flare*, a bimonthly report published by Strategies Unlimited, twenty-one more cities with 7.7 million inhabitants were considering the adoption of rate-based incentives programs. Buyback rates varied by city and PV system size. In Hamburg, for example, systems of 1 to 5 kilowatts in size would be reimbursed at the rate of 1.80 deutsche marks per kilowatt-hour, and large systems of 10 to 50 kilowatts could sell back their solar-generated electricity at 1.50 deutsche marks per kilowatt-hour. In addition, the city agreed to keep the program open to new subscribers until 1.6 megawatts of peak capacity has been installed. So far, installations in Hamburg total 150 kilowatts, with about thirty-eight homes and businesses participating. The reimbursement money comes from a 1-percent surcharge on all electric bills in Hamburg.[53] Household photovoltaics are so popular that

Greenpeace Germany has gotten into the business of selling PV systems.[54]

Another German city that is considering the adoption of rate-based incentive systems is Berlin. The movement for rate-based systems has been strongest in cities with Social Democrat/Green mayors and city councils, since these programs are often undertaken in alliance with the Green party. Conservative cities controlled by Christian Democrats have been slower to participate. According to Jensen, proposals in large cities such as Hamburg and Berlin, with privately owned electric utilities, have often faced strong utility opposition. In Berlin, for example, a few supporters of rate-based incentives, in order to pressure the utility, have started withholding 5 percent of their utility bills and paying them instead to an escrow fund, for release when the privately owned Berlin utility agrees to move forward on the issue. These organized protests against the utilities, which can rapidly grow if people are angry and committed enough, seem to have brought Berlin's electric utility to the negotiating table.[55] As a result of the rate-based incentives strategy in Germany, the amount of rooftop, grid-connected photovoltaic capacity in place nationwide leaped from 0.1 megawatts at the beginning of 1995 to a predicted 1.8 megawatts by the end of 1996.[56]

In the United States, the photovoltaic industry is divided in its opinion regarding rate-based incentives. Jim Trotter, a marketing executive with Solar Electric Specialities and president of the California Solar

Ökobanks and Öko-Crazies

In a "daredevil move by a few crazies," Germans from a variety of political backgrounds have founded a cooperative Ökobank ("Eco-bank"), now capitalized at over $100 million, to help finance green enterprises. Support has materialized from pension funds, as well as from the Catholic and Lutheran churches. Concerned individuals could buy shares for DM100 ($59) apiece. On May 1, 1993, the fifth anniversary of its founding, Ökobank announced that it had completed a year in the black for the first time—with a profit of about $7,500. Though similar banks have been founded in Belgium, Denmark, Spain, and Sweden, nothing comparable exists in the United States.

This Week in Germany, May 1993

Energy Industries Association in 1996, believes that rate-based incentives for photovoltaics are an essentially wasteful and unnecessary subsidy that distorts true market signals. People touting photovoltaics because of their concern with air quality, says Trotter, would contribute more by getting their car tuned up on a regular basis. If they are concerned with excessive greenhouse gases produced by fossil fuels such as coal, oil, and natural gas, they would get more results by starting a municipal program to spray attic insulation in older homes. Furthermore, argues Trotter, with a jump in U.S. production of almost 36 percent between 1994 and 1995, photovoltaic capacity is stretched to the limit, and there is no need to fund new markets.[57]

On the other hand, Thomas J. Starrs of Bellevue, Washington and Tim Townsend of Davis, California have written and presented a draft photovoltaic ordinance to the Natural Resources Commission of the city of Davis, California, a proposal that incorporates some of the Swiss and German rate-based incentives. As of yet, the plan has not been adopted. One obstacle is that taxing all residential electric bills 1 percent to redistribute the money to the few photovoltaic households would be a new tax, which is forbidden by California law without approval by a two-thirds local majority. Townsend believes that a system of rate-based incentives will be established in Davis only when a determined group of citizens organizes in support of the idea.[58]

"Sustained, Orderly Development" of Solar Energy

THE EXPERIENCE OF thousands of off-the-grid households and experiments proves that technical progress in photovoltaics has been very rapid in recent years. The first PV cells in the mid-1950s cost about $600 per peak watt, but the decline in price was startling: to $150 per peak watt by 1970, and as low as $1.75 per peak watt by 1996. (For total installation costs, these estimates should be doubled, since the "balance-of-system" costs for photovoltaics are usually assumed to equal the cost of the modules themselves.) And the price of a 10-foot by 10-foot PV array—enough to generate a year's electricity for one energy-efficient household—fell from about $16,000 in 1980 to $5,000 by 1996.

As emphasized throughout this book, when the noxious side effects of fossil fuels and the risk of fuel price hikes are factored in, photovoltaic electricity is already competitive today from an economic standpoint. The Rocky Mountain Institute (RMI) claims that electricity usage in the average American home could be reduced from 11 kilowatt-hours per day (4,015 kilowatt-hours per year) to 3 kilowatt-hours per day (1,095 kilowatt-hours per year), if inappropriate uses of electricity were remedied with efficient lighting and electrical motors; passive-solar space heating; more efficient, better-insulated refrigerators; and other strategies for effectively matching sources of energy with the most suitable "end-use." RMI is designing mounting-hardware that can be used to retrofit older homes with photovoltaic panels without drilling holes in the roof, and various companies are designing PV roof shingles and siding that will repel rain and also generate electricity.[59] The problem now is to get these renewable innovations into the mainstream. Such stunning advances have been made *despite* massive federal cuts in photovoltaic research and continued subsidies to fossil fuels and nuclear power.

Solar Hydrogen, by Joan M. Ogden and Robert H. Williams of Princeton University, takes the renewable-energy option a giant step forward.[60] The authors sketch a technical model for using community-produced hydrogen rather than household batteries as a storage medium. When the sun shines or wind blows, surplus electricity would power an electrolyzer to split water into its component elements of hydrogen and oxygen, and to compress and store the hydrogen for later combustion as a power source for heating, electricity, and transportation. Ogden believes that there is no particular economy of scale for electrolysis of water above 10 megawatts of capacity, which is enough electricity for a neighborhood of 3,500 to 6,000 households at present consumption levels.[61] The community would be immune from electrical rate increases because the renewable "fuel" (ordinary water) would be free, and the profits which would have been sent to wealthy absentee investors would stay within the community—à la Osage, Iowa, or Sacramento. A community-owned system would be under constant pressure to encourage conservation once the initial hardware was purchased and installed. Since the system's components are modular, it could be expanded on a house-by-house basis as the need arose, without mortgaging the community's economic health through huge bank loans. And, best of all, hydrogen power doesn't pollute once the equipment is in use.[62] The only by-products are carbon dioxide and water vapor.

If Ogden and Williams are right, all local communities could handle all of their electric power requirements using locally harvested solar energy and water, and thereby declare their energy independence from the private utilities. Such arrangements would turn even the most progressive of today's existing utility practices upside down. Instead of using renewable energy merely to shave peak-power demands on hot summer afternoons, photovoltaic and wind-generated electricity with hydrogen storage could become the centerpiece of the energy system, with fossil fuels perceived as a backup for rare or intermittent situations, mimicking present-day practice in remote stand-alone home energy systems powered by photovoltaics and wind or micro-hydro generators. If the financial support were provided, solar hydrogen projects could be initiated immediately in the desert Southwest and in tropical areas such as Brazil and Hawaii.[63]

The experience of California's North Coast is living proof that it is possible to live off-the-grid with PV power. Without giving up the modern conveniences that Americans have come to consider their birthright, off-the-grid homesteaders have learned to live with a quarter of average American electricity consumption. If the minor hassles and expense of battery maintenance could be avoided by socializing storage and distribution, or by developing maintenance-free batteries, solar electricity could become ubiquitous in a generation.

"Green" emporiums located in California's Solar Belt, including Real Goods Trading Corporation and Alternative Energy Engineering, as well as others in virtually every region of the country, have grown fabulously in the past five years. And yet, solar consumerism alone cannot ensure the widespread adoption of renewable-energy technologies. The danger is that solar consumerism will remain a dispersed, narcissistic preoccupation of a few well-heeled green consumers who have a technological bent. There is no reason why local planning departments shouldn't require solar hot water heating and even photovoltaic electricity, particularly in those areas not served by gas and electric lines, and local building contractors could be enlisted in the process of setting guidelines for customer service and quality control. To survive and expand when their original market is saturated, solar merchandizers must reach out and help create a social and political movement for clean energy and clean jobs at the local, state, and national levels. For example, John Schaeffer of Real Goods consistently includes solar advocacy in his catalogue notes, and publishes a magazine for activists called *Real Goods News*, which features a great deal of commentary from customers on environmental and social issues. In 1996, Schaeffer inaugurated the Real Goods Solar Living Center in Hopland, California, as a kind of solar theme park, designed and built in close collaboration with architect Sim Van der Ryn and David Arkin of the Ecological Design Institute, and incorporating innovative energy-generating and energy-efficient components as well as integral demonstrations of new approaches to ecological construction and landscaping.

Local Forms of Ownership and Control

Our type of economics says, "You mustn't ... produce anything unless you are quite sure you couldn't possibly buy it cheaper from outside." When I was in Puerto Rico, a luscious island, I found that the carrots were imported from Texas! You mustn't produce carrots in Puerto Rico if the Texas-produced carrots are cheaper. That's our system. The Chinese have turned this around. They have said, "You mustn't buy anything from outside unless you can be quite sure you can't make it yourself." It's as simple as that.

E.F. SCHUMACHER, *Good Work*

We have now come full circle: with the new viability of solar technologies, the clean dreams of the 1970s can finally become the realities of the new millennium. Now is the time to reprint the *Blueprint for a Solar America* and *Sun! A Handbook for the Solar Decade*, to assess how the programs of the major energy companies and environmental groups measure up, and to write a new *Citizens' Solar Blueprint*.[64] As the price of photovoltaic panels continues to fall, and as architects and designers figure out how to integrate PVs in beautiful, practical ways into new buildings, people with sunny backyards and roofs will be visibly generating more than enough electricity for their personal needs while feeding the excess back into the grid for storage and distribution. Experiments and demonstration projects are no longer sufficient. We need a reinvigorated, nationwide citizens' solar movement to insist that government and energy users replace increasing proportions of fossil and nuclear energy with renewable energy, so that the manufacturers of those technologies experience steady expansion in the market in order to reduce unit costs through economies of scale. Fifty years from now, the world's electrical systems could be organized as stand-alone systems presently are, utilizing strict efficiency, with integrated photovoltaic modules, solar water heaters, small wind and hydro turbines, and hydrogen-based storage systems—and with fossil-fuel generation used only as an occasional backup, saving the planet's petroleum reserves for use in processes where, unlike burning, the oil is rationally exploited and priced according to its scarcity.

The experiences of Osage Municipal Utilities and the Sacramento Municipal Utility District demonstrate that democratic empowerment, grassroots decision-making, and economic self-sufficiency are not only possible, but can be technically and administratively superior to private monopoly systems in running complex technologies such as electric power generation and distribution. The local generation and distribution of energy, if it is based on local control, has the ability to bring the center of economic gravity back to the local level. Yet, successful organizations such as those in Sacramento and Osage represent an open challenge to economic control and manipulation by multinational corporations and banks and their allies in big government.

Even in areas where electric power is now completely under the control of a private monopoly, the sale price of household solar electricity will become a bigger and bigger issue. Citizens who have invested in solar equiment connected with the grid will want to know why the utility sells them electricity at 12 cents per kilowatt-hour and credits them at 3 cents per kilowatt-hour, when their meter should simply run backward, crediting them one for one. Or why not pay them a premium of $1.24 to encourage solar electricity, like many cities in Germany and Switzerland? Citizens will want to know why they are not allowed a special nighttime rate to charge up their electric bicycles and cars, when the 100 to 600 square feet of solar modules on their roof generates enough electricity to power their house *and* vehicles. Since it is evident that many utilities will vehemently oppose locally owned and controlled energy systems, Congress itself, sensing a conflict of interest, has begun to question the wisdom of turning over total control of solar technology to the utilities.[65]

Although few people can afford to spend ten thousand dollars or more upfront for a stand-alone household photovoltaic or wind system connected to the electric grid, many people could afford $150 per month, particularly when they would own the systems free and clear in fifteen or twenty years. Yet the Federal Housing Administration and Veterans Administration—which have helped rehouse half of the American people since the end of World War II—will not at this time guarantee loans for independent solar installations. In a new solar economy, will building owners and tenants earn income-tax credits or deductions for their payments on such systems? A recent Pacific Gas & Electric/Bank of America proposal for providing loan money to purchasers of energy-efficient refrigerators and other appliances pointedly leaves solar water heating *off* the list of options, and participants would not necessarily be required to turn in their energy-wasting refrigerators in exchange for the new ones.[66] If PG&E truly wanted to encourage solar, it would promote a Solar Bank and offer premiums for solar and wind power over the retail electric rates, as the Europeans and Japanese do.

In the new solar economy, will the utilities universally buy back home-generated power at a price that makes an investment in PVs worthwhile, or will they pay a fraction of its true value and try to maintain a monopoly over the transmission equipment? Will building codes be modified to permit any qualified and bonded solar electricians to install the equipment, or will the utilities have the political power to control the installation of solar equipment, as they have controlled domestic electric meters? Will building codes, like those in San Francisco, still require builders to install gas or electric heating in San Francisco and in the other regions of California with a mild, Mediterranean climate? Or will building codes in the new solar economy *forbid* the installation of electric hot water heating in San Francisco except as backup?

The task of the new solar movement will be to put the democracy back in photovoltaics and the other solar technologies whose efficacy can no longer be denied or delayed on technological grounds. Solar entrepreneurs, socially responsible engineers, and other believers must break out of their old-hippie ghettos (like California's Mendocino County) and their technocratic ghettos (on the utility plantations) and confront the conventional wisdom about energy in small towns, suburbs, and cities across America. New institutions—solar banks, ecological businesses, builders' cooperatives—and new solar lobbies will need to be created. Without a vision and a new kind of political pressure, the U.S. government will continue to maintain the status quo, handing out big tax breaks in energy to oil and gas drillers and to monopoly utilities.

In this book, we argue that public ownership with local democratic control of utilities is a necessary if not a sufficient condition for a solar economy in the United States. We believe this for two simple reasons: unlike a large capitalist enterprise, a publicly owned utility is *not* impelled to constantly increase sales in order to increase profit margins and stock prices. Therefore, publicly owned utilities have no inherent drive to promote constantly increasing electricity sales. If well managed, a public entity can supply electricity more cheaply, because it doesn't pay inflated salaries and, most importantly, because stockholders don't siphon off 10 percent each year in dividends. As SMUD and Osage Municipal Utilities demonstrate, publicly owned and democratically governed utilities can be more efficient, in both a business and an environmental sense, than privately owned utilities. Commentators familiar with California utilities would be hard-pressed to claim that Pacific Gas & Electric is run more efficiently than SMUD, in part because the people of Sacramento have a democratic mechanism to allow them to help govern an extremely important institution in the life of the community. Sacramento's residents voted to shut down SMUD's nuclear plant when it proved to be a boondoggle; by contrast, PG&E used its extraordinary political clout to keep the Diablo Canyon nuclear power plant open, saddling ratepayers with billions of dollars in outlandish charges and interest over the decades. SMUD plants tens of thousands of shade trees each year; neighboring PG&E plants none. And SMUD's residential rates are 25 percent lower than PG&E's. As competition and restructuring make headway in the utility industry, people in many more cities and towns may find that municipal intervention may be a way to protect the average citizen's access to affordable, reliable electricity, and to assure continued utility promotion of renewable energy.

A new blueprint for a revitalized solar movement requires advocacy of the following:

- Public "ownership" of energy—just as with water or schools
- Access to loans for photovoltaics, solar water heating, wind and micro-hydro generators, and other forms of energy-generating and energy-saving technologies—just as those purchasing automobiles or homes have access to loans
- Reinstitution of tax credits and rebates for renewable-energy investments, with conscientious licensing and quality-control procedures to guarantee customer satisfaction
- Net metering and rate-based incentives, so that independent, home- and business-based electricity producers are paid the same price that they would be required to pay for the grid power they are not using
- Massive public- and private-sector investment in renewable-energy technologies and building techniques, to reestablish the U.S. as the preeminent leader in this economic domain
- Partnerships between industry, government, and local communities to oversee the new "green" industries, in order to make sure that the public knows what is being produced in a factory, by what means, and how the wastes and by-products will be managed
- Scholarships and retraining for displaced fossil-fuel and nuclear workers, and small-business loans to support new solar tradespeople
- Congressional hearings to examine why none of the above is federal or state policy

Bill Clinton made green sounds at appropriate times during the 1992 and 1996 presidential campaigns, and Clinton's choice of Al Gore as his two-time running mate might have presaged a new direction for the nation's energy policies. In his insightful book *Earth in the Balance: Ecology and the Human Spirit* (written before he knew he would be running for national office), Vice President Al Gore wrote movingly of the "dysfunctional" nature of modern technology and advocated a world "strategic environmental initiative" to deal with these problems. *Earth in the Balance* argued for the need to "accelerate" photovoltaics as the "most promising" of the new, clean, and decentralized solar technologies. It is impossible to disagree with his conclusion that "the technical barriers ... are ... less important than the political and institutional barriers ..." to photovoltaic and other renewable-energy technologies.[67] President Clinton was no less eloquent in his campaign advocacy of an energy policy that "lets Americans control America's energy future" rather than "coddling special interests whose fortunes depend on America's addiction to foreign oil," and he earnestly promised to hew to this vision of "national security, energy diversity, economic prosperity, and environmental protection" when it came to energy policy.[68]

Campaign promises notwithstanding, Clinton's most significant victory was the North American Free Trade Agreement, whose chief effect on the energy economy has been to open up Mexico's oil and gas reserves to drillers from the United States.[69]

In 1990, $84 billion of the country's $109-billion trade deficit was directly attributable to chronic negative trade balances in petroleum and in cars and trucks. The country may be facing another massive oil shock by the year 2000, if demand growth continues to climb faster than supply.[70] To pay for the deficit in oil and cars, average people must accept lower wages and living standards.[71] "The U.S. does not have an energy policy that discourages the growth of imports," said former Secretary of Energy James Schlesinger, at the end of the Bush presidency, and the same can be said of today.[72] In the fall of 1993, the Energy Department was "readying tax breaks for domestic oil and gas producers" rather than creating massive new incentives for solar and other renewables.[73] Rather than investing in the productivity of this country's workers, U.S. financiers and industrialists plan to compete on the world market by creating a secure, low-wage manufacturing base in Mexico and other Third World countries.

In his first Earth Day message after taking office in 1993, President Clinton issued a "clarion call ... for American ingenuity and creativity to produce the best and most energy-efficient technology" and to reduce greenhouse-gas emissions to 1990 levels by the year 2000.[74] But the president's Climate Change Action Plan is long on "voluntary compliance" and short on sanctions for those whose emissions of greenhouse gases have increased.[75] Under the Climate Change Action Plan's wide allowances, the utilities will be able to expand at will in the United States so long as they plant (or claim to be planting) carbon-dioxide–absorbing forests somewhere in the world. Perhaps it should not be surprising that the utilities were spared in the Climate Change Action Plan; after all, the retired chairman and CEO of Pacific Gas & Electric, Richard A. Clarke, was appointed to the President's Council on Sustainable Development.[76] Clinton has been unwilling and unable to stand up to the fossil-fuel interests in his own party and pass a Btu tax on fossil-fuel use.[77] A typical environmentalist charge was that Clinton's greenhouse plan was "one that George Bush could have produced . . . it has no hammers, no substantial requirements. . . ."[78]

John Schaeffer, founder and president of Real Goods, the largest of the renewable-energy retailers, has waxed enthusiastic about "the enlightened attitudes of a utility like SMUD," which operates 4.5 megawatts of photovoltaic power and has made a commitment to install PV modules on one hundred additional homes each year. Schaeffer also criticized the "fat-cat attitudes" of a New York utility executive "who said smugly that banks would finance only photovoltaic systems provided by utilities" to remote customers. These opportunistic utilities, wrote Schaeffer, "plan to deliver a 1-kilowatt photovoltaic system in a fully prepared box which costs them around $30,000 . . . [and] in turn, charge the customer 16 percent per year forever. While this solves the immediate problem of providing power to a remote customer, it misses entirely the appeal of independent living. The utilities . . . want to bring enslavement to the remote home market just as they have to the suburbs. . . . The revolution we are fighting is to maintain the values of sustainability, self-empowerment, and independence rather than to find a new way to place ourselves under the watchful eye of Big Brother."[79]

Along similar lines, Dr. Hermann Scheer, longtime member of Germany's Bundestag and president of the European Solar Energy Association (Eurosolar), argues that for energy activists to focus on energy efficiency to the exclusion of solar power is a fraud, though energy efficiency is an indispensable part of the sustainable-energy strategy.

"The oft-cited phrase 'Energy-saving is the greatest new energy source' is not only factually false—for 'saving energy' is not a real source—but it only reduces the energy demand to some extent. What is really necessary," Scheer argues in *Sun Strategy*, "is to replace [fossil] fuel with sun energy . . . the only energy which is really friendly to the environment." Scheer envisages solar energy as a new "people's energy," with solar equipment manufacturing and installation on the scale of today's automobile industry. This powerful solar industry would become the engine for a new and more decentralized economy, with well-paid, highly-skilled jobs that could provide the material basis for a more decentralized and more cooperative social system.[80] But most mainstream business thinkers couldn't care less. Indeed, the 150th anniversary issue of the *The Economist*, perhaps the world's most influential business magazine, published ten authoritative articles about possible developments for the period 1993 through 2143, with zero references to solar energy or the possible problems, such as the greenhouse effect, of building a society totally reliant upon fossil fuels.[81]

Though the debate over restructuring and retail wheeling was triggered by large industrial interests, it has created a forum where the needs of average electricity consumers can be discussed. The formal positions of almost all state public utilities commissions are rife with pious protestations of concern for the interests of the average consumer and the need to preserve and expand a clean-energy sector.[82] Whether those noble sentiments will become reality depends on a level of popular understanding and mobilization around the energy issue not seen in the United States since the 1970s-era oil shocks and the battles over nuclear energy. That mass movement of energy activism stopped nuclear power plant construction, launched the solar energy and building-efficiency movements, and for a short time managed to reverse the monotonic increase in petroleum consumption in the United States.[83]

Although most citizens have not yet recognized these truths, today's battles are about the social control of energy, including solar energy—about the struggle to shape a clean and decentralized, productive system that will live up to its democratic potential.

The program of the devotees of fossil-fuel is a formula for further concentration of wealth and power: keep society hooked on petroleum and uranium; delay the solar transition until the oil and gas are too costly to recover; then burn up the coal and slap a meter on the sun. The antidotes to perpetual fossil-fuel dependency are democracy and community self-reliance, along with respect for good design and skilled work.[84] If citizens are to transcend their dependency on fossil fuels, they must learn to use their intelligence to live within their means. We have the technical means now to provide all of the power we need for a sustainable world.

Democracy is a false promise if it does not include the power to steer the energy economy. In *Who Owns the Sun?* we have argued that the new solar technologies make it possible to maintain human society with constantly diminishing use of fossil fuels. However, to turn the tools of a solar transition over to utilities and fossil-fuel corporations, which is the present policy of the government and mainstream environmental organizations, is to guarantee that the coming Solar Age will arrive a century behind its time, and that it will be every bit as autocratic as today's fossil-fuel economy. We believe that a solar revolution will necessarily occur at the expense of the private energy monopolies, and that such a revolution will not take place without a passionate public fight for more democracy and more participation.

Since the sun shines everywhere and is free for the harvesting, solar energy is inherently democratic, if ordinary citizens are allowed to develop and control the tools that utilize sunlight where it falls. People have an unalienable right to the solar energy that they collect on their rooftops. In the process of establishing the political and economic means for widespread adoption of renewable technologies, solar energy can help bring back that taste of happiness that people feel when they control their own resources and their community's economic destiny. Just as the people have unequivocally said no to nuclear power and toxic wastes, they must now organize to say yes to solar. Public, local control of energy generation, storage, and distribution is the essential basis for a democratic solar economy. Just as war is too important to leave to the generals, energy is too important to leave to private monopolies. For the original vision of democracy to be realized at last, all of us must have our place in the sun.

Notes

Introduction

1. On the anthracite strike of 1902–1903, see Sidney Lens, *The Labor Wars* (Garden City, New York: Anchor Books, 1974), pp. 158–173. For the history of electric power, see Richard Rudolph and Scott Ridley's indispensable *Power Struggle: The Hundred-Year War Over Electricity* (New York: Harper & Row, 1986). Between 1897 and 1907 between 60 and 120 municipally owned power systems were created every year, badly frightening the privately owned power companies (pp. 36–38). See Richard Grossman and Gail Daneker's classic *Energy, Jobs, and the Economy* (Boston: Alyson Publications, 1979). Grossman went on to found the Labor-Energy Coalition to try to convince environmentalists and labor that renewable energy would mean more rather than fewer jobs. For an update of this discussion by Grossman and others, see *Environmental Action* (Spring 1992): 11–23. Environmental Action's address is 6930 Carroll Ave., 6th floor, Takoma Park, MD 20912. For an early 1970s look at these debates and the technical possibilities at the time, see Wilson Clark's superb, encyclopedic, and nearly forgotten book *Energy for Survival* (Garden City, New York: Anchor Books, 1974).

2. Nixon's 1973 speech was quoted in the one-hour *Frontline* documentary "The Politics of Power," written by Nick Kotz, October 8, 1992, KQED, San Francisco, script available from Journal Graphics, 1535 Grant St., Denver,

CO 80203, $5.00, which accurately portrays the deals and pressures behind the limp National Energy Security Act of 1992.

3. See, for example, Harvey Wasserman, *Energy War: Reports from the Front* (Westport, Connecticut: Lawrence Hill & Co., 1979). Green Mountain Post Films, Box 229, Turners Falls, MA 01376, distributes *Lovejoy's Nuclear War, The Last Resort* (about the Seabrook mass actions), and a few dozen other videos and films that vividly communicate the energy and peace insurgencies of the 1970s and early 1980s.

4. Anthony White, "It's Ba-a-a-ck! Duck and Cover! The TVA Is Doing Its Best To Raise Nuclear Power From the Dead," *Village Voice*, July 9, 1991.

Chapter 1 Solar America: A Dream Deferred

1. Kenneth E. Boulding, "The Economics of Spaceship Earth," in *The Environmental Handbook,* ed. Garret De Bell (New York: Ballantine/Friends of the Earth, 1970), pp. 96–101 (a collection prepared for the first national environmental teach-in, April 22, 1970).

2. A seductive "must read" is Ken Butti and John Perlin's *A Golden Thread: 2,500 Years of Solar Architecture and Technology* (New York: Cheshire Books; Palo Alto; Van Nostrand Reinhold, 1980). The pictures and illustrations are as good as the text. A new edition, with 3,500 years of solar history, will be available from the publisher Aatec in mid-1997 (call 800-995-1470). Reynold M. Wik, *Henry Ford and Grass-roots America* (Ann Arbor, Michigan: University of Michigan Press, 1972), esp. ch. 4, "Barnyard Inventors and the Model T." Ford himself read widely about different technologies, if his personal library is an indication (p. 239).

3. For an inspiring guide see Edward Mazria, *The Passive Solar Energy Book: A Complete Guide to Passive Solar Home, Greenhouse and Building Design* (Emmaus, Pennsylvania: Rodale Press, 1979). More recent editions exist.

4. *Reaching Up, Reaching Out: A Guide to Organizing Local Solar Events*, Project leader: Rebecca Vories, Solar Energy Research Institute (now National Renewable Energy Laboratory), Golden, Colorado, SERI/SP-62-326, U. S. Government Document No. 061-000-00345-2, c. 1981.

5. "The Harris Survey, June 1, 1978," (New York: Chicago Tribune-New York News Syndicate, Inc., 1978), quoted in *Blueprint For a Solar America*, Solar Lobby, January 1979, p. 2.

6. *A Time To Choose: America's Energy Future*, Energy Policy Project, Ford Foundation, S. David Freeman, Director (Cambridge, Mass.: Ballinger Publishing Co., 1974). The report was so controversial, says Freeman, that it was quietly shelved by Ford.

7. Ibid.

8. *No Time to Confuse* (San Francisco: Institute for Contemporary Studies, 1975).

9. See David Morris, *Self-Reliant Cities: Energy and the Transformation of American Cities* (San Francisco: Sierra Club Books, and St. Paul, Minnesota: Institute for Local Self-Reliance, 1982). Chapter 2 contains the best capsule history of the electrical grid.

10. Calvin Broomhead, long-time solar activist who now works in energy conservation for the Public Utilities Commission of the City of San Francisco, explained this concept to me. Barry Kelman in "De-Utilitizing the Energy Industry: Planning the Solar Transition," *UCLA Law Review* 28 (1980): 1–51, argued that utilities should be allowed into the naturally decentralized solar markets like water heating and photovoltaics only if they were not allowed to use their monopoly powers to keep out other competitors. No critique of Detroit's "insolent chariots" has improved on Lewis Mumford's 1958 essay "The Highway and the City," pp. 234–246 in his collection by the same name (New York: Harcourt, Brace & World, 1963). On the GM conspiracy to get rid of streetcars, see David J. St. Clair, *The Motorization of American Cities* (New York: Praeger, 1986), esp. ch. 3, "The Conspiracy Evidence."

11. Ray Reece, in *The Sun Betrayed* (Boston: South End Press, 1979), pp. 1–2, traces the strength of the initial surge of interest in renewable energy to a happy alliance of young entrepreneurs and counter-culture technologists.

12. Copyright Stuart Leiderman, *Songs of Conscience*, 1981; record and tape still available for $10.00 postpaid from Stuart Leiderman, PO Box 186, Woodstock, NY 12498.

13. Albert Bates, "Technological Innovation in a Rural Intentional Community, 1971–1987," *Bulletin of Science and Technology Society* 8 (1988): 183–199; available for $1 from Albert Bates, Box 90, Summertown, TN 38483.

14. Michael H. Brown, *Brown's Alcohol Motor Fuel Cookbook* (Cornville, Arizona: Desert Publications, 1979).

15. See *A Time To Choose*; Daniel Yergin, "Conservation: the Key Energy Source," in *Energy Future, Report of the Energy Project at the Harvard Business School*, ed. Robert Stobaugh and Daniel Yergin (New York: Ballantine Books, 1979), pp. 167–229.

16. In ten years, promised Rockefeller in 1976, the U.S. should conserve 5 percent more energy and increase domestic oil production by 50 percent, coal production by 100 percent, natural gas production by 10 percent, and nuclear power generation by 300 percent; see Jim Horn, "The National Energy Strategy: A Historical Perspective," *Northern California SUN* 17, no. 4 (1991): 14, published by the Northern California Solar Energy Asso-

ciation, PO Box 3008, Berkeley, CA 94703. Nelson, of course, was the grandson of the original John D. Rockefeller, founder of Standard Oil. On this theme see also Robert Engler, *The Brotherhood of Oil: Energy Policy and the Public Interest* (New York: New American Library, 1977), esp. the critique in ch. 8, "Planning for New Beginnings."

17. Denis Hayes, *Rays of Hope: The Transition to a Post-Petroleum World* (New York: W. W. Norton & Co., 1977).

18. See Eugene Frankel, "Technology, Politics and Ideology: The Vicissitudes of Federal Solar Energy Policy, 1974-1983," in *The Politics of Energy Research and Development*, vol. 3, John Byrne and Daniel Rich, eds. (New Brunswick, New Jersey: Transaction Books, 1986), esp. pp. 75–77.

19. Richard Rudolph and Scott Ridley, *Power Struggle: The Hundred Year War Over Electricity* (New York: Harper & Row, 1986), pp. 200–202; also John Howes, "The Politics of Electric Power Deregulation," *Regulation* (Winter 1992): 17–20.

20. Quote is from Frankel, "Technology, Politics and Ideology," p. 78.

21. Barry Commoner, *The Politics of Energy* (New York: Knopf, 1979), cover page.

22. Michael Tanzer and Stephen Zorn, *Energy Update: Oil in the Late Twentieth Century* (New York: Monthly Review Press, 1985), esp. pp. 13–39. Ernest Callenbach invented the word ecotopia in his famous novel *Ecotopia*, first published 1975, which has been translated into a dozen languages (New York: Bantam Books edition, 1977).

23. See Reece, *The Sun Betrayed*.

24. Ibid., p. 61.

25. Rudolph and Ridley, *Power Struggle*, p. 142.

26. Cited in Reece, *The Sun Betrayed*, p. 118.

27. Ibid., pp. 116–120.

28. Amory Lovins, "The Soft Energy Path," in *Political Ecology: An Activist's Reader on Energy, Land, Food, Technology, Health and the Economic and Politics of Social Change*, ed. Alexander Cockburn and James Ridgeway (New York: Times Books, 1979), pp. 76–88. See "The Future of Planetary Exploration," an interview with NASA administrator Daniel Goldin and Carl Sagan, in *Engineering & Science* (Winter 1993), published by the Alumni Association of the California Institute of Technology.

29. Reece, *The Sun Betrayed*, p. 168.

30. Ibid., pp. 116–120.

31. Ibid., p. 120.

32. This account is based on the personal experiences of Denver businessman Dr. Jerry Plunkett, founder and owner of Materials Consultants, Inc., and a director of the American Association of Small Research Companies, at

1975 hearings of the Senate Select Committee on Small Business, quoted in ibid., pp. 71–74, 81.

33. Wilson Gonzalez, an economist with Columbia Gas, Columbus, Ohio, notes that the electric utility industry has "vigorously attacked" proposed regulations by the Office of Energy Efficiency and Renewable Energy in the Department of Energy which would ban electric resistance water heaters (letter to Dan Berman, Oct. 4, 1994).

34. Reduction depending on the size of the solar hot water heater, the climate, and personal differences in hot water usage. The *SMUD SOLAR PRO-GRAM PLAN* (April 1992, p. 11) estimates that solar hot water heaters cut gas or electric energy used for water heating by 60 to 70 percent in Sacramento.

35. Gigi Coe, "California's Experience in Promoting Renewable Energy Development," in *State Energy Policy: Current Issues, Future Directions*, Stephen W. Sawyer and John R. Armstrong, eds. (Boulder, Colorado: Westview Special Studies, 1985), pp. 193–211.

36. See *Building Energy*, which represents the proceedings of three conferences: the First International Conference on Solar Electric Buildings, cosponsored by the Solar Heating and Cooling Programs Division of the International Energy Agency; RENEW '96; and the Twelfth Annual Quality Building Conference, March 4–6, 1996, Boston, Massachusetts. Two volumes of proceedings available from the Northeast Sustainable Energy Association, Greenfield, Massachusetts.

37. Robert Engler's indispensable book *The Politics of Oil: A Study of Private Power and Democratic Institutions* (New York: MacMillan, 1951) calls this the "private government of oil."

CHAPTER 2 Keeping Solar Culture Alive

1. *Oxford English Dictionary* (1928), republished as *The Compact Edition of the Oxford English Dictionary* (Oxford: Oxford University Press, 1971).

2. Ken Butti and John Perlin, *A Golden Thread: 2,500 Years of Solar Architecture and Technology* (New York: Cheshire Books; Palo Alto: Van Nostrand Reinhold, 1980), ch. 11. See note 2 on p. 247 for reprint information.

3. T. Hisada and I. Oshide, "Use of Solar Energy for Water Heating," *United Nations Conference on New Sources of Energy*, E 35 Gr-S13 (Rome, 1961), cited in ch. 6, "Heating Water," of Farrington Daniels' pioneer book *Direct Use of the Sun's Energy* (New York: Ballantine Books, 1974, first printed by Yale University Press, 1964).

4. Louis Peck, "Whatever Happened to Solar?" *Garbage* 3 (January/February 1991): 25–31.

5. *Power to Change: Case Studies in Energy Efficiency and Renewable Energy*, Energy Policy Research Unit, Greenpeace International, Keisersgracht 176, 1016 dw Amsterdam, Netherlands (December 1993), pp. 7, 8.

6. Butti and Perlin, *Golden Thread*, ch. 18.

7. Arthur Shavit is a professor in the Department of Mechanical Engineering at the Technion, in Haifa. He spent part of a sabbatical year at the Florida Solar Energy Center in Cape Canaveral. His views are recounted in John Blackburn's excellent little book: *Solar Florida: A Sustainable Energy Future* (Winter Park, Florida: Florida Conservation Foundation, 1993), p. 129.

8. A worthy successor to *A Golden Thread* is Norwegian architect Harald N. Roestvik's colorfully illustrated book *The Sunshine Revolution* (1992), available from SUN-LAB publishers, Steingaten 87, 4024 Stavanger, Norway. For more information on Israel and other national programs, see pp. 152–164, "National Solar Programs."

9. For a brief description of eight types of common solar water heaters, see Mary Freen, John Harrison, and Colleen McCann Kettles, *Florida's Energy Future Is Here Today: A Report to the Florida Solar Energy Industries Association on the Status of Solar Water Heating in Florida*, Florida Solar Energy Center, April 1990, pp. 3–10.

10. Butti and Perlin, *A Golden Thread*, ch. 12.

11. Judy Stark, "Florida House Shows How Earth-friendly a Building Can Be," *San Francisco Sunday Examiner & Chronicle*, June 26, 1994, Zone 1, p. 12.

12. Assuming electricity costs of 10 cents per kilowatt-hour. Estimates of an average of 300 kWh/household/month are from the *Staff Report on Energy Efficiency Issues*, prepared for the Subcommittee on Public Utilities, Committee on Regulated Services and Technology, Florida State Legislature, December 11, 1991, pp. 3, 4.

13. Estimated total demand for 1991 was 144,986 GWh (billion watt-hours). Five percent of Florida houses already have solar hot water heaters, which reduce electric demand by roughly 2,520 kilowatt-hour per year apiece (70 percent of the 3,600 kilowatt-hours per year expended on electric water heating). Only 16 percent of Florida's residences are supplied with natural gas. If solar water heaters were installed in all the 4 million residences which lack them, the savings would total nearly 10,000 GWh, or about seven percent of total electricity consumption. Estimates on total electricity demand in Florida for 1991 are from Howard Geller and Steven Nadel, "Analysis of Electricity Savings Potential in Florida From the Adoption of Equipment Efficiency Standards and Aggressive Utility DSM Programs," American Council for an Energy-Efficient Economy, Washington, D.C., July 1991; estimates of average kilowatt-hours used for water heating are from the *Staff Report on Energy Issues*, pp. 3, 4.

14. Freen, et al., *Florida's Energy Future*, pp. 34, 52.
15. See Florida Energy Office, *Florida Energy Data Report 1990*. Almost all of the "renewable" energy comes from wood waste and sugar cane bagasse burned for process steam and cogenerated electricity.
16. Freen, et al., *Florida's Energy Future* , esp. "Summary of Florida Energy Legislation," pp. APP-20–APP-23.
17. *Staff Report on Energy Efficiency Issues*, p. 4.
18. Reference to Florida Light and Power's DSM plan for the nineties is from Freen, et al., *Florida's Energy Future*, p. 57.
19. Don Kazimir, telephone interview, March 5, 1993.
20. Telephone interview, March 4, 1993. Kazimir is the president of Solar Development Inc. (SDI) of Riviera Beach, Florida.
21. David L. Block, Ph.D., P.E., "Solar Energy and Energy Efficiency: Jobs For Florida," February 1993, typescript copy in authors' possession. Dr. Block is the Director, Florida Solar Energy Center, Cape Canaveral, Florida, tel. 407-783-0300.
22. See "Financing for Solar: Florida Firm Focuses on Affordability," *Florida Energy Reporter* 1 (July/August 1993): 12.
23. Interviews with Independent Savings Plan Company representative, September 20 and October 11, 1993.
24. See Colleen Kettles, "Solar System Sale Activity: Solar Water and Pool Heating Systems, Survey of the Florida Solar Industry," typescript, 2pp., based on an April 1994 survey by the Florida Solar Energy Research & Education Foundation, administered by Kettles, who is the foundation's director.
25. Dr. David Block, telephone interview; July 31, 1996.
26. Freen, et al., *Florida's Energy Future*, Appendix VIII, Consumer Survey, p. APP-43.
27. Robert Blackburn to Dan Berman, October 9, 1993.
28. Dr. David Block estimates that 11,850 solar water heating systems were installed in 1991, generating 135 jobs in manufacturing and 1,088 jobs in installation. Increasing that total to 50,000 installations a year would generate 1,230 new jobs; if 200,000 a year were installed, it is easy to imagine 5,000 new jobs created in new domestic installation alone, until the initial demand was satisfied in 20 years. See Block's "Solar Energy and Energy Efficiency."
29. The Energy Foundation, *Annual Report, 1992*, p. 18.
30. Steve Warn, Energy Policy Analyst, Florida Chapter, American Planning Association, telephone interview, March 4, 1993. The project has published the *Florida Energy Reporter* every two months since early 1993.
31. *Staff Report on Energy Efficiency Issues*, pp. 3, 4.

32. Colleen Kettles, telephone interview, July 31, 1996.

33. "Energy Efficiency Saves Dollars, Makes Sense," *Consumer Action News*, Florida Consumer Action Network (Spring/Summer, 1992): 1, 6, 7. See *Florida's Energy Future*, p. APP-21; also telephone interview with Steve Warn, March 4, 1993.

34. Brian Kerwin, Solar Coordinator at Utility Services, City Hall, Tallahassee, Florida. Telephone conversations, July 30 and 31, 1996.

35. Telephone interview with Carilyn Shon, Program Manager, Energy Division, Department of Business, Economic Development, and Tourism, July 1996. The converse would be to ask whether computers exist which function well under varying heat and humidity. Surely the U.S. military must have investigated this point.

36. Government and utility informants agree on this figure.

37. "Hawaii and Energy," Department of Planning and Economic Development, 1984, quoted in Jan Hamrin and Nancy Rader, *Investing in the Future: A Regulator's Guide to Renewables* (Washington, D.C.: National Association of Regulatory Utility Commissioners, February 1993), pp. F9–F13. For examples of what can be done in Hawaii with solar energy, see Jim Pearson, *The Hawaii Home Energy Book* (Honolulu: The University Press of Hawaii, 1978), which includes an appendix on meter reading.

38. See, for example, J.W. Shupe and J.W. Weingart, "Emerging Energy Technologies in an Island Environment," *Annual Review of Energy*, 1980, pp. 293–333, cited in Amory B. Lovins and L. Hunter Lovins, *Brittle Power: Energy Strategy for National Security* (Andover, Massachusetts: Brick House Publishing Company, 1982).

39. Docket No. 94-0226, Order No. 13441, Before the Public Utilities Commission of the State of Hawaii, *Instituting a Proceeding on Renewable Energy Resources, Including the Development and Use of Renewable Energy Resources in the State of Hawaii.*

40. Keith Block, "Energy Savings Will Benefit Oahu," *Hawaii Remodeling* (July 1996): 22. Mr. Block is the Program Manager, Energy Services Department, Hawaiian Electric Company, Inc.

41. Robert J. Mowris, Energy Efficiency and Least-cost Planning: the Best Way to Save Money and *Reduce Energy Use in Hawaii*, Revised draft, August 20, 1990.

42. Retail electricity is three times as expensive as in Nebraska and 1.5 times as expensive as in most of California. Gas is not an option, since most offices and residences outside of downtown Honolulu are not hooked up to gas pipelines. Synthetic natural gas for the few who burn it is a refinery byproduct which sells for over twice the price of piped natural gas in California.

43. Solar water heaters cost $4,000 in Hawaii and save about $500 year on a person's electric bill. With an $800 HECO rebate and a 35 percent state income tax rebate on the remaining $3,200, the out-of-pocket cost comes down to $2,080, which can be repaid in four years, under the new Hawaii PUC regulations. For the following 11 years, the expected life of the system, hot water bills should be negligible, since the systems are designed to provide 100 percent of hot water needs 90 percent of the time; see Rolf Christ, "The Smartest Home Improvement You Can Make: Solar," *Hawaii Remodeling* (July 1996): 19–20.

44. Hawaiian Electric Industries (HEI) is an independent holding company that owns 100 percent of the shares of the Hawaiian Electric Company, Inc. (HECO) on Oahu; the Hawaii Electric Light Co. (HELCO) on the Big Island of Hawaii; the Maui Electric Co., Ltd.; and a number of other companies. Three-quarters of HEI's 1994 sales of $1.2 billion came from its three utilities. Profits that year came to $73 million. Kawai Electric is is the only electric company on the islands that is not owned by HEI.

45. Telephone interview with Clyde Murley, Natural Resources Defense Council, San Francisco, April 5, 1993.

46. Telephone interview with Clyde Murley, July 14, 1993. Saddled with "lemon turbines" and a skeptical utility, Hawaiian Electric Industries are rapidly souring on wind power, according to Robert W. Righter, *Wind Energy in America: A History* (Norman, Oklahoma: University of Oklahoma Press, 1996), pp. 260–63.

47. Decision and Order No. 14709, Docket No. 94-0216, "For Approval of a Residential New Construction Program, Recovery of Program Costs and Lost Revenues, and Consideration for Shareholder Incentives," May 29, 1996 Whether this order, filed by the Hawaii Public Utilities Commission and intended to encourage solar and energy-efficient water heating in new construction, raises this percentage remains to be seen.

48. Telephone interview with Keith Block, Program Manager, Energy Services Department, HECO, July 19, 1996.

49. Compared to about 40 percent on the mainland.

50. Around 1915, William J. Bailey would hide the water tank of his popular Day and Night solar heater in a false chimney, according to Butti and Perlin, in *A Golden Thread*. The new trend in photovoltaics is to integrate the solar electric systems into the building's roofs, walls, and windows, according to *Proceedings of the First International Conference on Solar Electric Buildings*, Boston, Massachusetts, March 4-6. 1996, available from the Northeast Sustainable Energy Association, Greenfield, Massachusetts.

51. The 35-percent tax credit is scheduled to lapse on December 31, 1998 if it is not renewed. Governor Cayatano, worried about a state budget deficit of

over $100 million in 1995, has often stated that he wants to end all tax credits. The political pressure to cut the deficit could make the tax rebates for solar water heating short-lived.

52. Telephone interviews with Rolf Christ, July 1996.

53. Mowris, in *Energy Efficiency*, Table 1, p. 12, estimates the number of solar water heaters as 40,000; Mowris assumes that the water heater would save an average of 3,320 kilowatt-hours per year of electricity, worth about $500. Nancy Rader estimated the total number of solar water heaters in Hawaii at 50,000 in *The Power of the States* (Washington, D.C.: Critical Mass Energy Project, Public Citizen, June 1990), p. 80.

54. Formerly the Department of Planning and Economic Development.

55. Financial and ownership information is from *Who Owns Whom? 1995 Directory of Corporate Affilations*, vol. III, *U.S. Public Companies* (New Providence, NJ: National Register Publishing, 1995); and *Value Line*, May 24, 1996, available in good public libraries.

56. Decision and Order No. 14709.

57. Decision and Order No. 14730, Docket No. 94-0206, "For Approval of a Residential Efficient Water Heating Program, Recovery of Program Costs and Lost Revenues, and Consideration for Shareholder Incentives," June 5, 1996.

58. Also including 8,000 heat pump water heaters, 2,500 energy efficient electric water heaters, and 38,000 low-flow shower heads.

59. Building codes require systems that earn tax rebates to meet all of a household's water heating requirements 90 percent of the time. The financial benefits work as follows: $4,000 minus $800 times .35 equals $1,120; $4,000 minus $1,120 equals a net system cost of $2,080, which means a payback time of four years at $518 per year in electric bill savings. After the system is paid for, of course, water heating will be almost free for the rest of the life of the system, minus do-it-yourself annual maintenance, and a five-year professional tune-up. See Rolf Christ, "The Smartest Home Improvement."

60. Decision and Order No. 14730.

61. During the five years through the year 2000, if HECO's plans work out, 16,000 new solar water heaters at $4,000 apiece (with $800 tax credits), 8,000 new heat pumps at $1,900 apiece (with $344 in tax credits), and 2,500 efficient electric water heaters at $700 apiece (with $138 in tax credits) will be installed on the island of Oahu alone. State-wide, plans call for the installation of 20,000 new solar water heaters altogether through the year 2000.

62. Some of the administrative activities are listed in Decision and Order No. 14730, pp. 6 and 7, but the document does not specify the cost breakdown by activity.

63. "Recovery of lost revenue margin," in HECO's utility-speak.

64. "Shareholder incentives," in HECO utility-speak.

65. The authors did not evaluate how the figure of 82.5 million kilowatt-hours was calculated, so we do not feel comfortable assigning a cost-per-kilowatt-hours figure to the electricity saved.

66. The May 24, 1996, *Value Line* survey (available at any good library) was "unenthusiastic" about Hawaiian Electric, and called its dividend-growth potential "minimal" until 1999-2001.

67. Telephone interview with Jay Hansen, April 7, 1993.

68. Telephone inteview with Moanikeala Akaka, July 24, 1996.

69. *Native Hawaiian Position Paper on Geothermal Development*, Pele Defense Fund, P.O. Box 404, Volcano, Hawaii 96785, 5pp.

70. The quote from Lovins is from Joan Conrow, "Geothermal: Pele's Last Stand," *Honolulu* (June 1990): 56-59, 78-86. Mission Energy, the go-go subsidiary of a Southern California utility holding company then called SCEcorp and now called Edison International, also made a bid, ultimately unsuccessful, to build a controversial geothermal plant on the island of Hawaii in a tropical rain forest sacred to the Hawaiian people, according to Hamrin and Rader in *Investing in the Future*, p. F12.

71. See *Proposed Marine Mineral Lease Sale in the Hawaiian Archipelago and Johnston Island Exclusive Economic Zones, Draft Environmental Impact Statement*, A Joint Effort by the U. S. Department of the Interior, Minteral Management Service and the Department of Planning and Economic Development, January 1987, p. 216. See also Timothy Egan, "Hawaii Debates Peril to Rain Forest as an Energy Project Taps a Volcano," *New York Times*, January 26, 1990.

72. "True Victory in Hawaiian Rainforest," *Action Alert* #90, Rainforest Action Network, San Francisco, April 1995.

73. Telephone interview with Moanikeala Akaka, July 24, 1996.

74. Telephone interviews with Shirley Casuga, August 15, 1993 and June 19, 1996, who can be reached at PO Box 6101, Captain Cook, HI 96704.

75. Quotation is from George Stricker, former president of the California Wind Energy Association, in "The California Covenant," *Windpower Monthly* 5 (July 1989): 18-20; cited in Richard W. Righter, *Wind Energy in America: A History* (Norman, Oklahoma: University of Oklahoma Press, 1996), p. 196.

76. Gigi Coe, "California's Experience in Promoting Renewable Energy Development," in *State Energy Policy: Current Issues, Future Directions*, ed. Stephen W. Sawyer and John R. Armstrong (Boulder, Colorado: Westview Special Studies, 1985), pp. 193-211.

77. Ibid.

78. See Paul Gipe, *Wind Energy Comes of Age* (New York: Wiley, 1995), esp. ch. 1; and Righter, *Wind Energy in America*, esp. ch. 10, "California Takes the Lead." See also Thomas A. Starrs, "Legislative Incentive and Energy Technologies: Government's Role in the Development of the California Wind Energy Industry," *Ecology Law Quarterly* 15, no. 1 (1988): 103–158.

79. Peter Barnes, phone interview with Dan Berman, February 13, 1992.

80. Richard Grossman and Gail Daneker, *Energy, Jobs, and the Economy* (Boston: Alyson Publications, 1979), ch. 3, "Jobs From Conservation" and ch. 4, "Jobs From the Sun."

81. Telephone interview with Dick Behm, February 13, 1992.

82. Information supplied by Ester Epperson from her family's May/June 1993 propane bill.

83. William Greider, "The Education of David Stockman," *Atlantic Monthly* (December 1981): 51, 52.

84. For 1984 estimate see Coe, "California's Experience," p. 194. See graph "Annual Solar Investment, California 1973-1991," prepared by Alec Jenkins, California Energy Commission. Jenkins estimates the solar water heating business at just under $250 million in 1984.

85. The best source for the level of renewable energy activity is Nancy Rader, et al., *The Power of the States: a Fifty-State Survey of Renewable Energy*, Critical Mass Energy Project, Public Citizen, 215 Pennsylvania Avenue, SE, Washington, DC 20005, June 1990, which includes "A Critique of the Federal Government's Collection of Data on Renewable Energy" on p. 82.

86. Alec Jenkins, graph, "Annual Wind Turbine Investment, California, 1980–1991," California Energy Commission, 1992; employment figures from Coe, "California's Experience," p. 194.

87. U.S. Department of Energy, Energy Information Administration, *Energy Facts 1989*, DOE/EIA-0469(89), p. 95. *Energy Facts* is available free from any EIA office.

88. "Solar Thermal Collector Shipments by Type," *The 1992 Information Please Almanac*, p. 668, based on data from the federal Energy Information Administration publication, *Solar Collector Manufacturing Activity*.

89. Michael Lotker, *Barriers to Commercialization of Large-Scale Solar Electricity: Lessons Learned from the LUZ Experience*, SAND91-7014, Sandia National Laboratories, November 1991, p. 16, where he estimates that the real price of natural gas, in 1980 dollars, fell from $6.61 to $1.45 per million British Thermal Units (MMBTU) between 1982 and 1991.

90. See Mike O'Keeffe, "Power Failure," *Westword* (February 27–March 5, 1990), pp. 20–25.

91. On the cooperation between the U. S. and Saudis in setting oil prices, see Edwin S. Rothschild, "The Roots of Bush's Oil Policy," *The Texas Observer*,

February 14, 1992, pp. 10–14. George Minter, a lobbyist for Southern California Gas, estimated that 60 percent of natural gas comes from oil companies (chat in Sacramento, October 15, 1991, at a SoCal Gas meeting designed to enlist environmentalist support for the gas industry's $250 million federal research agenda).

92. *Energy Facts 1989*, p. 95. New capacity installed in California wind plants crashed from 398 megawatts (1985) to 19 megawatts (1992 projected); see "Wind Energy: A Resource for the 1990s and Beyond," American Wind Energy Association, 777 N. Capitol St., NE, Suite 805, Washington, DC, 20002.

93. According to Bob Aldrich, California Energy Commission spokesperson, quoted in Ricardo Sandoval, "Riding Winds of Energy Change," *San Francisco Examiner*, August 31, 1992, p. B1.

94. The deal went through, according to one insider, because Mitsubishi offered a 10-year warranty and a 90 percent performance guarantee for the new machines.

95. See *Independent Power: A California Success Story*, prepared by Edson & Modisette, sponsored by Independent Energy Producers Association, October 11, 1993, tables and charts on pp. 7 and 15.

96. Jonathan Marshall, "California OKs Electricity Sales to 3 Big Utilities," *San Francisco Chronicle*, June 23, 1994, p. D1.

97. Jeff Pelline, "S.F. Energy Firm Wins Go-Ahead for Wind Projects," *San Francisco Chronicle*, December 11, 1993, p. D1.

98. Coe, "California's Experience." Barnes helped found Working Assets, a "socially conscious" investment firm.

99. Denis Hayes, Herb Epstein, Susannah Lawrence, and Janet Dierker (Solar Lobby); Tom Cohen (Common Cause); Grant Thompson (Conservation Foundation); Pam Duel (Environmental Action); Garry Deloss (Environmental Policy Center); Jonathan Lash (Natural Resources Defense Council); and Jonathan Gibson (Sierra Club), *Blueprint for a Solar America*, January 1979, pp. 31, 32.

100. For a good summary of the lively range of opinion of pro-solar forces, see *Sun! A Handbook for the Solar Decade*, ed. Stephen Lyons (San Francisco: Friends of the Earth and Solar Action, 1978). *Sun!* was billed as "The Official Book of the First International Sun Day."

101. Daniel Yergin, *The Prize, The Epic Quest for Oil, Money, and Power* (New York: Simon & Shuster, 1991), p. 753.

102. Maureen Dowd, "War Introduces Nation to a Tougher Bush," *New York Times*, March 2, 1991, p. 1.

103. See, for example, the evidence about the Banco Nazionale de Lavoro's role in the Iraqgate scandal, presented in floor speeches by Representative

Henry Gonzales of Texas, the chair of the House Banking Committee, *Congressional Record- House*, September 21, 1992, pp. H8820–H8829.

104. Richard Grossman to Dan Berman, October 28, 1992.

105. Telephone interview with Peter Barnes, February 13, 1992.

106. David Morris, *Self-Reliant Cities: Energy and the Transformation of American Cities* (San Francisco: Sierra Club Books; St. Paul, Minnesota: Institute for Local Self-Reliance, 1982), p. 196, apparently quoting from a 1980 study by the California Public Utilities Commission: *Demonstration Solar Financing Program*. The Continental Federal Savings and Loan Association let investors earmark deposits for solar projects in single or multifamily homes.

107. See Eugene Frankel, "Technology, Politics, and Ideology: The Vicissitudes of Federal Solar Energy Policy, 1974–1983," in *The Politics of Energy Research and Development*, vol. 3, ed. John Byrne and Daniel Rich (New Brunswick, New Jersey: Transaction Books, 1986), esp. pp. 75–79, for the ins and outs of federal policy under President Carter, which supports the contention of solar radicals like Richard Grossman that the Department of Energy never had any intention of seriously supporting a solar economy.

108. Consensus of NRDC and EDF representatives meeting at NRDC office, San Francisco, March 19, 1992, called by Southern California Gas Company; Christopher Flavin and Nicholas Lenssen, *Beyond the Petroleum Age: Designing a Solar Economy*, Worldwatch Paper 100 (Washington, D.C.: Worldwatch, December 1990) note that 25 to 65 percent of the homes in these countries have solar water heaters (p. 19).

109. The bill was AB 1328, introduced by Assembly member Sam Farr, according to *Solar Industry Journal* (fourth quarter, 1991): 5.

110. Regarding Davis and Village Homes, see *Those Who Make Memories: A Century of the Davis Enterprise* (Davis, California: The Davis Enterprise, 1996), pp. 64–67 and 130–132; Michael N. Corbett (with Judy Corbett and John Klein), *A Better Place To Live* (Emmaus, Pennsylvania: Rodale Press, 1981); also David Bainbridge, Judy Corbett, and John Hofacre, *Village Homes' Solar House Designs* (Emmaus, Pennsylvania: Rodale Press, 1979); and Edward L. Vine, *Solarizing America: The Davis Experience*, Public Policy Report N. 5, Lawrence Berkeley Laboratory, University of California, Berkeley, California, August 1981.

111. Amory Lovins, "Forward," *Real Goods Sourcebook 1991*, Real Goods Trading Corporation, Ukiah, California.

112. *Consumer Reports* (February 1994): 74, 80–84, disagrees.

113. Timothy Ziegler, "Living Beyond the Power Lines," *San Francisco Chronicle*, October 7, 1994, pp. A21, A23.

114. *PV-4U-Connections* 4 (Spring 1992): 2, published by Photovoltaics for Utilities, Boston, Massachusetts.

115. See *Real Goods News*, June 1994, p. 12.

116. Interview with Keith Rutledge of the Bank of Willits, August 24, 1992.

117. "David Katz: Alternative Energy Engineering," *The New Settler Interview* 49 (May 1990): 29.

118. For the way of life and economics of Garberville, the center of sinsemilla (no seeds) marijuana growing, see Ray Raphael, *Cash Crop: An American Dream* (Mendocino, California: The Ridge Times Press, 1985). The ferocious (depending on your point of view) federal/state Campaign Against Mariajuana Planting (CAMP) has made marijuana trade too dangerous for most in Northern California. (The "devil weed" is still the region's largest cash crop, though a North Coast agricultural commissioner was nearly "run out of town on a rail" for admitting it publicly. Despite a decade of search and destroy operations, rumors were still rife of "city slickers" with suitcases full of $100 bills offering up to $5,000 a pound for good clean bud from the Emerald Triangle, according to *Lookout* 36 (Winter 1992), "Around the Emerald Triangle."

119. Quote is from p. 2 of the *Alternative Energy Engineering 1991–92 Catalogue and Design Guide*, available for $3 from Alternative Energy Engineering, PO Box 339, Redway, CA 95560, or call toll-free 800-777-6609 "for ordering and technical support." The monthly *New Settler Interview* (PO Box 730, Willits, CA 95490) seems to distill as well as any source the lives and culture of the "ecotopian" alternative community in Mendocino and Humboldt counties. The interview with David Katz appeared in *New Settler Interview* 49 (May 1990), which has sold out since it was published together with an interview with Earth First! leader Judi Bari the same month she and Darryl Cherney were bombed and nearly murdered in Oakland.

120. *Alternative Energy Engineering 1991–92 Catalogue & Design Guide*, p. 2.

121. Telephone interview with Robert Klayman, head of the catalogue division, Real Goods Trading Corporation, July 26, 1996. See the cover story on the Solar Living Center in *Solar Today* (May/June 1996).

122. Ibid., which for $14 claims to "walk you through all the steps toward cutting your powerline." Write Real Goods Trading Corporation, 555 Leslie St., Ukiah, CA 95482; toll-free telephone, 800-762-7325.

123. Katz is honest enough to point out that AEE's $10,000 solar systems make it possible to bring in the appliances and televisions that earlier, more purist "new settlers" had struggled to escape. On the other hand, without AEE's solar tools, those same people would be running generators all day to feed their electronic habits.

124. Real Goods Trading Corporation, *Annual Report, 1995,* "Environmental Audit," which also claims to have prevented the release of 838 million pounds of carbon dioxide into the atmosphere. Calculations are based on a Natural Resources Defense Council formula. My figure of almost 10,000 houses for two years is based on an estimated annual energy consumption of 6,000 kWh per single-family house. He also claims to have saved 306,000 trees, and prevented the production of 183 million pounds of carbon dioxide.

125. Though it is primarily a how-to magazine à la *Popular Mechanics, Home Power*'s publishers Richard and Karen Perez have provided a forum for activists such as Don Loweburg of Independent Power Providers.

126. Bill Clinton and Al Gore, *Putting People First: How We Can All Change America* (New York: Times Books, 1992), pp. 89-92, 98.

127. Albert Gore, *Earth in the Balance: Ecology and the Human Spirit* (Boston: Houghton Mifflin, 1992), especially ch. 15, " A Global Marshall Plan."

128. See the three-volume (with appendices) report of the *Task Force on Strategic Energy Research and Development,* Secretary of Energy Advisory Board, U.S. Department of Energy, June 1995, chaired by Dr. Daniel Yergin, President, Cambridge Energy Research Associates, Cambridge, Massachusetts.

129. See Don Loweburg, "National PV Production Statistics," *Home Power* 51 (February/March 1996); and also his "Independent Power Providers," *Home Power* 52 (April/May 1996), Ashland, Oregon.

130. *Task Force on Strategic Energy Research and Development,* Annexes 2–4, pp. 39 and 61.

131. See Joseph J. Romm and Charles B. Curtis, "Mideast Oil Forever?" cover story in *Atlantic Monthly* (April 1996). Mr. Romm is the acting principal deputy assistant secretary for energy efficiency and renewable energy in the U.S. Department of Energy, and Mr. Curtis is the deputy secretary of the Department of Energy.

132. See the masterful analysis by Carol Alexander and Ken Stump, *The North American Free Trade Agreement and the Energy Trade* (Washington, D.C.: Greenpeace, 1992).

133. Geri Smith, "The Remaking of an Oil Giant, 1993," *Business Week,* August 16, 1993, pp. 84, 85.

134. Tom Raum, "Congress Did Business As Usual with Deficit Bill," *San Francisco Examiner,* August 7, 1993, p. A10.

135. Hazel O'Leary, Secretary of Energy, as quoted in Jeff Pelline, "In Visit to S.F., Energy Chief Talks of Supplies and Demands," *San Francisco Chronicle,* July 28, 1993, p. A6.

136. *Energy Efficiency,* California Energy Commission, October 1990, P400-90-003, esp. ch. 3, "Moving Forward: An Overview."

137. Charles Imbrecht, California Public Utility Commission, en banc hearings, San Francisco, February 25, 1993. On the sources of California electricity savings, see *Electricity: 1990 Report*, California Energy Commission, October 1990, P106-90-002, pp. 2-9 through 2-15.

138. See Amory Lovins' brilliant epitaph, "The Origins of the Nuclear Power Fiasco," in Byrne and Rich, eds., *The Politics of Energy Research and Development*, pp. 7-34.

139. Telephone interview with Monte Belote, Executive Director, Florida Consumer Action Network, December 24, 1992. For on-going coverage of the nuclear power issue, contact the Nuclear Information and Resource Center (NIRS), 1424 16th St. NW, Suite 601, Washington, DC 20036.

140. See Richard Rudolph and Scott Ridley, *Power Struggle: The Hundred Year War Over Electricity* (New York: Harper & Row, 1986), pp. 151, 152; also Charles Komanoff, "Dismal Science Meets Dismal Subject: The (Mal)practice of Nuclear Power Economics," *The New England Journal of Public Policy* (1985): 47–59. For a good general discussion on why Rancho Seco, owned by the Sacramento Municipal Utility District, was shut down in 1989, see interview with SMUD director Ed Smeloff, published in *Northwest Conservation Act Report*, August 19 and September 2, 1991, and extensive coverage in the *Sacramento Bee*. The *Los Angeles Times* (November 1, 1992) and *The Oregonian* (December 20, 1992 and January 10, 1993) have covered the grassroots opposition to and cost and safety problems of the Trojan Nuclear Plant, owned by Portland General Electric.

141. *Nuclear Electricity and Energy Independence*, a pamphlet from the U.S. Council on Energy Awareness, Washington, D.C., 1991.

142. Telephone interview with Steve Warn, March 1993, regarding the push for a new nuclear plant in Florida. See also Ed Lane, "Westinghouse, GE Win Reactor Design Contracts," *The Energy Daily*, January 12, 1993, p. 1.

143. O'Leary, "Responses to Questions from the Senate Energy and Natural Resources Committee," January 14, 1993, pp. 25, 30. See "Clinton Budget Sets Shift in Energy Priorities," *The Nuclear Monitor*, March 1, 1993, Nuclear Information and Resource Service, Michael Mariotte, Executive Director and Editor.

144. Matthew L. Wald, "A Utility's Strategy for Life in a Nonnuclear Age," *New York Times*, September 11, 1992, p. C1. See also *Statistical Abstract of the United States, 1992*, U.S. Department of Commerce, Bureau of the Census, p. 575.

145. Pamela Russell, "Republicans Would Require PMAs to Charge 'Market' Rates," *California Energy Markets*, No. 250, March 11, 1994, p. 12.

146. A. B. Lovins, L. H. Lovins, and F. Krause, *Least-Cost Energy—Solving the CO2 Problem* (Snowmass, Colorado: Rocky Mountain Institute, reprinted

from 1981 Brick House edition, with a new preface, April 1989); also A. B. Lovins and L. H. Lovins, "Make Fuel Efficiency Our Gulf Strategy," rewritten from *New York Times* op-ed piece from December 3, 1990, with copious footnotes, available from RMI.

CHAPTER 3 The Empire of Oil

1. See Daniel Yergin, *The Prize: The Epic Quest for Oil, Money, and Power* (New York: Simon & Schuster, 1991), especially ch. 2, "'Our Plan': John D. Rockefeller and the Combination of American Oil." See also the snapshot history of oil in Wilson Clark's prescient and nearly forgotten book *Energy for Survival* (New York: Anchor Books, 1974), pp. 29–33.

2. Robert Engler, *The Politics of Oil: A Study of Private Power and Democratic Institutions* (New York: MacMillan, 1951), pp. 151–152.

3. Yergin, *The Prize*, p. 254.

4. Harvey O'Connor, *The Empire of Oil: A Study of Private Power and Democratic Institutions* (New York: Monthly Review Press, 1955), ch. 6, "Conservation."

5. See O'Connor, *Empire of Oil*, pp. 62–67; also Yergin, *The Prize*, pp. 248–259, for the account of Governor Sterling's invasion of East Texas. The ins and outs of private/public oil price fixing has spawned a voluminous literature in economics. The result, of course, is the same: continued hegemony by the majors. See, for example, David Glasner, *Politics, Prices and Petroleum: The Political Economy of Energy* (Cambridge: Harvard University Press, 1985).

6. Yergin, *The Prize*, p. 250.

7. The phrase comes from Engler, *Politics of Oil*. For an early post-oil shock analysis, see also Engler's *The Brotherhood of Oil: Energy Policy and the Public Interest* (Chicago: University of Chicago Press, 1977).

8. See Yergin, *The Prize*, ch. 16, "Japan's Road to War." See also Jonathan Marshall, *To Have and Have Not: Southeast Asian Raw Materials and the Origins of the Pacific War* (Berkeley: University of California Press, 1994).

9. See Joseph Borkin, *The Crime and Punishment of I.G. Farben* (New York: Simon & Schuster, 1978), chs. 2 and 3.

10. Yergin, *The Prize*, pp. 344–346.

11. See John W. Frey and H. Chandler Ide, *A History of the Petroleum Administration for War, 1941–1945*, Petroleum Administration for War, U.S. Government Printing Office, 1946, p. 1.

12. See especially Charles A.S. Hall, Cutler J. Cleveland, and Robert Kaufmann, *Energy and Resource Quality: The Ecology of the Economic Process* (New York: John Wiley & Sons, 1986), esp. ch. 8, "Imported Oil," for the best estimates of the long-term availability of oil and other fossil fuels.

13. Yergin, *The Prize,* pp. 422–425; figures on percent oil use come from Tony Judt, "Europe: The Grand Illusion," *New York Review of Books,* July 11, 1996, pp. 6–9, an excerpt from *A Grand Illusion: An Essay on Europe,* to be published in late 1996.

14. Yergin, *The Prize,* p. 424.

15. Ibid., p. 544.

16. See William L. Shirer, *The Rise and Fall of the Third Reich* (New York: Simon & Schuster, 1959), p. 266; 1985 figure for cars in Germany was calculated from figures in *The Bicycle: Vehicle for a Small Planet,* Worldwatch Paper 90, September 1989, pp. 11, 12.

17. Thomas W. Lippman, "Remembering the First Great Oil Panic of 20 Years Ago," *San Francisco Chronicle,* November 27, 1993, p. D4.

18. On Carter's energy policy, which was strongly pro-nuclear, see Barry Commoner's *The Politics of Energy* (New York: Knopf, 1979), especially ch. 5, "Solar Versus Nuclear Energy: The Politics of Choice."

19. See Linda R. Cohen and Roger G. Noll, "Synthetic Fuels from Coal," in *The Technology Pork Barrel,* (Washington, D.C.: The Brookings Institution, 1991), pp. 259–320.

20. Ibid., p. 297.

21. David A. Stockman, *The Triumph of Politics: How the Reagan Revolution Failed* (New York: Harper & Row, 1986), p. 131.

22. See Yergin, *The Prize,* ch. 34.

23. This account of the manipulation of oil prices is from Edwin S. Rothschild, "The Roots of Bush's Oil Policy," *Texas Observer,* February 14, 1992, pp. 1–14; Thomas W. Lippman, "U.S. Tries to Influence Oil Prices, Papers Show," *The Washington Post,* July 21, 1992, p. 1; and Robert D. Hershey, Jr., "U.S., In Shift, Seems to View Fall In Oil Prices as a Risk, Not a Boon," *New York Times,* April 3, 1986, p. A1.

24. Michael Parenti, *Against Empire* (San Francisco: City Lights Books, 1995), pp. 45, 48, 49.

25. "Veterans Tell of Being Unwitting Medical Subjects," *San Francisco Chronicle,* May 7, 1994, p. A4.

26. "Stability in Oil Prices is Likely Result of War," *San Francisco Examiner,* March 1, 1991, p. B1.

27. Louis Peck, "Whatever Happened to Solar?" *Garbage* 3 (January/February 1991): 25–31; Walter Karp, "Who Decides What is News? (Hint: It's Not Journalists)," *Utne Reader* (November/December 1989): 60–68, exerpted from *Harper's Magazine,* July 1989.

28. Alexander Cockburn (*Anderson Valley Advertiser,* March 6, 1991) reported on a survey of 250 Denver residents—mostly strong war supporters—by academics Sut Jhally, Justin Lewis and Michael Morgan which

concluded that the more people watched TV, the less they knew about the true issues of the war. For example, 32 percent of the heavy viewers but only 15 percent of the light viewers thought Kuwait was a democracy.

29. Youssef M. Ibrahim, "Money Scandals Rock Kuwait's Ruling Family," *San Francisco Chronicle*, January 11, 1993.

30. See Seymour Hersh, "A Reporter At Large: The Spoils of the Gulf War," *The New Yorker*, September 6, 1993. In the same article General Norman Schwartzkopf says he rejected "what would have been hundreds of millions of dollars in commissions for me personally" to do business in Kuwait.

31. See Leslie Cockburn and Andrew Cockburn, "Royal Mess," *The New Yorker*, November 28, 1994, p. 60.

32. Ibid.

33. On June 25, 1996, a giant truck bomb exploded thirty-five yards from an apartment complex in Dhahran, Saudi Arabia, housing hundreds of American military personnel, killing 23 and injuring 345 more. A month before the explosion, Saudi authorities had beheaded four Saudi religious militants who had been convicted of detonating a car bomb that had killed five Americans and two Indians at a U.S.-run military training facility. Pentagon officials immediately assumed there was a connection between the executions and the Dhahran truck bombing. "Bomb Kills 23 at U.S. Compound," *Sacramento Bee,* June 26, 1996, p. A1.

34. Quote is cited in Cockburn and Cockburn, "Royal Mess."

35. "Senate Would Ban Arms Sales to Nations Insisting on Israel Boycott," *San Francisco Chronicle*, January 29, 1994, p. A11; "U.S., Saudis Renegotiate Arms Deal," *San Francisco Chronicle*, February 11, 1994, p. A8.

36. "McDonnell Douglas: Unfasten the Seat Belts," *Business Week*, February 14, 1994, p. 36.

37. S.S. Marsden, Petroleum Engineering Department, Stanford University, letter to the editor, *San Francisco Chronicle*, "Sunday Punch," October 3, 1993, p. 3.

38. Mark Fineman, "U.S. Oil Firms Have Big Stake In Somalia Effort: Industry Giants Hold Right to Promising Field," *San Francisco Chronicle* (excerpted from *Los Angeles Times* article), January 19, 1993, p. A9. Amoco, Chevron, and Philips also have "promising concessions" in the East African country, despite oil company denials that the U.S. intervention was occasioned by those concessions.

39. See *Energy Facts 1989*, Energy Information Administration, U.S. Department of Energy, 1990, available free from any regional office of the EIA, esp. pp. 60, 71.

40. The best critique of this fact is by Senator Paul Wellstone (D-Mn), pp.

432–459. See also Clifford Krauss, "Energy Bill Is Derailed In Senate," *New York Times,* November 2, 1991, p. 19. See Andrea Ricci, "Analysts Say Imported Oil Has Crucial Role in Trade Deficit," *San Francisco Chronicle,* January 13, 1992, p. B1. Three-fourths of the trade deficit with Japan stems from automobiles.

41. For the dollar figures on their historical dominance of public research and development monies see Public Citizen's *Renewable-Energy Research and Development: An Alternative Budget Proposal for FY 1993-1995,* Third Edition, March 24, 1992, Washington, D.C., esp. pp. 47, 48. For the energy barons' comprehensive wish list, see the *National Energy Security Act of 1991,* introduced by Senator Bennett Johnston, D-La., Chairman, Senate Committee on Energy and Natural Resources, Report 102-72.

42. "Senate OKs Landmark Energy Policy Legislation," *San Francisco Chronicle,* July 31, 1992, p. A3, (from *Los Angeles Times*). Daniel Becker of the Sierra Club said of an earlier version, "This is half a good energy bill. It has a gaping hole where more efficient cars should be. Since half of our oil goes into cars, that is an enormous hole." (Clifford Krauss, "Energy Measure Passed by House With New Offshore Drilling Ban," *New York Times,* May 28, 1992, pp. A1, C10.

43. Bill Clinton and Al Gore, *Putting People First: How We Can All Change America* (New York: Times Books, 1992), p. 98.

44. "Clinton Energy Policy Promotes Natural Gas Over Oil," *San Francisco Chronicle,* December 10, 1993, p. B2.

45. Commoner's books include *Science and Survival, The Closing Circle, The Poverty of Power, The Politics of Energy,* and *Making Peace With the Planet.*

46. Stephen D. Ban, President and CEO, Gas Research Institute, "Focus for the Nineties," 1990 Annual Meeting, Gas Research Institute, April 5, 1990. Discussions with staffers at SoCal Gas and the Gas Research Institute in October 1991 failed to uncover anyone who had done any thinking or planning about the possibility of converting the natural gas infrastructure to hydrogen.

47. American Petroleum Institute, *Basic Petroleum Data Book: Petroleum Industry Statistics* 11, no. 3 (September 1991): 93, note 46, as cited in Carol Alexander, *Natural Gas: Bridging Fuel or Roadblock to Clean Energy?* (Washington, D.C.: Greenpeace, 1993), p. 32 and note 115, p. 58; George Minter, lobbyist for Southern California Gas, chat with Dan Berman in Sacramento, October 15, 1991.

48. "It's become a cliché that natural gas is the U.S. fuel of the future, but now there are dollar estimates to support that prediction," *The Oil and Gas Journal,* August 30, 1993, p. 3. According to the article, the Department of Energy predicted that two-thirds of gross domestic wellhead revenue

would come from natural gas by 2002—a reverse of the usual ratio between 1970 and 1990.

49. *Energy Facts 1989*, Department of Energy, DOE/EIA-0469(89), p. 74.

50. Barry Commoner, *The Politics of Energy* (New York: Knopf, 1979), ch. 6, "The Solar Transition," esp. pp. 58–65.

51. Proposed by the South Coast Air Quality Management District. The plan slapped emission limits on gasoline lawn mowers and restricted the use of barbecue starter fluids, among other novelties.

52. See Dan McCosh and Stuart F. Brown, "The Alternative Fuel Follies," *Popular Science*, July 1992, beginning p. 54. Anne Chen Smith, Southern California Gas Company, special press briefing for the California environmental movement, Sacramento, October 14, 1991.

53. See *Oil and Gas Journal*, December 28, 1992, p. 45, and *State of the World 1992*, Worldwatch Institute, Washington, D.C., 1992, p. 39, both quoted in Alexander, *Natural Gas*, ch. 7, "Projected Supply and Demand of Natural Gas," footnotes 127 and 129.

54. Beth Miller, *Information Sheet*, United States Advanced Battery Consortium (USABC), FAX transmission from U.S. Department of Energy, Washington, D.C., October 25, 1991.

55. California Energy Commission, *Electricity 1990 Report*, October 1990, P106-90-002, p. 3-2.

56. Telephone interview, August 6, 1993.

57. Robert Johnson and Caleb Solomon, "Power Source: Coal Quietly Regains a Dominant Chunk of Generating Market," *Wall Street Journal*, August 20, 1992, p. 1.

58. See Carol Alexander and Ken Stump, *The North American Free Trade Agreement and the Energy Trade* (Washington, D.C.: Greenpeace, 1992), p. 36. For copies send $5 to Greenpeace, 1436 U St. NW, Washington, DC 20009.

59. Carol Alexander, *Natural Gas*; *Energy Facts 1989*. The estimate regarding natural gas reserves comes from Daniel A. Dreyfuss, Vice President, Strategic Planning and Analysis, Gas Research Institute, "GRI Baseline Projection of U.S. Energy Supply and Demand, 1992 Edition," published August 1991.

60. Mainstream think tanks and organs of opinion are just beginning to pay attention to potential oil shortages again. See, for example, James J. MacKenzie, "Oil as Finite Resource: When Is Global Production Likely to Peak?" (Washington, D.C.: World Resources Institute, March 1996); and Joseph J. Romm and Charles B. Curtis, "Mideast Oil Forever?" *Atlantic Monthly* (April 1996), cover story.

61. See Curtis Moore and Alan Miller, *Green Gold: Japan, Germany, the United States and the Race for Environmental Technology* (Boston: Beacon Press, 1994).

CHAPTER 4 Public Power

1. From Tom Johnson, *My Story* (New York: D.W. Huebsch, 1951), cited in *Power Struggle: The Hundred-Year War Over Electricity*, ed. Richard Rudolph and Scott Ridley (New York, Harper & Row, 1986), p. 22.
2. See Scott Ridley, "Seeing the Forest from the Trees: Emergence of the Competitive Franchise," *Electricity Journal* (May 1995): 39–49. See Ilana DeBare, "Power Play," *Sacramento Bee*, June 25, 1995, p. G1, for a summary of electricity deregulation in California.
3. See Nancy Rader, *Power Surge: The Status and Near-Term Potential of Renewable Energy Technologies*, Critical Mass Energy Project, Public Citizen, Washington, D.C., May 1989; see also *Towards a Fossil-Free Energy Future: A Technical Analysis for Greenpeace International*, Michael Lazarus, Principal Investigator, Tellus Institute, Boston, April 1993, esp. pp. 144–158, "Electricity Generation."
4. See Charles M. Coleman's company history, *P.G. and E. of California* (New York: McGraw-Hill, 1952), esp. ch. 6, "Dawn of the Electric Day"; and David Roe, *Dynamos and Virgins* (New York: Random House, 1984), pp. 3–4 for early accounts of the commerce in electricity.
5. The Wabash reaction is cited in Vic Reinemer, "Public Power's Roots," *Public Power*, centennial issue, September-October 1982, pp. 22–23, as quoted in Rudolph and Ridley, *Power Struggle*, pp. 22–27, endnote, p. 267.
6. Rudolph and Ridley, *Power Struggle*, p. 32.
7. Coleman, *P. G. and E.*, pp. 81, 82, 343.
8. Olney was quoted in 1893, six years after the formation of the ICC, according to Richard Hofstadter, *The Age of Reform* (New York: Knopf, 1955), pp. 178, 179.
9. See discussion by Rudolph and Ridley in *Power Struggle*, pp. 38–40.
10. Ibid., pp. 40, 41.
11. The discussion of Tom Johnson is taken from Ibid., pp. 22–25 and 37, 38. Direct quotes are from Johnson, *My Story*, as quoted in *Power Struggle*.
12. Rudolph and Ridley, *Power Struggle*, p. 38. Much of this material is from the *Cleveland Plain Dealer* and from Eugene C. Murdock's 1951 doctoral dissertation at Columbia University, *Life of Tom Johnson*.
13. See Rudolph and Ridley, *Power Struggle*, pp. 41–56. See also Gifford Pinchot, *The Power Monopoly: Its Makeup and Its Menace* (Milford, Pennsylvania, 1928); and *Fighting Liberal: The Autobiography of George Norris* (New York: Macmillan, 1945).
14. Rudolph and Ridley, *Power Struggle*, p. 70.
15. From a famous book on utility propaganda, *The Public Pays . . . and still*

pays: A Study of Power Propaganda, by Ernest Gruening, Senator from Alaska (New York: Vanguard Press, 1931, reprinted with a new introduction, 1964), pp. 164, 165.

16. Rudolph and Ridley, *Power Struggle*, chs. 2 and 3.
17. Ibid., p. 65.
18. Speech of September 15, 1932 in Portland, referred to in Ibid., pp. 67, 68.
19. Ibid., p. 73.
20. Introduced by Senator Burton K. Wheeler (D-Montana) and Congressman (and future Speaker) Sam Rayburn (D-Texas).
21. The story is told in "The First Co-op in the Valley: An Experiment Triumphs," *Rural Electrification*, July 1935, pp. 19, 20; cited in Rudolph and Ridley, *Power Struggle*, p. 82.
22. Section 6, the Raker Act, cited in J. B. Neillands, "How PG&E Robs San Francisco of Cheap Power," excerpt from a March 22, 1969 article in the *San Francisco Bay Guardian*, reprinted in the Sixteenth Anniversary Edition of the *San Francisco Bay Guardian*, October 6, 1982, pp. 7, 8. The following quotation by John Muir is from the Neillands article as well. A summary of the same material can be found in Rudolph and Ridley, *Power Struggle*, pp. 253–257.
23. Neillands, "How PG&E Robs San Francisco."
24. Craig McLaughlin, "Vote Yes on a PG&E Feasibility Study," *San Francisco Bay Guardian*, January 23, 1991, p. 17.
25. Interview with mayoral candidate Richard Hongisto, October 8, 1991. Hongisto believes home electric bills could be cut by at least 15 percent with a municipal electric company. During the campaign Hongisto leased a battery-powered Ford Escort for $1,800 a year, rebuilt by Solar Electric Engineering, Santa Rosa, California.
26. "Agnos' Last Dirty Trick," editorial in the *San Francisco Bay Guardian*, January 8, 1992, p. 6. The *Bay Guardian* story has been corroborated by Nancy Walker.
27. Tim Redmond, "How Many Light Bulbs . . . ?" *San Francisco Bay Guardian*, September 22, 1993, p. 9. "A City Held Hostage: How San Francisco Could Lose Its Most Valuable Asset: The Hetch Hetchy Dam," Redmond's cover story in the same issue is the best single account of the public power issue in San Francisco. More recently, see editorial "GOP Eyes Hetch Hetchy and it's S.F.'s Own Damn Fault," *San Francisco Bay Guardian*, December 21, 1994, p. 6.
28. David Behler, *The Presidio Electrical System: Analysis of Basic Options*, December 14, 1992.
29. Telephone interview with Joel Ventresca, chair, San Franciscans for Public

Power, late July 1996. In 1993 and 1994, author Dan Berman helped fight PG&E's attempt to secure the Presidio power contract, and in the latter year served as co-chair of the Preserve the Presidio Campaign.

30. San Francisco's Public Utilities Commission awarded the feasibility study contract to Economic & Technical Analysis Group (ETAG), a San Francisco-based gropu whose members conceded that it had done "about $140,000" of consulting work for PG&E in the last decade. See memo by Ron Knecht, President, ETAG, to Lawrence Klein, General Manager, Hetch-Hetchy Water & Power, June 4, 1996, in which he minimizes the importance of ETAG's PG&E work in recent years. Ventrusca's comments are from telephone interviews in late July 1996.

31. Quoted in Robert M. Fogelson's masterful book *The Fragmented Metropolis: Los Angeles, 1850-1930* (Cambridge: Harvard University Press, 1967), esp. ch. 11, "The Municipal Ownership Movement"; quote is from p. 233.

32. Rudolph and Ridley, *Power Struggle*, pp. 254, 255.

33. Ibid., p. 211. See also ch. 8, "Wall Street: the Stall of the Dividend Machines."

34. *Moody's Public Utility Manual, 1990*, pp. 4523, 4524.

35. Rudolph and Ridley, *Power Struggle*, p. 229.

36. Roe, *Dynamos and Virgins*, p. 68, quoting an anonymous friend who almost lost his job in the mid-seventies by trying to arrange a presentation by Environmental Defense Fund lawyers on the debt overhang of the public utilities industry because of overbuilding. David Roe's book, to which we will return, is a lucid and charming insider's account of the attempts to get Pacific Gas & Electric to accept least-cost energy accounting and energy conservation, rather than continually building new power plants.

37. Ibid., p. 127.

38. Paul Loeb, *Nuclear Culture: Living and Working in the World's Largest Atomic Complex* (Philadelphia: New Society Publishers, 1986), pp. 115, 116.

39. Ibid., pp. 114–116.

40. Ibid., p. 116.

41. Rudolph and Ridley, *Power Struggle*, pp. 1–8.

42. James T. Madore, four part series in the *Buffalo News*, Business section, January 12–15, 1992.

43. Richard Miller, Policy Analyst, Labor Institute, New York City, and Policy Analyst, Oil, Chemical and Atomic Workers International Union, "Economic Development in the Niagara Falls, New York Region (Why Cheap Power Isn't Saving Jobs)," about December 1991.

44. Ibid.

45. Ibid.

46. *The Bee Through 100 Years: The Bee's Centennial Album*, Part the Ninth, *The Sacramento Bee*, February 3, 1958, pp. CD14, CD15.

47. For PG&E's company history see Coleman, *P.G.&E. of California*, pp. 320–322.

48. Most of the historical material before 1972 is from Ruth Sutherland Ward, "*. . . For the People*": *The Story of the Sacramento Municipal Utility District*, available from SMUD, Sacramento, 1973.

49. Harold E. Stassen, "Atoms for Peace," *Ladies' Home Journal* (August 1955): 48, cited in Stephen L. Del Sesto, "Wasn"t the Future of Nuclear Energy Wonderful?" in *Imagining Tomorrow: History, Technology and the American Future*, ed. Joseph J. Corn (Cambridge, Massachusetts: MIT Press, 1986), p. 58.

50. *San Francisco Chronicle*, August 13, 1993.

51. The early history of Rancho Seco is from Ward, "*. . . For the People*," ch. 15, "Nuclear Power Emerges: Rancho Seco."

52. Matthew L. Wald, "A Utility's Strategy for Life in a Nonnuclear Age," *New York Times*, September 11, 1992, p. C1.

53. These early anti–Rancho Seco activists formed a group called Citizens for Safe Energy, which never acquired a mass following.

54. Edward Smeloff, "Building a Solar Utility," address to the Florida Solar Energy Industry Association, about November 1992.

55. Edward Smeloff, "SMUD Puts Eggs in the Conservation Basket: *Report* Interview with Edward Smeloff," *Northwest Conservation Act Report* 10, no. 16 (August 19, 1991).

56. Telephone interview with Edward Smeloff, August 4, 1993. Smeloff remembers that about 20 permanent workers refused to accept the severance package and were laid off.

57. Telephone interview with Edward Smeloff, August 11, 1993.

58. See, for example, S. David Freeman, "Is There An Energy Crisis: An Overview," *The Annals of the the American Academy of Political and Social Science*, vol. 410, November 1973.

59. Edward Smeloff, "Advanced and Renewable Conversion Technologies: One Utility's Approach to Commercialization," presentation at the Conference on Opportunities for Ecologically Clean Power and Energy Conservation, Minsk, Belarus, May 25–27, 1993. Figures from NRDC on efficiency investments are cited in Wald, "A Utility's Strategy for Life," p. C1.

60. *Sacramento Municipal Utility District, Annual Report, 1991*, p. 5. According to an August 29, 1996 memo from Ginger Salmon, Shade Program manager at SMUD, and Joanna Jullien, SMUD spends an annual total of $2 million on shade trees.

61. Based on a telephone call on September 3, 1993 to 1-800 TREE GEO.

62. Telephone interview with Sharon Dezurick, August 3, 1992. See Serena Herr, "Saving Energy with Shade Trees," *California Releaf Remarks* 2, no. 2 (Fall 1991).

63. Telephone interview with Heidi Lian, August 22, 1996.

64. Telephone interviews with Edward Smeloff, September 22, 1993, and Cliff Murley, June 27, 1996.

65. Based on telephone interviews with Terry Parks, Harvest Sun Energy, and Professor Jim Berquam, Berquam Energy, November 14, 1992; also *SMUD Solar Program Plan*, April 1992, available for $5 from Donald E. Osborn, Project Manager, Solar Program, SMUD, PO Box 15830, MS #75, Sacramento, CA 95852-1830, tel.: 916-732-6679.

66. Telephone interview with Peter Lowenthal, domestic solar water heating expert, Solar Energy Industries Association, Washington, D.C., August 23, 1994. Apparently SMUD had overestimated the number of electric resistence water heaters still to be replaced. See also Donald E. Osborn, "The Sustained, Orderly Development of Utility, Grid-Connected Photovoltaics," First International Conference on Solar Electric Buildings, March 1996, Proceedings, vol. 1, pp. 99–107.

67. *Annual Report, SMUD, 1991*, p. 51.

68. Interviews with Cliff Murley, solar water heating program head, SMUD, late June and early July 1996.

69. Sacramento Municipal Utility District, *1995 Annual Report*, pp. 10, 11.

70. Edward Smeloff to Dan Berman, December 20, 1993; write for *Today's Solar Power Towers*, Solar Power, Sandia National Laboratories, Albuquerque, NM 87185-5800, December 1991, c/o Dan Alpert, for a lucid description of the progress and problems of molten salt power towers. See the DOE's strategy in *Solar 2000: A Collaborative Strategy*, vol. II, Office of Solar Energy Conversion, U.S. DOE, Washington, D.C., February 1992, pp. 6, 7, which predicts that the costs of solar thermal systems can be brought down between 5-10 cents/kWh by the year 2000. See also Smeloff, "Advanced and Renewable Energy Conversion Technologies"; *SMUD Solar Program Plan*.

71. Edward Smeloff, telephone interview, December 21, 1992. In 1993 SMUD finally got lucky with the weather. After six years of drought, it snowed and snowed and snowed in the watershed of the upper American River, providing 26 percent more hydroelectric power than in normal years, and 2.5 times more than in 1992. As a result, a quarter rather than a tenth of SMUD generation came from inexpensive water power in 1993.

72. Public Information Department, *SMUD News*, "SMUD Bond Rating Upgraded to A- by Standard & Poor's," press release, March 30, 1993.

73. Mark Ryan, et al., *Standard & Poor's Utility*, "S&P Raises Sacramento Muni Util Revs to A-," typescript, March 31, 1993.

74. Donald Osborn, chief of solar programs, Sacramento Municipal Utility District, talk in Sacramento, September 10, 1992. See Osborn's "Using Solar Energy at the Sacramento Municipal Utility District," *Solar Today* (July/August 1992): 11–14, where he mentions the figures regarding capacity to be replaced.

75. See Daniel M. Berman, *Death On the Job* (New York: Monthly Review Press, 1978), esp. ch. 3, "How Cheap Is a Life?" The Ohio State Fund returns 93 percent of its premium income in income maintenance and medical benefits to injured workers, compared to about 65 percent for private insurance companies nation-wide. In Ohio, business associations have resisted private insurance companies' attempts to open the state to their business.

76. See Martha Groves, "Bellwether Nuke Vote: Oregon Effort Could Set a Precedent for Who Will Pay for Plant Closures," *Los Angeles Times*, November 1, 1992, p. D1; also Spencer Heinz, "Closure Fails to Keep Lid on new World of Trojan Ills," *The Oregonian*, January 10, 1993, p. 1. On PG&E's problems with interruptions of service caused by maintenance cutbacks, see *San Jose Mercury News*, May 19–21, 1996.

77. See *The Public Benefits of Public Power*, pamphlet from the American Public Power Association, Washington, D.C.

78. See Diane Moody, "Public Power Costs Less," *Public Power* (January–February 1992), pp. 24, 25.

79. Cited in Rader, *Power Surge*, p. 9.

80. See Andrew White, "It's Ba-a-a-ck!: Duck and Cover! The TVA is Doing Its Best To Raise Nuclear Power From the Dead," *Village Voice*, July 9, 1991; also issues of the *The Energy Monitor*, a quarterly published by the Tennessee Valley Energy Coalition, PO Box 1842, Knoxville, TN 37901.

81. All amounts regarding TVA are expressed in 1995 dollars.

82. *Tennessee Valley Authority: Financial Problems Raise Questions About Long-Term Viability*, GAO/AOMD/RCED-95-134 (Washington, D.C.: United States General Accounting Office, August 1995), p. 20.

83. Telephone interview with Michelle Neal, August 27, 1996.

84. Letter of July 1996 to authors.

85. Compare, for example, the term used to refer to utilities in the *Statistical Abstract of the United States* for 1975 and 1991. In 1975 they were termed "private;" in 1991 the new term was "investor-owned."

86. Richard L. Grossman and Frank T. Adams, *Taking Care of Business: Citizenship and the Charter of Incorporation*. Available from Charter, Inc., PO Box 806, Cambridge, MA 02140, January 1993.

87. Telephone interview with Terry Bundy, June 22, 1992.
88. *Lincoln Electric System 1991, Facts About the Lincoln Electric System*, LES brochure.
89. Carl Weinberg supervised many of these experiments while he was research director at PG&E. He is now a renewables consultant based in Walnut Creek, California. See a study a study carried out under Weinberg's direction, "Photovoltaics as a Demand-Side Management Option: Benefits of a Utility-Customer Partnership," presented at the World Energy Engineering Congress, Atlanta, Georgia, October 1992, by Howard Wenger, PG&E, San Francisco; Tom Hoff, Palo Alto, California; and Richard Perez, AWS Scientific, Albany, New York.
90. According to a handout from the American Public Power Association. For a discussion of recent campaigns, see Clinton A. Vince, "Lesson Learned by Cities Looking at the Public Power Option," paper presented at the APPA Legal Seminar, Hilton Head, South Carolina, October 16, 1990.
91. *Citizen Action*, publication of Ohio Citizen Action, vol. 1, 1993, p. 4.
92. John Nichols, "79% of Toledans in Survey Want City to Pressure Edison," *The Toledo Blade*, July 28, 1993, p. 1.
93. In *Taking Care of Business,* Grossman and Adams give a short history of corporate charters and argue that what the state legislature giveth the state legislature can take away.
94. Paula Ross, phone interview with Dan Berman, August 13, 1993; Nancy Hirsch, "Toledo Residents Advocate Public Power," *Power Line*, July/August 1992, pp. 10–11.
95. Hirsch, "Toledo Residents Advocate Public Power," pp. 10, 11.
96. Paula Ross, telephone interview, November 3, 1993; also James Drew, "A Progressive Coalition in the Heartland," *In These Times*, March 21, 1994, pp. 8, 9.

Chapter 5 Green Capitalism and Wall Street Environmentalism

1. "Twenty Years Ago," reminiscences of Stephen P. Duggan, cofounder, first chairman, and chairman emeritus, Board of Trustees, Natural Resources Defense Council; also a partner in Simpson Thacher & Bartlett, New York City; in *Twenty Years Defending the Environment, NRDC 1970–1990* (New York: NRDC, 1990), p. 10.
2. The EDF office is across San Francisco Bay in Berkeley.
3. David Roe, *Dynamos and Virgins* (New York: Random House, 1984), pp. 70, 71.
4. Ibid., p. 121.

5. Ibid., pp. 12, 13.
6. Leslie Lamarre, "Shaping DSM as a Resource," *EPRI Journal* (October/November 1991): 4–15.
7. Carl Blumstein, Jeffrey P. Harris, Arthur H. Rosenfeld, and John P. Millhone, *Energy Efficiency RD&D: New Roles for States,* Program on Workable Energy Regulation (POWER), University-wide Energy Research Group, University of California, Berkeley, March 1993, p. 16.
8. "Conservation Power," *Business Week* cover story, September 16, 1991; also telephone interview with National Audubon Society building manager Ken Hamilton, May 14, 1996.
9. The "five gurus" are not mentioned in Duggan's *Twenty Years Defending the Environment,* but are mentioned in Robert Gottlieb's book *Forcing the Spring: The Transformation of the American Environmental Movement* (Washington, D.C.: Island Press, 1993).
10. Roe, *Virgins and Dynamos,* p. 27.
11. Ibid., p. 28; also John E. Bryson and Jon F. Elliott, "California's Best Energy Supply Investment: Interest-free Loans for Conservation," *Public Utilities Fortnightly,* November 5, 1981.
12. Roe, *Dynamos and Virgins,* ch. 5, "Dangerous Radicals"; Wilson Clark, *Energy for Survival* (New York: Anchor Books, 1974), pp. 129, 130, citing Senator Lee Metcalfe of Montana, pointed out that the 193 largest utilities spent seven times as much for advertising as for research in 1969; in 1973 advertising expenditures had declined to "only" three times as much.
13. Interview with Calvin Broomhead, June 6, 1991.
14. Chris J. Calwell and Ralph C. Cavanagh, *The Decline of Conservation at California Utilities: Causes, Costs and Remedies* (San Francisco: Natural Resources Defense Council, 1989).
15. Public Utilities Commission of the State of California, Decision 90-08-068, August 29, 1990, p. 29.
16. Calwell and Cavanagh, *The Decline of Conservation,* p. 19.
17. Ibid.
18. *An Energy Efficiency Blueprint for California,* Report of the Statewide Collaborative Process, January 1990, p. 5.
19. Annual Report, The Energy Foundation, 1994.
20. See Ralph Cavanagh, Director, Energy Program, Natural Resources Defense Council, letter to the editor, *San Francisco Examiner,* February 16, 1993, p. A12. For the clearest expression of the theory and practice of the "collaborative process" see "'Transforming a Mega-Utility: How California's Power Giant Turned from Fighting Environmentalists to Cooperating on Energy Conservation," an interview with Ralph Cavanagh, by Sarah Van Gelder, *In Context* (Fall 1992): pp. 50-55.

21. The advertisement appeared in the *San Francisco Chronicle*, November 1, 1991, p. A17. For a complete list of the new coal-fired plants planned for the U.S. see Ralph Cavanagh, Ashok Gupta, Dan Lashof, and Marika Tatsutani, "Utilities and CO_2 Emissions: Who Bears the Risks of Future Regulation?" *The Electricity Journal* (March 1993): 64–75. All the authors are on the staff of the Natural Resources Defense Council. Inset 3 (p. 72) shows how to calculate estimated carbon dioxide emissions from new coal-fired power plants. The basic data for the article come from the *Inventory of Power Plants in the United States* (about $250), published annually by the Utility Data Institute in Washington, D.C. The article misleadingly assigns wholly-owned subsidiaries like Mission Energy Co. (100 percent owned by SCEcorp, the parent holding company of Southern California Edison, which shares the same top management) and U.S. Generating Company (50 percent PG&E, 50 percent Bechtel) to the category of "non-utility generators," which they are in name only. For more information, see *1992 Summary Annual Report, PG&E*, p. 4; and Donna Soper, "Latest Cedar Bay Player Steps Forward: U.S. Generating Co. to Buy Half-finished Cogenerator Plant," *Jacksonville Business Journal*, Jacksonville, Florida, August 21, 1992, pp. 1, 2. U.S. Generating is headquartered in Bethesda, Maryland.
22. "Natural Gas: Roadblock to Clean Energy," Greenpeace, Atmosphere & Energy Campaign, flyer, October 1992.
23. Mark Fineman, "U.S. Oil Firms Have Big Stake in Somalia Field," reprinted in the *San Francisco Chronicle*, January 19, 1993, p. A9.
24. *San Francisco Chronicle*, November 1, 1991, p. A17.
25. *The Energy Foundation, 1994*, p. 33.
26. Alexander Cockburn and Ken Silverstein, *Washington Babylon* (New York: Verso, 1996), pp. 210, 211. "Notable exceptions include the Turner Foundation and smaller opponents of the Pew Cartel such as Levinson and Patagonia," say the authors.
27. Gro Harlem Brundtland, Prime Minister of Norway and chair, The World Commission on Environment and Development, *Our Common Future* (Oxford, England: Oxford University Press, 1987).
28. See Charles Komanoff's exhaustive *Power Plant Cost Escalation: Nuclear and Coal Capital Costs, Regulation, and Economics* (New York: Komanoff Energy Associates, 1981).
29. Testimony of Charles Imbrecht, Chairman, California Energy Commission, Full Panel Hearing Agenda, California Public Utilities Commission, San Francisco, February 25, 1993.
30. *Electricity 1990 Report*, California Energy Commission, October 1990, P106-90-002, pp. 2-9 through 2-16.

31. From *Efficient Electricity Use: Estimates of Maximum Energy Savings*, prepared for EPRI by Barakat & Chamberlain, Inc., CU-6746, March 1990, and cited in *Energy Efficiency Report*, California Energy Commission, October 1990, P400-90-003, footnote on p. C6.

32. Personal communication between Rosenfeld and A. A Faruqui of Barakat & Chamberlain, Inc., June 4, 1990, cited in *Energy Efficiency Report*, October 1990, Appendix, p. C6.

33. Sarah Van Gelder, "Transforming A Mega-Utility." For a glimpse at the dense and questionable literature on the measurement of induced demand-side management, see *Impact Evaluation of Demand-Side Management Programs: Volume 1: A Guide to Current Practice* and *Volume 2: Case Studies and Applications (Revision 2)*, prepared by RCG/Hagler, Bailly, Inc., Boulder, Colorado, EPRI CU-7179, Palo Alto, California, 1991. List price $500, quote is from vol. 1, p. 1-1.

34. Ricardo Sandoval, "Debate Over Pay for Utility Execs," *San Francisco Examiner*, September 5, 1993, pp. E1, E4.

35. "The Role of NRDC in Negotiating Energy Conservation Investments in California: the Demand Side Management (DSM) Single Text," prepared for the William and Flora Hewlett Foundation, 1990, pp. 54, 55.

36. Wendy Tanaka, "PG&E Unit to Buy Gas Company," *San Francisco Examiner*, July 7, 1994, p. F1.

37. See John Bryson, Chairman and Chief Executive Officer, SCEcorp and Southern California Edison, speech at California Institute of Technology, January 24, 1994. See also *SCE News*, February 25, 1994, a company weekly, for ownership shares. Supposedly the Paiton plant will meet Indonesia's "stringent" air quality standards, which are "comparable" to those of the U.S. Environmental Protection Agency. See also Paul Blustein, "In Asia, an Eldorado of Infrastructure," *Washington Post*, June 4, 1995.

38. Charles Komanoff, letter to John Bryson, September 9, 1993. For a detailed technical critique of the Carbon II plant, see Komanoff's letter to the editor, *Electricity Journal*, January 1994. New evidence from the EPA suggests that ultra-small soot particles 10 microns in diameter or under (a human hair is 75 microns thick) may be responsible for over 50,000 extra deaths a year in the United States. See Philip J. Hilts, "Studies Say Soot Kills up to 60,000 in U.S. Each Year," *New York Times*, July 19, 1993, p. A16.

39. Co-authored by Jon F. Elliott and published in *Public Utilities Fortnightly*, November 5, 1981, pp. 19–23.

40. California Energy Commission, *Energy Efficiency Report*, p. C6. The California Energy Commission estimates are from the industry-sponsored Electric Power Research Institute (EPRI) estimates based on a study for

the entire U.S. and scaled to California: *Efficient Energy Use: Estimates of Maximum Energy Savings,* prepared by Barakat & Chamberlain, Inc., CU-6746, March 1990.

41. See California PUC, Decision 90-08-068, August 29, 1990, p. 25. The "net life-cycle benefits" constitute the total cost of transmission, distribution, and energy systems not built, minus PG&E's expenses for encouragement of conservation. Under the Collaborative Agreement, PG&E would keep 15 percent of the net life-cycle benefits as profits.

42. Based on figures from Martha Groves, "PG&E Launches $2-Billion Energy-Saving Program," *Los Angeles Times,* first page, Section D, January 31, 1991.

43. Dale Sartor, "*A Better Mouse Trap? or Can There Be a Better Utility Rebate?*" President's Message, San Francisco Bay Area Chapter, Association of Energy Engineers, May 1992.

44. Eugene Coyle, *Prepared Direct Testimony on Shareholder Incentives for DSM Program, Investigation No. 91-08-002,* TURN, May 7, 1993, p. 3. For a published summary of the argument see *The Quad Report,* "TURN Analyst Gene Coyle Looks At Utility Incentives To Build or Save," August 1993. The NRDC strongly disagreed with TURN; see Peter Miller, Natural Resources Defense Council, "Response to the Prepared Testimony of Eugene P. Coyle on Shareholder Incentives for DSM Programs," 91-08-002, June 17, 1993, before the California PUC.

45. Toward Utility Rate Normalization (TURN), "Prepared Direct Testimony of Eugene P. Coyle on Shareholder Incentives for DSM Programs," May 7, 1993, pp. 3, 7.

46. "Response of Peter Miller, Natural Resources Defense Council to the Prepared Testimony of Eugene P. Coyle on Shareholder Incentives for DSM Programs," R. 91-08-003, June 17, 1993, p. 1.

47. Jeff Pelline, "Utilities Turn to Power-Selling," *San Francisco Chronicle,* September 20, 1993, p. B1.

48. See *Facts on ActSquare,* Issue 5, September 1991; telephone interview with Merwin Brown, February 18, 1992.

49. Telephone interview with Merwin Brown, February 18, 1992.

50. Telephone interview with Lance Elberling, August 20, 1992, who reported that Merwin Brown is now employed at the Pacific Northwest Laboratory in Richland, Washington.

51. *A Time to Choose: America's Energy Future,* Final Report of the Energy Policy Project, Ford Foundation (Cambridge, Massachusetts: Ballinger, 1974), esp. ch. 5.

52. Adapted from Ronald Sutherland, Argonne National Laboratory, "1990 Households Benefitting from Residential Utility Conservation Programs,

by Income," from David Lapp, "The Demanding Side of Utility Conservation Programs," *Environmental Action*, Summer 1994, pp. 27–31.

53. The Building Industry Association of Superior California.

54. Ad in *San Francisco Chronicle*, February 2, 1992, p. F3.

55. Telephone interview with Grant Brohard, August 22, 1996.

56. *Facts on ActSquare*, Issue 8, May 1992. On the Davis plans see *Facts on Act-Square* for November 1992 and July/August 1993.

57. See "Building Energy," Proceedings of the First International Solar Electric Buildings Conference, Boston, Massachusetts, March 4–6, 1996; available from the Northeast Sustainable Energy Association (NESEA), Greenfield, Massachusetts. According to the NESEA staff, no one who identified himself or herself as PG&E staff attended the conference.

58. Jim Chace, Technical Coordinator, Pacific Energy Center, Pacific Gas and Electric, during tour conducted for author Dan Berman and a group of architects, March 5, 1992.

59. Talk by David Altsher, Pacific Gas & Electric, April 28, 1994, San Francisco. The program was scheduled to begin on June 1, 1994, with estimated annual interest rates of 9 percent and up to 60 months for payback.

60. Merwin Brown, Director of ActSquare in the Research and Development Department, PG&E, "PG&E's Supply and Demand Bridging Strategy for the 1990s," October 29, 1991.

61. Christopher Flavin and Nicholas Lenssen, *Beyond the Petroleum Age: Designing a Solar Economy*, Worldwatch Paper 100 (December 1990), p. 19, based on U.S. DOE analyses, estimate that PV generating capacity will fall below existing gas turbine generation costs by 2030.

62. For California see *Electricity 1990 Report*, California Energy Commission, October 1990, p. 2-1. Peak demand was expected to grow even faster. Natural gas is predicted to supply an increasing percentage of power supply (p. 4-90).

63. See "Wind Energy: A Resource for the 1990s," American Wind Energy Association, Washington, D.C., February 15, 1991, p. 4, on the decline of new-installed wind capacity in California; on the decline of new solar hot water heaters in the U.S., see Energy Information Administration, *Energy Facts 1989*, DOE/EIA-0469(89), p. 95.

64. Robert Johnson and Cales Solomon, "Coal Quietly Regains a Dominant Chunk of Generating Market," *Wall Street Journal*, August 20, 1992.

65. See Richard L. Grossman and Frank T. Adams, *Taking Care of Business: Citizenship and the Charter of Incorporation*, 1992, available from Charter, Inc., PO Box 806, Cambridge, MA 02140.

66. Richard W. Stevenson, "To Help Itself, Utility Aids Riot Area," *New York Times* (West Coast Edition), May 30, 1992, p. 17.

67. According to a telephone interview with Vallee Bunting, Communications Manager, RLA, July 11, 1994.

68. *Bringing Solar Electricity to Earth,* Electric Power Research Institute, 1991, produced by Ideas In Motion, San Francisco.

69. See for example, Carl Weinberg's poignant talk "Enoughness and Sustainability," at the Soltech '92 Conference in Albuquerque, New Mexico, February 19, 1992, reprinted in *Solar Today,* May/June 1992, p. 19.

70. See, for example, "Photovoltaics as a Demand-Side Management Option: Benefits of a Utility-Customer Partnership," by Howard Wenger (PG&E), Tom Hoff (Palo Alto, California), and Richard Perez (AWS Scientific, Albany, New York), presented at the World Energy Engineering Congress, Atlanta, Georgia, October 1992.

71. Remarks by Larry Simi, as recorded by Tim Redmond, "The Boys in the Bag," *San Francisco Bay Guardian,* January 22, 1992, p. 12.

72. Merwin Brown, telelphone interview with Dan Berman, February 18, 1992.

73. Eugene Coyle, Ph. D., *Report on the California Photovoltaic Industry for the California Commission on Industrial Innovation,* Eugene P. Coyle and Associates, June 1982, p. 71.

74. Keith Coyne, memo of July 2, 1993.

75. "PG&E Emphasizes Commitment to Keep Electric Rates in Check," News Department, Immediate Release, July 14, 1993.

76. Brigid Schulte, "More Power to Small Utilities," *San Francisco Examiner,* February 24, 1992, p. D1.

77. Philip Shabecoff, "Apple Sales Strong Despite Scare in '89 About Chemical Use," *New York Times,* November 13, 1990, p. A1. See William Greider, *Who Will Tell the People: The Betrayal of American Democracy* (New York: Simon & Schuster, 1992), pp. 320–323, for a few words on how NRDC won the Alar issue.

78. Sylvia Siegel, telephone interview, June 6, 1991.

79. Martin Espinoza, "SF Fails to Counter PG&E's Ratehike Bid," *The San Francisco Bay Guardian,* August 12, 1992, p. 9.

80. Martha Groves, "Bellwether Nuke Vote: Oregon Effort Could Set a Precedent for Who Will Pay for Plant Closures," *Los Angeles Times,* November 1, 1992, pp. D1. D8, D9.

81. Spencer Heinz, "Two Leaks Stir Doubts on Safety of Trojan," *The Sunday Oregonian,* December 20, 1992, pp. A1, A20.

82. Heinz, "Closure Fails to Keep Lid on New World of Trojan Ills," pp. A1, A14.

83. *Toward a Sustainable Energy Future,* The Energy Foundation, 1991 Annual Report, San Francisco, 1992. For 1992 and 1993 The John D. and Catherine

T. MacArthur Foundation, The Pew Charitable Trusts, and The Rockfeller Foundation combined had already pledged $15,875,000 in grants to the Energy Foundation.

84. Mark Worth, "Utility Officials and Environmentalists Agree to Close Nuclear Plant," *In Context* (Fall, 1992): 50–55; see also letter by Ralph Cavanagh to the *San Francisco Examiner*, February 16, 1993, p. A12.

85. Sarah Van Gelder, "Transforming a Mega-Utility," p. 51.

86. See Grossman and Adams, *Taking Care of Business.*

87. Telephone interview with Carl Weinberg, August 6, 1993.

88. See annual reports of The Energy Foundation for 1991 through 1995.

89. Greider, *Who Will Tell the People,* p. 44.

90. See *Newsweek,* August 5, 1991, for example.

91. SEER Official Program, August 1992, p. 3.

92. *Solar Hot Water Systems for the Homeowners,* Pacific Gas & Electric publication 62-8844, revised October 1985.

93. *Blueprint for a Solar America,* Solar Lobby, January 1979, p. 12.

94. Full name: Solar Energy and Conservation Mortage Corporation, a name a lot longer than its reputation. For an account of its brief life and untimely death, see Virginia Coe, "California's Experience in Promoting Renewable Enrgy Development," in *State Energy Policy, Current Issues, Future Directions,* ed. Stephen W. Sawyer and John R. Armstrong (Boulder, Colorado: Westview Special Studies, 1985), pp. 201–203.

95. Telephone interview with Paul Erickson, Director of Energy Efficiency Services, Iowa Association of Municipal Utilities, June 22, 1992.

96. The high point of the celebrations was the all day en banc Demand Side Management hearing of the California Public Utility Commission in San Francisco, February 25, 1993.

CHAPTER 6 Labor, Solar, and the Energy Economy

1. A Washington-based group founded by by Hazel Henderson of the Princeton Center for Alternative Futures.

2. Leonard Rifas, editor/publisher, *Energy Comics No.1* (San Francisco: Educomics, January 1980). The comic book form belied the serious intent of the authors and editor. Crammed into 34 pages are 10 strips about "hard" vs. "soft" energy strategies; "ecocorrecto" lifestyles and their limits; attacks on cars, nuclear power in the Phillipines, uranium mining in the Black Hills, and a proposed orbiting solar power plant; safety tips for wood stoves; jobs and energy; and a capsule history of how PG&E was able to snatch the sales of cheap hydroelectric power from the citizens of San Francisco who had built the Hetch Hetchy dam which supplied it.

3. Ibid., "Introduction."

4. Richard Grossman and Gail Daneker, *Energy, Jobs, and the Economy* (Boston: Alyson Publications, Inc., 1979); also October 28, 1992 letter to Dan Berman from Richard Grossman. *Energy, Jobs, and the Economy* has been recently republished. For more information contact Grossman at Charter, Inc., PO Box 806, Cambridge, MA 02140.

5. Harvey Wasserman, "Two Movements, One Goal," Introduction to *Energy, Jobs, and the Economy*.

6. *Statistical Abstract of the United States* (Washington, D.C.: U.S. Department of Commerce, 1976), Table 906, p. 548.

7. "Solar," as we use the term here, includes windpower as well as solar water and space heating, photovoltaics, solar power towers, and all energy forms which have come to be called "renewable" energy in the technical literature.

8. *Employment Impact of the Solar Transition: A Study prepared for the Subcommittee on Energy of the Joint Economic Committee, Congress of the United States*, April 6, 1979, pp. 1, 2.

9. For a cogent update see Leonard S. Rodberg, "Employment Impact of Alternative Energy Demand/Supply Options," Department of Urban Studies, Queens College/CUNY, Flushing, New York, prepared for the Coalition of Environmental Groups for a Sustainable Energy Future, Toronto, December 1992. For a recent sampling see Howard Geller, John DeCicco, and Skip Laitner, *Energy Efficiency and Jobs Creation: The Employment and Income Benefits from Investing in Energy Conserving Technologies* (Washington, D.C.: American Council for an Energy Efficient Economy; and Eugene, Oregon: Economic Research Associates, October 1992); Frank Muller, Skip Laitner, Alan Miller, and Lyuba Zarsky, *Job Benefits of Expanding Investment in Solar Energy* (College Park, Maryland: Center for Global Change, University of Maryland; and Eugene, Oregon: Economic Research Associates, 4th quarter 1992). In the 1990s, Ian Goodman, Matthew Clark, Michael Anthony, and Peter Kelly-Detwiler of The Goodman Group in Boston have prepared a number of studies that demonstrate that energy efficiency measures that reduce energy demand create more jobs than highly centralized energy production facilities such as hydroelectric, coal-fired, and nuclear power plants.

10. *Our Jobs, Our Health, Our Lives, Our Fight: Report of the First National Labor Conference for Safe Energy & Full Employment*, October 10-12, 1980, Pittsburgh, Pennsylvania, Labor Committee for Safe Energy and Full Employment, Washington, D.C., March 1981.

11. See Christopher Flavin and Nicholas Lenssen, *Beyond the Petroleum Age:*

Designing a Solar Economy, Worldwatch Paper 100 (December 1990), p. 43, for the estimated employment in nuclear and solar thermal plants.

12. "Section 13A: Elimination of Restructions on Production," *Laborers' Master Agreement between Associated General Contractors of California, Inc. and Northern California District Council of Laborers, affiliation with the Laborers' International Union of North America AFL-CIO, 1989–1993*, p. 24.

13. *Laborers' Master Labor Agreement, 11 Counties of Southern California: Wages Scales Working Rules, and By-Laws of the Southern California District Council of Laborers affiliated with the Laborers' International Union of North America, as Revised July 1, 1984*, pp. 13, 14.

14. Jonathan Marshall, "Public Works Costs in State Under Review: Assembly Panel to Debate 'Prevailing Wage' Laws," *San Francisco Chronicle*, March 16, 1993, p. A13.

15. "GM Plans Huge Layoffs," *San Francisco Examiner*, December 18, 1991, banner headline, p. 1.

16. "Detroit South: Mexico's Auto Boom: Who Wins, Who Loses," *Business Week* cover story, March 16, 1993, pp. 98–103.

17. According to Ron Blum, economist with the Research Department, United Automobile Workers, Detroit, telephone interview, May 6, 1994. Output per worker-hour increased by at least 5 percent and average of production and nonsupervisory personnel increased 5 percent over that period. Figures are for "light vehicle output." It is common knowledge that the automobile companies prefer increased overtime to hiring new workers, so they don't have to incur high costs of non-wage benefits, especially health insurance.

18. "11,000 Go on Strike At General Motors," *San Francisco Chronicle*, September 28, 1994, p. A3.

19. "Rumble in Buick City," *Business Week*, October 10, 1994, pp. 42, 43.

20. Johnson and Solomon, "Power Source," pp. A1–A4; *Inventory of Power Plants in the United States, 1991*, Energy Information Administration, U. S. Department of Energy, 1992, esp. Table 4, "Fossil-Fueled Steam-Electric and Nuclear Steam-Electric Operable Capacity and Planned Capacity Additions, as of December 31, 1991," p. 13.

21. See Charles A. S. Hall, Cutler J. Cleveland, and Robert Kaufmann, *Energy and Resource Quality: The Ecology of the Economic Process* (New York: Wiley, 1986), ch. 11, "Coal." Harry Caudill's classic *Night Comes to the Cumberlands* describes the devastation the absentee coal barons can wreak on a region.

22. Martha Brannigan, "U.S. Electric Companies See Promised Land Elsewhere," *Wall Street Journal*, March 2, 1993.

23. Ralph Cavanagh, Ashok Gupta, Dan Lashof, and Marika Tatsutani, "Utilities and CO_2 Emissions: Who Bears the Risks of Future Regulation?" *The Electricity Journal*, March 1993, esp. pp. 70, 71. See also Brannigan, "U.S. Electric Companies See Promised Land Elsewhere."

24. See, for example, Greenpeace International, *The Climate Time Bomb: Signs of Climate Change from the Greenpeace Database* (Amsterdam: Greenpeace International, 1994).

25. On the world-wide impact of "structural adjustment" policies, see *Dark Victory: The United States, Structural Adjustment, and Global Policy*, available from the Institute for Food and Development Policy, Oakland, California, 1994; also Walden Bello and Shea Cunningham, "The World Bank and the IMF: The Reaganite and the Subordination of the Third World," *Z Magazine*, July/August 1994, pp. 66–69. *Z Magazine* is published by South End Press in Boston, Massachusetts.

26. Quoted in "Caliifornia's Growth," *Engineering & Science*, Fall 1992, California Institute of Technology, pp. 14–16, a panel moderated by Bryson.

27. Edward Abbey, *The Journey Home* (New York: E.P. Dutton, 1977), p. 163. The "at least 11 percent" figure comes from SCEcorp's Form 10K for the fiscal year ending 1994, filed with the Securities and Exchange Commission.

28. See Marla Cone, "Consensus Could Clear Air Over Grand Canyon," *Washington Post*, June 20, 1996, p. A20.

29. "Judge: One for Citizens' Rights, Zero for Peabody and OSM," and editorial, "Beyond Shame: Peabody in Navaholand," from *The Reporter*, Citizens Coal Council, Washington, D.C., Spring 1996. "OSM" stands for the Office of Surface Mining in the Department of the Interior; its mandate is to enforce the cleanup of abandoned mines and ensure that new mines are operated so they don't damage the environment.

30. Ward Churchill and Winona LaDuke, "The Political Economy of Radioactive Colonialism," in *The State of Native America: Genocide, Colonization, and Resistance*, ed. M. Annette Jaimes (Boston: South End Press), p. 241, cited in Winona LaDuke, "A Society Based on Conquest Cannot be Sustained," in *Toxic Struggles* (Philadelphia: New Society Publishers, 1993), p. 101.

31. *1992 Annual Report*, SCEcorp, pp. 17, 21; Cavanagh, et. al. in "Utilities and CO_2 Emissions," note that Mission Energy is planning to build 442 megawatts of new coal-fired capacity in the United States.

32. Tom Lent, Greenpeace Climate Change Campaign, "Southern California Edison & CO_2: Greenwashing the Corporate Record," Internal Draft, May 24, 1993. At the time new capacity additions planned or under construction included 1,400 megawatts in Mexico, 1,220 megawatts in Indonesia, 1,000 megawatts in Australia, and 442 megawatts in the U.S. Since then

Mission Energy has withdrawn from the CARBON II project in Mexico, perhaps forever. Mission Energy's chief Indonesian partner in the $2.6 billion, 1,230 megawatt Paiton coal-fired power plant in eastern Java is Hashim Djojohadikusumo, the brother of a son-in-law of Indonesian President Suharto. Mission Energy secured a $200-million loan from the Overseas Private Insurance Corporation to help pay for its share of the Paiton project, and it pays $2 million for $200 million in OPIC insurance to hedge against the risk of civil war, nationalization, or currency transfer problems. OPIC is an agency financed by U.S. taxpayers. For more on the Paiton project, see Paul Blustein, "In Asia, an Eldorado of Infrastructure," *Washington Post*, June 4, 1995. Pratap Chatterjee, a reporter for Inter Press Service based in Berkeley, California, tracked down the OPIC loan and insurance information.

33. Energan is an abbreviation for Energía del Norte, S.A.C.V.

34. See John MacCormack, "Plant is Going On Line," *San Antonio Express-News*, June 6, 1993, pp. 1B, 7B.

35. Ted Bardacke, "The Hunger for Power: Government Electricity Privatization In the Works," *El Financiero International* (Mexico City), May 31, 1993.

36. Andy Pasztor, "Power Plants In Mexico Cast Pall Over NAFTA," *Wall Street Journal*, September 8, 1993; see letter to SCEcorp Chairman and CEO John Bryson from Nicholas Lenssen, Senior Researcher, and Christopher Flavin, Vice President for Research, Worldwatch, July 8, 1993.

37. Miguel Flores, U. S. National Park Service, Denver, memo to Jim Yarbrough, EPA Region 6, Dallas, in regard to Rio Escondido (Mexico) Electric Generating Station, April 30, 1993. See, especially, Keith A. Yarborough, William C. Malm, and Jarvis Moyers, "Visibility In Southwest National Park Service Areas"; and Malm, Kristi A. Gebhart, and Ronald C. Henry, "An Investigation of the Dominant Source Regions of Fine Sulfur in the Western United States and Their Areas of Influence," *Atmospheric Environment* 24A, no. 12 (1990): 3047–60.

38. "Talking Points for Alan Hecht RE: Carboelectrica Power Stations, Coahuila, Mexico," EPA internal memo, early June 1993.

39. Pasztor, "Power Plants In Mexico Cast Pall Over NAFTA."

40. "Fact Sheet" regarding Carbon II, Mission Energy Company, April 7, 1993; Bardacke, "The Hunger for Power." See also front-page story by Tod Robberson and Ted Bardacke, "Cloud Over Trade Pact—Texas Too," *Washington Post*, June 22, 1993.

41. See Michael Stremfel, "SCEcorp Finds its 'Green' Reputation in Jeopardy: Mexico Plant Purchase Raises Environmentalists' Ire," *Los Angeles Business Journal*, week of July 12-18, 1993, pp. 3–5. See letters from Congresswoman

Cardiss Collins, D.-Illinois, to Carol M. Browner, Administration of EPA, June 25, 1993; and from Congressman John D. Dingell, D.-Michigan, to Secretary of Commerce Ronald H. Brown, Secretary of Energy Hazel R. O'Leary, and EPA Administration Carol M. Browner, July 13, 1993.

42. "NAFTA Side Accords: A Border Polluters' Pact," editorial signed by Louis Dubose, *Texas Observer*, October 1, 1993, p. 2.

43. The most comprehensive account of the controversy is by Chris Green and Mary Kelley, *The Carbon II Dilemma: A Case Study of the Failings of U.S./Mexico Environmental Management in the Border Region*, preliminary draft, August 1993, Texas Center for Policy Studies. Mary Kelley is the executive director of TCPS. See Pasztor, "Power Plants In Mexico Cast Pall Over NAFTA."

44. Reuters, "SCEcorp Downgraded by Natwest," *Executive News Service*, October 6, 1993. Other brokerage giants which published reports critical Mission Energy the week of October 4 included Merrill, Lynch; Prudential, and Kemper.

45. Michael Parrish, "Edison Scuttles Plan to Invest in Massive Coal-Fired Power Plant," *Los Angeles Times*, October 12, 1993, p. D1; see also Andy Pasztor, "SCEcorp Drops Mexican Electric Plant Under Political, Environmental Pressure," *Wall Street Journal*, October 12, 1993, p. A6; Sanford Cohen and Morgan Stanley, "SCEcorp (SCE): Mission Impossible Dos?" October 12, 1993, for example.

46. On November 3, 1993.

47. John Bryson, letter to the editor, *Wall Street Journal*, September 14, 1993; finally published October 13, 1993, after SCEcorp had pulled out of Carbon II.

48. Paul Klebnikov, "Mission's Mission," *Forbes*, December 6, 1993, pp. 90, 91.

49. Kathleen Lally, "SECcorp: Mexican Project Terminated," *Salomon Brothers*, October 12, 1993.

50. Klebnikov, "Mission's Mission."

51. Ricardo Sandoval, "PG&E Near Work-force Reduction Goal," *San Francisco Examiner*, October 12, 1993, p. B1.

52. John Bryson, speech on the future of California at the California Institute of Technology, January 24, 1994.

53. "Women Tunnel On Despite Ban," *Sunday Morning Post*, August 20, 1994, p. 8.

54. On July 7, 1994 SCEcorp's stock price closed at 12 3/4, less than half the high of 25 3/4 in the preceding twelve months.

55. Tim Madigan, "Pollution Clouds Big Bend's Future," *Fort Worth Star-Telegram*, September 10, 1995, p. 1; also Tod Robberson, "Haze at the Border," *Washington Post*, August 11, 1995, p. A25.

56. Joe Nick Patoski, "Big Bend, R.I.P.?" *Texas Monthly,* March 1996.

57. Ralph K.M. Haurwitz, "Mexico Smokestacks Mar View at Big Bend," *Austin American-Statesman,* December 18, 1995, p. A5; also Sam Howe Verhovek, "A Diplomatic Haze Pervades Park's Air Pollution Dispute," *New York Times,* June 3, 1996, p. A-1.

58. Hauritz, "Mexico Smokestacks Mar View at Big Bend."

59. Telephone interview with Jim Yarbrough, EPA Region 6, Houston, late March 1996.

60. For some recent figures on jobs, energy, and employment, see Michael Renner, *Jobs in a Sustainable Economy,* Worldwatch Paper 104, Washington, D.C., September 1991, pp. 13, 14; and *Statistical Abstract of the United States, 1991,* Tables 669, 695–697 for a comparison of wage rates in different economic spheres.

61. Harry M. Caudill, *Night Comes to the Cumberlands* (Boston: Little, Brown & Co., 1962), p. 372.

62. See Sidney Lens, *The Labor Wars: From the Molly Maguires to the Sitdowns* (New York: Doubleday-Anchor edition, 1974), pp. 162-168. Blanket recognition of the union was not part of the deal.

63. *Historical Statistics of the United States, Colonial Times to 1970* (Washington, D.C.: U.S. Department of Commerce, 1975), p. 592 (anthracite production and employment), p. 589 (bituminous employment), p. 589 (bituminous production). In 1923, 98 percent of the bituminous coal came from underground mines, compared to only one-half today.

64. Ibid., pp. 587, 588 for proportion of BTUs contributed by coal; p. 608 for number of coal miners.

65. Caudill, *Night Comes to the Cumberlands,* ch. 14, "The Union Drives."

66. See Lens, *The Labor Wars,* ch. 14, "Clearing the Cobwebs."

67. Telephone interview with Mike Buckner, Research Director, United Mine Workers of America, March 9, 1993.

68. The best accessible accounts of the events surrounding the Pittston strike are in *Labor Notes,* a hard-hitting monthly dedicated to "Putting the Movement Back in the Labor Movement," June 1989–February 1990, available from 7435 Michigan Avenue, Detroit, MI 48210.

69. Paul Wellstone and Barry Casper, *Powerline: The First Battle of America's Energy War* (Amherst: University of Massachusetts Press, 1981). Paul Wellstone is now U.S. Senator from Minnesota. The issue of the relationship between exposure to electromagnetic fields (EMF) and leukemia and other forms of cancer is too complicated to explore here. For an introduction to the subject, see the books of *New Yorker* writer Paul Brodeur: *Currents of Death: Power Lines, Computer Terminals, and the Attempt to Cover Up their Threat to Your Health* (New York: Simon & Schuster, 1989); and

The Great Power-Line Cover-Up: How the Utilities and the Government are Trying to Hide the Cancer Hazards Posed by Electromagnetic Fields (Boston: Little, Brown and Co., 1993). The bi-monthly *Microwave News,* published by Louis Slesin, covers the new research on EMFs and microwaves from the point of view of the activists. Both Slesin and Brodeur believe the evidence for a causal link between EMFs and cancer is overwhelming. Slesin can be reached at *Microwave News,* PO Box 1799, Grand Central Station, New York, NY 10163. Dr. Patricia Buffler, Dean of the School of Public Health at the University of California, Berkeley, and a long-time consultant at the Electric Power Research Institute, believes there is no connection between EMFs and cancer.

70. Telephone interview with Leslie Coleman, National Mining Association, August 21, 1996. Source for Campbell County is *Coal Industry Annual, 1994* (Energy Information Administration, October 1995), p. 11, table 4. The figure for Germany, says Coleman, comes from a UN publication.

71. Telephone interview with Leslie Springer, Marketing Department, Arco Coal, Denver, Colorado, August 21, 1996.

72. Telephone interview with Terry O'Connor, Arco Coal, Denver Colorado, April 26, 1993.

73. Sheryl DuWinn, "Chinese Suffer From Rising Pollution As Byproduct of the Industrial Boom," *New York Times,* February 28, 1993, p. 11. See Hazel R. O'Leary, Secretary-Designate of Energy, "Responses to Questions from the Senate Energy and Natural Resources Committee," January 14, 1993, p. 1.

74. Peter Britton, "Clean Up Coal: the 21st Century Imperative," *Popular Science,* April 1993, pp. 78-83.

75. From a video produced by the United Mine Workers, cited by L.A. Kauffman, "Union Blues," *San Francisco Weekly,* October 13, 1993, p. 9, reporting on the AFL-CIO convention in San Francisco in early October.

76. Arthur Gottschalk, "Workers, Management Benefit In Coal Strike Deal, Analysts Say," *Journal of Commerce,* December 16, 1993.

77. News coverage of the 1993 strike was typical in this regard; see "Coal Strike Claims First Victim," *San Francisco Chronicle,* July 24, 1993, p. A7.

78. "60,000 Ruhr Workers Protest Mine Closings," *This Week In Germany,* September 24, 1993, p. 4. The strikers were members of the Mining and Energy Union.

79. Hermann Scheer, *Sonnen Strategie: Politik ohne Alternative* (Munich: Piper, 1993), p. 234.

80. See Dennis Normile, "Science Gets the CO_2 Out," *Popular Science,* February 1994, pp. 65-70.

81. Robert Johnson and Caleb Solomon, "Power Source: Coal Quietly Regains

A Dominant Chunk Of Generating Market," *Wall Street Journal*, August 20, 1992, p. 1.

82. See Union of Concerned Scientists, "World Scientists' Warning to Humanity," *Nucleus* (Cambridge, Massachusetts) 14, no. 4 (Winter 1992–1993): 2.

83. See Gro Harlem Brundtland, Prime Minister of Norway, *Our Common Future*, The World Commission on Environment and Development (Oxford: Oxford University Press, 1987).

84. The Union of Concerned Scientists, "Executive Summary," in *America's Energy Choices: Investing in a Strong Economy and a Clean Environment* (Cambridge, Masachusetts: The Union of Concerned Scientists, 1991), with the equal endorsement of the Alliance to Save Energy, the American Council for an Energy-Efficient Economy, the Natural Resources Defense Council and the Union of Concerned Scientists in consultation with the Tellus Institute, Boston.

85. Britton, "Clean Up Coal."

86. Tom Lent, "Energy for Employment: How to Heat Up the Economy, Not the Planet," Greenpeace, Atmosphere & Energy Campaign, San Francisco, July 1992.

87. *OCAW Reporter*, January–February 1993, p. 19. For a discussion of OCAW's Superfund for Workers proposal, see *Jobs and the Environment*, January 1994, available from The Public Health Institute, 853 Broadway, Rm. 2014, New York, NY 10003.

88. Anthony Mazzocchi, telephone interview with Dan Berman, August 10, 1993.

89. Greenpeace, *Chlorine Free* 2, no. 1 (Fall 1993).

90. The best recent summary of the status of relations between unions, workers, and the environmental movements is in *Environmental Action* (Spring 1992), with articles by Halwey Truax, Laura McClure, Eric Mann, Ruth Caplan, and Richard Grossman, available from Environmental Action, 6930 Carroll Avenue, 6th floor, Takoma Park, MD 20912; see also controversy between Lin Kaatz Chary and Mike Merril, *New Solutions: A Journal of Environmental and Occupational Health Policy* 3, no. 1 (1991), available from OCAW, PO Box 2812, Denver, CO 80201; also Ann Rabe, Director, Citizens' Environmental Coalition, New York State, "Building a Movement for Labor and Environmental Justice," *Everyone's Backyard*, published by the Citizen's Clearinghouse for Hazardous Wastes in Falls Church, Virginia, October 1992, pp. 3, 10.

91. Interview with Dick Leonard, Special Projects Director, Oil, Chemical and Atomic Workers, Denver, Colorado, November 11, 1989. The best single article on the union's successful battle to fight the lockout is Zack Nauth's "BASF: Bhopal on the Bayou?" *In These Times*, January 24–30, 1990.

92. See Zack Nauth, *The Great Louisiana Tax Giveaway, 1980–1989*, published by the Louisiana Coalition for Tax Justice, a project of the Louisiana Coalition, Inc., Baton Rouge, Louisiana, 1992. For a copy of *The Observer*, published by OCAW Local 4-620 and the Louisiana Coalition for Tax Justice, write the Louisiana Labor/Neighbor Fund, OCAW Local 4-620, 8841 Bluebonnet Road, Baton Rouge, LA 70810.

93. Telephone interview with Richard Miller, August 21, 1996.

94. Mentioned by Dan Nicolai of the Jobs and Environment Campaign, *Founding Seminar*, September 12, 1993, Washington, New Hampshire.

95. Lin Kaatz Chary, "The Great Lakes United Labor/Environmental Task Force," *New Solutions* 3, no. 1 (1992): 13, 14; for information about *New Solutions*, a quarterly, contact the Oil, Chemical, and Atomic Workers International Union in Lakewood, Colorado, USA.

96. See Rabe, "Building a Movement for Labor and Environmental Justice," pp. 3, 10.

97. Michael Merrill, "Accepting the Challenge: A Response to Lin Kaatz Chary," *New Solutions* 3, no. 1 (1992): 14, 15.

98. For a sense of the broad sweep of labor history, see Richard O. Boyer and Herbert M. Morais, *Labor's Untold Story*, 3d edition (New York: United Electrical Workers, 1971), esp. chapters 11 and 12, for the glorious triumphs of the CIO in the thirties and early forties, and the damage to labor of the Red Scare and the Taft-Hartley Act in the late forties and fifties. To get a sense of labor's hard times today, see Thomas Geoghegan, *Which Side Are You On? Trying to Be for Labor When it's Flat on Its Back* (New York: Farrar, Straus & Giroux, 1991).

99. See "Adequacy of OSHA Protections for Chemical Workers," Hearing before the Employment and Housing Subcommittee, Committee on Government Operations, U.S. House of Representatives, November 6, 1989; Caleb Solomon, "Volatile Situation: Rash of Fires at Oil and Chemical Plant Sparks Growing Alarm," *Wall Street Journal*, November 7, 1989, p. 1; John Gray Institute, Lamar University System, *A Study of Safety and Health Practices as They Pertain to the Reliance Upon Contractors in Selected Petrochemical Industries: A Preliminary Report*, April 20, 1990; also Robert E. Wages, President, Oil, Chemical and Atomic Workers Union, "Concerning the Final Report of the John Gray Institute," submitted to the Environmental and Housing Subcommittee of the Committee on Government Operations, U.S. House of Representatives, October 2, 1991.

100. Richard Miller, Presidential Assistant, Oil, Chemical, and Atomic Workers Union, presentation to founding seminar, Jobs and Environment Campaign, Washington, New Hampshire, September 12, 1993. Miller estimates that existing refineries represent about $30 billion in sunk capital costs.

According to the *San Francisco Chronicle*, "Some Oil Refineries Are White Elephants," November 28, 1991, p. B-1, retooling for "clean" gasoline will cost $50 billion.

101. See Ray Davidson, *Peril On the Job* (Washington, D.C.: Public Affairs Press, 1970); Richard Engler, *Oil Refinery Health and Safety Hazards: Their Causes and the Struggle to End Them* (Philadelphia: PhilaPOSH, 1975); and Daniel M. Berman, *Death On the Job: Occupational Health and Safety Struggles in the United States* (New York: Monthly Review Press, 1978), ch. 5, "The Workers and the Unions," esp. pp. 120–125.

102. "An OCAW Political Action Plan for 1993–1994," *OCAW Reporter* 48 (January–February 1993): 7.

103. *Labor Party Advocate* 5, no. 2 (March–April 1996), Washington, D.C.

104. Leo Seidlitz, Organizer, Labor Party Advocates, Berkeley, California, telephone interview, November 3, 1993. For accounts of the Labor Party's founding convention write for the *Labor Party Press* 1, no. 1 (Summer 1996), available from Labor Party, PO Box 53177, Washington, D.C. 20009-3177. Included is a description of a party platform advocating a "just transition movement" for workers in industries shut down for environmental reasons.

105. The best account of the M&M agreement is Louis Goldblatt's *Men and Machines* (San Francisco: International Longshoremen's and Warehousemen's Union and Pacific Maritime Association, 1963), which gives an eloquent photographic history of longshoring work without slighting the differences between the shippers and the union. Lincoln Fairly, *Facing Mechanization: the West Coast Longshore Plan* (Los Angeles: Institute of Industrial Relations, UCLA, 1979), pp. xii, 16–20, 249.

106. "New Technology, New Bargaining Shape, New Local 9 Contract," *The Dispatcher*, December 20, 1994, p. 7.

107. Goldblatt, *Men and Machines*, pp. 33, 41.

108. See John T. O'Connor and Gary Cohen, eds., *Fighting Toxics* (Washington, D.C.: Island Press, 1990), and Jane Nogaki, "Community Inspection Leads to Good Neighbor Agreement with Dynasil Corporation of American," typescript, 1988.

109. See Tom Abate, "The Scully Years," *San Francisco Examiner*, April 4, 1993, pp. E1, E4, where second generation Apple CEO John Scully conceded that it was "the nature of [the computer] industry that it's taking fewer and fewer people to turn out more and more productivity. That's just an overall trend...." Three months later Apple announced layoffs of 2,500 employees worldwide, mirroring similar cuts in 1985 and 1990.

110. Edited by Sanford Lewis and others, available to community and public interest groups for $25 from The Good Neighbor Project, 42 Davis Road,

No. 3B, Acton, MA 01720, tel: 508-264-4060; FAX: 508-263-0068. The 1992 edition was updated through September 8, 1993.

111. See, for example, *U.S. Citizens' Analysis of the North American Free Trade Agreement*, December 1992, available from The Development GAP, 1400 I Street, NW, Suite 520, Washington, DC 20005, and The Institute for Agriculture and Trade Policy, 1313 5th Street, SE, Suite 303, Minneapolis, MN 55414. Organizational co-sponsors of the document and participants in the Analysis Team include the Sierra Club, United Auto Workers, Fair Trade Campaign, Institute for Policy Studies, National Lawyers Guild Free Trade Task Force, Greenpeace, The Development Group for Alternative Policies, International Labor Rights Education & Research Fund, Economic Policy Institute, Institute for Agricultural and Trade Policy, and Public Citizen.

112. Speech on October 6, 1993, Union Square, San Francisco. See also Sierra Club, *An Analysis of the North American Free Trade Agreement and the North American Agreement on Environment Cooperation* (Washington, D.C.: Sierra Club, October 6, 1993), handed out as a press packet at the anti-NAFTA rally sponsored by the AFL-CIO at its annual convention, held in San Francisco for the first time.

113. Patrick Crow, "NAFTA on Shaky Ground," *Oil and Gas Journal*, September 27, 1993, p. 39. For a penetrating analysis of the implications of NAFTA for energy, see Carol Alexander and Ken Stump, *The North American Free Trade Agreement and Energy Trade*, (Washington, D.C.: Greenpeace, Atmosphere and Energy Campaign, 1992).

114. Keith Schneider, "Environmentalists Fight Each Other Over Trade Accord," *New York Times*, September 16, 1993, p. A1.

115. Charles Lewis, et al., *The Trading Game: Inside Lobbying for the North American Free Trade Agreement* (Center for Public Integrity, 1993); "Sweet Victory: the NAFTA War Is Won. Now Clinton Must Mend Fences," *Business Week*, November 29, 1993.

116. "EEMs and the Climate Change Action Plan (C-CAP)," Memo to Mark Chupka, The White House, Office of Environmental Policy (OEP), from Jim Curtis of the Ad-Hoc Task Force for a Workable Existing-Home EEM Program, November 26, 1993. See Sandia National Laboratories, *Today's Solar Power Towers* (Albuquerque: Sandia National Laboratories, December 1991), for a lucid description of this new technology.

117. Figures are from the Confederation of German Employers Associations, reported in *This Week in Germany*, April 2, 1993, p. 5. See also "Germany Still Second-Largest Creditor Nation With DM210 Billion [$323 billion] in Net Foreign Assets in Mid-1992," *This Week In Germany*, January 29, 1993. Figures are from the Bundesbank's monthly report for January 1993.

118. "Volkswagen, Metalworkers' Union Agree to Introduce Four-Day Week, Cut Personnel Costs 20%," *This Week in Germany*, December 3, 1993.

119. In California, 44 percent mentioned "Jobs/economy" among the "issues that matter most"; "U.S. deficit" was second with 27 percent and "Environment" was eighth with 7 percent, according to an exit poll of 2,269 voters conducted by Voter Research & Surveys, in *San Francisco Examiner*, November 4, 1992, p. A10. In the two races for U.S. Senate "Economy" was a close second to "Abortion," and "Environment" didn't even make the list.

120. Mark Dowie, "American Environmentalism: a Movement Courting Irrelevence," *World Policy Journal* (Winter 1991–1992): 67–92.

121. Wiener quote is from David F. Noble, *Forces of Production: A Social History of Industrial Production* (New York: Knopf, 1984), pp. 74–76. Indispensable is Wiener's pioneering book *The Human Use of Human Beings: Cybernetics and Society* (Boston: Houghton-Mifflin, 1954), a brilliant book in which he tried to alert the general public about the social costs of automation. For more on struggles over time and leisure in the United States, see David R. Roediger and Philip S. Foner, *Our Own Time: A History of American Labor and the Working Day* (London & New York: Verso, 1989); Roy Rosenzweig, *Eight Hours For What We Will: Workers and Leisure in an Industrial City, 1870–1920* (Cambridge, England and New York: Cambridge University Press, 1983). For a detailed analysis of where the hours go in France, see Adret, *Travailler Deux Heures Par Jour* (Paris: Editions du Seuil, 1977); and the delightful utopian book by Andre Gorz, *Paths to Paradise: On the Liberation from Work* (Boston: South End Press, 1985), translated from the French. See, more recently, the excellent books by Stanley Aronowitz and William DeFazio, *The Jobless Future: Sci-Tech and the Dogma of Work* (Minneapolis: University of Minnesota Press, 1994), especially ch. 11, "The Jobless Future?"; and Jeremy Rifkin's *The End of Work: The Decline of the Global Labor Force and the Dawn of the Post-Market Era* (New York: G.P. Putnam's Sons, 1995), especially ch. 17, "Empowering the Third Sector," and ch. 18, "Globalizing the Social Economy."

CHAPTER 7 Solar Homesteaders or Solar Sharecroppers?

1. Quoted in L. M. Boy, "Grab Bag," *San Francisco Chronicle*, September 25, 1993, p. C1.

2. *Experiences and Lessons Learned With Residential Photovoltaic Systems*, prepared by M. C. Russell and E. C. Kern, Jr. of Ascension Technology, Inc., Lincoln Center, Massachusetts, July 1991, GS-7227, EPRI Research Project 1607-15.

3. See Gail Robinson, "Microinverter Finds Place in the Sun," *Electronic Engineering News,* February 5, 1996,p. 39.

4. William Young and Kirk Collier, *Evaluation of Roof-integrated PV Module Designs and Systems, Final Report,* Florida Solar Energy Center, University of Central Florida, Cape Canaveral, Florida, for the National Renewable Energy Laboratory, July 15, 1992. The NAHB Research Center is in Upper Marlboro, Maryland. See also Sandy Fritz, "Solar Power: Photovoltaic Shingles," *Popular Science,* December 1993, p. 43. On Swiss practice, see Thomas Nordmann, President, Solar Industry Association of Switzerland, "Photovoltaic Applications in Switzerland," paper presented at the International Solar Energy Symposium, February 16, 1993, at University of Delaware, Newark.

5. EPRI, 1991.

6. Rob Roy, "The Solar Powered Home," An Earthwood Building School/Chevalier-Thurling Production, 1996. Distributed by Chelsea Green Publishing Co. (800-634-4099).

7. See James E. Rannels, Director, National Photovoltaic Program, U.S. Department of Energy, "Photovoltaics at the Crossroads," *Real Goods News,* June 1994, beginning p. 4; also Don Loweburg, phone conversation with Dan Berman, September 15, 1996.

8. *Photovoltaic Energy Program Overview,* National Renewable Energy Laboratory, Golden, Colorado, February 1992.

9. See *Solar Energy Uses in the Utility Sector,* Solar Energy Industries Association, September, 1990, for a state-by-state list of innovative projects.

10. Michael Lotker, *Barriers to Commericialization of Large-Scale Solar Electricity: Lessons Learned from the LUZ Experience,* SAND91-7014, Sandia National Laboratories, November 1991.

11. See Thomas J. Starrs, "Overcoming Legal and Insititutional Obstacles to Private Investment in Grid-Integrated Solar Electric Buildings: The Case for 'Net Metering,'" Proceedings of the First International Solar Electric Buildings Conference, Boston, Massachusetts, March 4–6, 1996; pp. 252–259, for a summary of state regulations for customer-located systems, different metering and billing systems and how they pay off for the customer. Two-volume Proceedings may be purchased from the Northeast Sustainable Energy Association, 50 Miles Street, Greenfield, Massachusetts 01301. The small municipal utility of Palo Alto, California allows net periodic billing with credits for "net excess production" as the retail rate. Iowa, though not on Starrs's list, is also a net-metering state.

12. Section 216B.164, kindly supplied by Dan Ahrens of the Minnesota Department of Public Service. Also see *1996 Energy Policy and Conservation Report,* Draft, Minnesota Department of Public Services, July 1996.

13. See *Energy: Minnesota's Options for the 1990s*, Minnesota Department of Public Service, November 1988. Also, Paul D. Wellstone and Barry M. Casper, "Politics and Policy: The Minnesota Energy Program," in *State Energy Policy: Current Issues, Future Directions*, Stephen W. Sawyer and John R. Armstrong, editors (Boulder, Colo.: Westview Press, 1985), pp. 131-145.

14. Steve Coonen, telephone interview, June 8, 1992.

15. "Southern California Edison and Texas Instruments Develop a Low-Cost Solar Cell"; "Photovoltaics Primer: How Photovoltaics Cells Work"; "Background on Photovoltaics at Edison"; glossy public relations handouts, Southern California Edison, no dates, received October 1991. Also, telephone conversations with Don Loweberg, Offline Independent Energy Systems, North Folk, California, September 15, 1996.

16. James Clopton, Senior Research Engineer, Research and Electric Transporation, Southern California Edison, September 10, 1992, interview in Sacramento.

17. "RMI Generates Its Electricity From the Sun," *Rocky Mountain Institute Newsletter*, spring 1991, pp. 5-7. Address is RMI, 1739 Snowmass Creek Road, Snowmass, CO 81654-9199. Telephone interview with John Bigger, then of the Electric Power Research Institute, September 24, 1992.

18. SCE Schedule D-PG, from Thomas A. Starrs, "Net Metering of Customer-Owned, Utility-Integrated Rooftop Photovoltaic Systems," paper presented at the National Solar Energy Conference of the American Solar Energy Society, June 25-30, 1994, San Jose, California.

19. Telephone interviews with Ken Koch, October 3, 1991, April 16, 1992, July 12, 1994, and August 29, 1994. The Electric Vehicle Association of Southern California has 180 paid members and draws an average of 50 people to its monthly meetings. It was formed in the mid-seventies, mostly among steelworkers from the Kaiser steel mill in Fontana who were trying to invent their own response to the gas crisis. A dozen members drive electric vehicles, and nine more are under construction. Information is from article by Mitch Boretz, "July Speaker, Thomas A. Burhenn, Southern Calif. Edison," *EVAOSC News*, Electric Vehicle Association of Southern California, August 1994.

20. Based on August 30, 1993 interview with Dale R. Foster, electric-assisted bicycle project coordinator, AeroVironment, Monrovia, California.

21. See, for example, "Building Energy," Proceedings of the First International Conference on Solar Electric Buildings, Boston, Massachusetts, March 4-6, 1996.

22. Nick Patapoff, telephone interview, September 30, 1991.

23. John Bigger, Electric Power Research Institute, telephone interview with Dan Berman, September 10, 1992.

24. John Bryson, quoted in William J. Cook, "Solar Energy Hits the Market," *U.S. News & World Report*, December 30, 1991/January 6, 1992, p. 60.

25. See for example Joan M. Ogden and Robert H. Williams, *Solar Hydrogen: Moving Beyond Fossil Fuels* (Washington, D.C.: World Resources Institute, October 1989). EPRI and PG&E are quite aware of the possibilities of solar hydrogen.

26. "Southern California Edison and Texas Instruments Develop a Low-Cost Solar Cell," information flyer from Southern California Edison.

27. "SoCal Edison Unveils Solar Strategy," *California Energy Markets*, March 12, 1993, p. 16.

28. "PV for Utilities: California State Working Group," *Meeting Agenda*, September 10, 1992.

29. Telephone interview with Don Loweburg, founder and acting director of Independent Power Producers, January 11, 1995. The IPP position is shared by the Division of Ratepayer Advocates of the California Public Utility Commission, and TURN and UCAN, ratepayer groups in Northern California and San Diego, respectively.

30. Offline Independent Energy Systems is located in North Fork, California. Loweburg and Independent Power Producers are leading the resistance to the Edison proposal.

31. Independent contractors claim they can beat Edison's price; handy homeowners who do the work themselves can bring the cost of a 1-kilowatt system down to $15,000, or $135 a month for fifteen years at 8 percent interest. Edison argues that the figure of $320 a month includes maintenance, but small contractors note that PV requires almost no maintenance once it is installed. They also call attention to the fact that the normal lifespan of the batteries Edison would install is about fifteen years.

32. Telephone interview with Sam Vanderhoof, Photocom, Grass Valley, California, who has been in the PV business for fifteen years. Several loan programs exclude consideration of houses that are not connected to the electric grid: Farmers Home Administration (FMA), Veterans Administration (VA), Government National Mortgage Association (Ginny Mae), and the Federal Home Loan Mortgage Corporation (Fanny Mae), among others.

33. Don Loweburg, RE: Advice No. 1027-E, "Request for Denial," November 23, 1994. For more information write IPP, PO Box 231, North Fork, CA 93643. See also Michael Stremfel, "Solar System Firms Afraid Edison Could Overshadow Them," *Los Angeles Business Journal*, April 4, 1994, p. 1.

34. Telephone interview with Ray Paz, August 26, 1996. The monthly charge for Edison's off-grid PV service is 1.6 percent of the total cost of the installation, according to Experimental Schedule PVS, "Off-Grid Photovoltaic

Service," and the buyer gets to select a contractor from among three lowest bids.

35. Rich Ferguson of the Sierra Club floated this trial balloon, which was quickly shot down. See "Should California Retire its Nuclear Plants?" Center for Energy Efficiency and Renewable Technologies, Sacramento, *Coalition Energy News* (Fall 1993): 1-4.

36. Cost estimates are from Ronald W. Larson, Frank Vignola, and Ron West, *Economics of Solar Energy Technologies* (Boulder, Colorado: American Solar Energy Society, 1992), cited in Keith Lee Kozloff and Roger C. Dower, *A New Power Base: Renewable Energy Policies for the Nineties and Beyond* (Washington, D.C.: World Resources Institute, December 1993), p. 8. *A New Power Base* is a good technical read, but it never asks about the political implications of leaving photovoltaics to the tender mercies of the private utilities. World Resources' "Renewable Energy Project Advisory Panel" is heavily salted with representives from the Electric Power Research Institute, Southern California Edison, the New England Electrical System, and the Institute of Gas Technology.

37. Strempel, "Solar System Firms Afraid Edison Could Overshadow Them."

38. Gary Beckwith, "Welcome Back, Solar Tax Credits," *Real Goods News*, June 1994, pp. 14, 15. Beckwith reports that the Solar Energy Industries Association in Washington, D.C. is compiling a national list of incentives for customer-located solar and renewable energy systems. For $35 the Alliance to Save Energy, 1725 K Street, Suite 509, Washington, D.C. 20006 makes available its summer 1994 report *State and Local Taxation: Energy Policy by Accident.*

39. Beckwith, "Welcome Back, Solar Tax Credits."

40. The federally-owned Salt River Project, which supplies Solar One, called it the "largest grid-connected residential photovoltaic development in the nation to use a centralized PV system rather than separate roof-managed units on each home," *Salt River Project, R&D Focus*, Winter 1992.

41. Telephone interview with Magnus Jolayemi, May 16, 1992.

42. According to Williams, the transformer fed 25,560 kilowatt-hours into the Salt River Project system in August 1991, but SRP paid for only 6,300 kilowatt-hours, according to the monthy bill. In March 1992, after an extended series of meetings with SRP officials, the company paid for 15,840 kilowatt-hours of the 22,960 apparently generated.

43. The inverter which changed the electric current from direct to alternating broke down in July 1989.

44. The *In Concert with the Environment* teaching packet includes a 13-minute video, an "Energy Survey" that students are expected to perform at their own houses, an extensive "Teacher's Guide," all apparently produced by

EcoGroup, Inc. Though two EcoGroup authors have advanced degrees behind their names, no address is furnished for EcoGroup itself. The packet dates from 1991.

45. Victor Dricks, "Taking Special Interest: Utility Windfall Shows Lobbyists' Clout," and "Passing the Bucks: Energy Panelists Allegedly Favored Employers When Doling Out Millions," *The Phoenix Gazette*, April 27 and 28, 1990, respectively.

46. Victor Dricks, "Solar Village Frustrated With Problems," part of a six-part series, "The Sun Sets On Solar," *The Phoenix Gazette*, about the destruction of solar in the sunniest state in the Union, republished in its entirety, March 2, 1993.

47. Interview with Shimon Awerbuch, June 26, 1992. "Measuring the Costs and Benefits of New Technology: A Framework for Photovoltaics," prepared for Solar World Congress, 1991, available from Shimon Awerbuch, College of Management, University of Lowell, Lowell, Massachusetts.

48. See Awerbuch's "Valuing PV Technology: The Application of Finance Theory to IRP Resource Selection," in *PV for Utilities: Developing a National Photovoltaic Strategy for Utilities*, Tucson Meeting Report, December 8-10, 1991, published February 1992, by *PV for Utilities*, Boston, Massachusetts, February 1992, Appendix II.

49. For a more conventional theory of how photovoltaics will spread, see Joe Iannucci, Carl Weinberg, and Richard Sellers, "A Diffusion Model For the Entry of Photovoltaics into Utilities," also from *PV for Utilities*, Appendix II.

50. John Douglas, "Renewables on the Rise," *EPRI Journal*, June 1991. See also *The Potential of Renewable Energy*, an Interlaboratory White Paper, prepared for the Office of Policy, Planning and Analysis, U.S. Department of Energy, published by the Solar Energy Research Institute, March 1990, p. 36.

51. See M.J. Grub, "The Integration of Renewable Electricity Sources," *Energy Policy* 19, no. 7 (1991): 670–688; Thomas B. Johansson, Henry Kelly, and Amulya D. N. Reddy, *Renewables Energy: Sources for Fuels and Electricity* (Washington D.C.: Island Press, 1992); both cited in Kozloff and Dower, *A New Power Base*, p. 91.

52. Telephone interview with Don Loweburg, Janaury 11, 1995. For a utility technical rationale, see Howard Wenger (Pacific Gas & Electric), Tom Hoff (Innotative Analysis, Palo Alto), and Richard Perez (AWS Scientific, Albany, NY), "Photovoltaics as a Demand-Side Management Option: Benefits of a Utility-Customer Partnership," presented at the World Energy Engineering Conference, Atlanta, Georgia, October 1992.

53. Office of the Press Secretary, "Solar Power Lights Up Rural Villages in Brazil," *DOE News*, January 19, 1993. Another source on these projects is

PV News, July and December 1992, c/o PV Energy Systems, PO Box 290, Casanova, VA 22017. See also Rannels, "Photovoltaics at the Crossroads."

54. C. M. Fortmann, M. V. Farley, M. A. Smoot, and B. F. Fieselmann, "Safety Gas Handling and System Design for the Large Scale Production of Amorphous Silicon based Solar Cells," in *Photovoltaic Safety,* AIP Conference Proceedings, Denver, January 19–20, 1988, ed. Werner Luft (New York: American Institute of Physics, 1988), pp. 129–137. Available from Solarex Thin Film Divison, 876 Newtown-Yardley Rd., Newtown, PA 18940. The proceedings of a January 15–17, 1986, conference on photovoltaic safety were published in *Solar Cells* 19, nos. 3–4 (January 1987).

55. The most comprehensive single source on occupational and environmental health and safety is Luft, ed., *Photovoltaic Safety.* See R.P. Gale, J.P. Salerno, P.M. Zavrecky, and W.P. Brisette, "Interaction of Safety and the Facility for Photovoltaic R & D" (Taunton, Massachusetts: Kopin Corporation), for this summary of the steps in PV production.

56. Chris Keavy, personal communication, August 8, 1994.

57. Telephone interviews, November 1992, with Paul Moscowitz, Environmental Health Scientist and Principal Investigator, Photovoltaic Environmental Health and Safety Assistance Center, Brookhaven National Laboratory, Building 475, Upton, NY 11973, tel: 516-282-2017. Mr. Moscowitz encourages interested people to call or write. See also *Bibliography, National Photovoltaic Environmental, Health and Safety Information Center,* Brookhaven National Laboratory, May 1990, for a 78-item list of articles produced by the Brookhaven Center's associates.

58. For a short summary of the SVTC's work, write for the *Silicon Valley Toxics News* 10, no. 1 (Winter 1992), esp. the article "Pollution in Paradise: The Legacy of High-Tech Development," by Ted Smith, Executive Director, Silicon Valley Toxics Coalition (SVTC), 760 North 1st Street, San Jose, CA 95112. The Santa Clara Center on Occupational Safety and Health/Injured Workers United (SCCOSH/IWU), 304 West Hedding St., San Jose, CA 95110, deals primarily with in-plant problems of workers in a largely non-union industry. See also Lenny Siegel and John Markoff, *The High Cost of High Tech* (New York: Harper & Row, 1985), esp. ch. 8, "The Toxic Time Bomb"; and Robin Baker and Sharon Woodrow, "The Clean, Light Image of the Electronics Industry," in *Double Exposure: Women's Health Hazards on the Job and at Home,* ed. Wendy Chavkin, M.D. (New York: Monthly Review Press, 1984), pp. 21–36.

59. Lenny Siegel, Ted Smith, and Rand Wilson, "Sematech, Toxics and U.S. Industrial Policy," by the Campaign for Responsible Technology, 1990, available from the Silicon Valley Toxics Campaign, San Jose, California.

60. Chris Keavny, personal communication, August 8, 1994.

61. *Toxics Release Inventory, 1991.*

62. An unanswered question is whether the inverter (which changes direct current from the PV arrays into alternating current which can be fed into the electric grid) generates a dangerous electromagnetic field. For a quick summary of the potential hazards of electromagnetic fields, see Flora W.F. Chu, "Electromagnetic Fields and Your Health," *Silicon Valley Toxics Action* (Spring 1993): 6. Ms. Chu is the Director of the Santa Clara Center for Occupational Safety and Health (SCCOSH) and of the Asian Workers Health project in San Jose, California.

63. See Brian O'Regan and Michael Graetzel, "A Low-cost, High-efficiency Solar Cell Based on Dye-sensitized Colloidal TiO2 Films," *Nature* 353 (October 24, 1991), pp. 737–740; and Thomas E. Mallouk, "Bettering Nature's Solar Cells," Ibid.: 698–699. The National Renewable Energy Laboratory believes that O'Regan and Graetzel have made a real breakthrough in keeping the colloidal film process going for many cycles, and as of October 1992 it was considering funding O'Regan.

64. Andy Zipser, "Solar Eclipse: Will the Mideast Crisis Make It a Hot Item Again?" *Barron's*, August 20, 1990, pp. 16–31.

65. Information on the research grant is from *Photovoltaic Energy Contract List Fiscal Year 1990*, U.S. Department of Energy, Programs in Utility Technology.

66. Julie Edelson Halpert, "Harnessing the Sun and Selling It Abroad: U.S. Solar Industry in Export Boom," *New York Times*, June 15, 1996, p. C1.

67. Ibid.; "ASE GmbH of Germany Takes Over Mobil Solar Energy Corporation," *Press Release*, ASE Americas, Billerica, Massachusetts, August 1, 1994.

68. *Renewable Energy for the World, 1990,* U.S. Export Council for Renewable Energy, p. 8; cited in H. Richard Heede, Richard E. Morgan, and Scott Ridley, *The Hidden Costs of Energy: How Taxpayers Subsidize Energy Development* (Washington, D.C.: Center for Renewable Resources, October 1985), p. 26. The unpublished 1991 DOE study was cited in Kozloff and Dower, *A New Power Base,* p. 71.

69. See Ogden and Williams, *Solar Hydrogen,* p. 95.

70. *Renewable Energy for the World, 1990,* p. 9, and footnotes 34–37.

71. Ibid.; Zipser, "Solar Eclipse," pp. 16, 17, 29–31.

72. Green Seal press packet, summer 1991.

73. David Roe, *Dynamos and Virgins* (New York: Random House, 1984), pp. 59–62.

74. "David Katz," *New Settler Interview* 49 (May 1990): 33.

75. Letter from Bill Maag to Daniel M. Berman, Fabrimex Solar, Erlenbach,

Switzerland, July 14, 1992. See Public Citizen, *Renewable-Energy Research and Development: An Alternative Budget Proposal for FY 1993-1995*, 3d ed., March 24, 1992, p. 47.

76. Steven J. Strong, "An Overview of Worldwide Development Activity in Building-Integrated Photovoltaics," Proceedings of the First International Conference on Solar Electric Buildings, March 4–6, 1996, Boston, Massachusetts, pp. 14–38. Strong's presentation includes accounts of activity in solar electric architecture in thirteen countries: Austria, Canada, Finland, Germany, Italy, Japan, The Netherlands, Norway, Switzerland, the United Kingdom, the United States, Spain, and Sweden.

77. See also Dr. M. Niederberger, "Die Elektrizitaetswirtschaft und 'Energie 2000'," *Bulletin SEV/VSE*, August 21, 1991, pp. 33, 34. Niederberger is president of VSE Electrizitaetsgesellschaft, an electrical utility company.

78. Nordmann, "Photovoltaic Applications in Switzerland."

79. Retail rate is for baseline electricity use under 300 kWh/month for large private utilities, various sources.

80. Southern California Edison, DOMESTIC—*Parallel Generation, Schedule D-PG, Revised California PUC Sheet No. 16737-E, Effective June 7, 1992*. A careful reading shows that "net energy transmitted" from a large residential PV array (for example) would be purchased by SCE at a rate equal to "the Company's applicable standard offer energy payment rate" filed with the PUC. This "avoided cost" was about 4 cents, compared to SCE's charge of 10.8 cents to 14.5 cents/kWh.

81. Information supplied by Rosemarie Williams, Treasurer, Solar One Homeowners Association, applicable May 15–October 14, 1991.

82. Regulations of the Minnesota Department of Public Service, Section 216B.164.

83. Starrs, "Net Metering of Customer-Owned, Utility-Integrated Rooftop Photovoltaic Systems."

84. B. Decker, U. Jahn, U. Rindelhardt, and W. Vaassen, "The German 1000-Roof-Photovoltaic-Programme: System Design and Energy Balance," apparently 1993, typescript, available from Ulrike Jahn, Institut für Solarenergie-Forschung BMBH, Sokelanstrasse 5, D-3000 Hannover 1, Germany. The 1,500 small grid-connected PV installations on one- and two-family houses earn a 70 percent subsidy of installation costs from the Federal Ministry for Research and Technology (BMFT), and the utilities are obliged to buy back the excess electric power at 90 percent of retail rates. Some utilities in Germany, often municipally-owned, have even more favorable terms for the installation of PV and wind systems.

85. Letter from Dr. O. Hohmeyer, April 22, 1993.

86. Oeko-Institut e.V., Freiburg, FAX, August 11, 1992.

87. Jin Ohara, "First Step to Commercializing Alternative Energy," *Nuke Info Tokyo*, May/June 1992, pp. 6,7.

88. Ricardo Sandoval, "Riding Winds of Energy Change: Bay Area Firm Puts Windmills in Wales," *San Francisco Examiner*, August 31, 1992, p. B1.

89. Letter to Daniel M. Berman, April 23, 1993, from Thomas Nordmann, President, Solar Industry Association of Switzerland, c/o TNC Consulting AG, CH-7000, Rheinfelsstrasse 1, CH-7000, Switzerland. The buyback rates represent the marginal utility cost of expansion of conventional grid capacity in Switzerland. See also Nordmann, "Stand der PV-Anwendung in der Scheiz," paper presented at the session on "Photovoltaics in Daily Life," February 19, 1993, BEA Expo, Bern, Switzerland.

90. Hubert A. Aulich, Siemens Solar GmbH, "Introduction of Photovoltaic Systems in the Near and Intermediate Future," paper presented at the International Solar Energy Symposium, February 1993, Newark, Delaware.

91. Letter from Dr. Olav Hohmeyer, April 22, 1993. Dr. Hohmeyer is the Director, Fraunhofer-Institut für Solare Energiesysteme, Freiburg, Germany.

92. Letter from Professor A. Goetzberger, Fraunhofer-Institut für Solare Energiesysteme, April 30, 1993. See Steven J. Strong, "An Overview of Worldwide Development Activity.

93. Strong, "An Overview of Worldwide Development."

94. In Harvard, Massachusetts.

95. Alan Paradis and Daniel S. Shugar, "Photovoltaic Building Materials," *Solar Today* (May/June 1994): 34–37. BIPV is illustrated at the Fairfield factory of Advanced Photovoltaics Systems, Inc., in Fairfield, California, where Shugar works. See also Gail Robinson, "Micro-inverter Finds Place in the Sun," *Electronic Engineering Times,* February 5, 1996, pp. 39, 42.

96. See Larry E. Shirley and Jodie D. Sholar, "State and Utility Financial Incentives for Solar Applications," *Solar Today* (July/August 1993): 11–14.

97. In constant dollars, according to Public Citizen's *Renewable-Energy Research and Development: An Alternative Budget Proposal*, 3d ed., March 24, 1992, p. 48.

98. Larry E. Shirley, "Drowning in Debate" (editorial), *Solar Today* (March/April 1996): 5.

99. Figures supplied by Patrick Summers, Public Affairs, National Renewable Energy Laboratory, April 17, 1996.

100. Shirley, "Drowning in Debate."

101. Ralph Nader, "Government Buying Can Help Agenda on Environment," *Monroe Evening News*, Monroe, Michigan, May 24, 1992, p. 5B. In July 1992, the Center for Responsive Law, founded by Nader, began to publish

the monthly *Energy Ideas*, on lighting technologies the first month and solar the second. Free to government employees, a subscription to *Energy Ideas* is available from the Government Purchasing Project, Center for the Study of Responsive Law, PO Box 19367, Washington, DC 20036, which also sells a book called *More than 40 Ways to Make Government Purchasing Green*.

102. Holger Eisl and Barry Commoner, *Photovoltaic Cells: Converting Government Purchasing Power Into Solar Power* (Flushing, New York: Center for the Biology of Natural Systems, July 1993).
103. Ralph Nader, Eleanor J. Lewis, and Eric Weltman, "Shopping for Innovation: The Government as Smart Consumer," *The American Prospect* 11 (Fall 1992): 71–78. See also Matthew L. Wald, "Jump-Starting the Electric Car's Future," *New York Times*, June 28, 1993, p. D1; Mark Cohen and Barry Commoner, *How Government Purchase Programs Can Get Electric Vehicles on the Road* (Flushing, New York: Center for the Biology of Natural Systems, June 1993).
104. Such as the VA, FHA, Fannie Mae, and Freddie Mac.
105. Kenneth R. Harney, "Group Pushes Energy-efficient Mortgages," *San Francisco Examiner*, April 25, 1993, p. F11. For more information write Jim Curtis, National Association of Energy-Efficient Mortgage Companies, Bay Area Energy Consultants, 3121 David Ave., Palo Alto, CA 94303, tel: 415-858-0890. For a dollar Curtis will mail you the report: "The Energy-Efficient Mortgage of 1993." See also Shirley and Sholar, "State and Utility Financial Incentives for Solar Applications."
106. President Bill Clinton, and Vice President Al Gore, Jr., *The Climate Change Action Plan*, The White House, October 1993.
107. Interview with Jim Curtis, September 27, 1993. Katy McGinty, who used be the environmental aide to then-Senator Al Gore, is the head of the Office of Environmental Policy; in September 1993 the person working on EEM at the White House was Mark Chupka. Other information is from a letter from Jim Curtis, April 11, 1994, and a telephone interview with Jim Curtis on July 12, 1994.
108. On the issue of technology and control, Harry Braverman argued convincingly in *Labor and Monopoly Capital: The Degradation of Work in America* (New York: Monthly Review Press, 1974) that a major purpose of new technology is to increase control over workers by appropriating their skills and incorporating them into the new machinery. David Noble makes the same point in his exhaustive (and exhausting) study of the development and introduction of automatic machine tools by the Department of Defense and General Electric, *Forces of Production: A Social History of Industrial Production* (New York: Knopf, 1984). Noble's most brilliant

short treatment of this issue is his Introduction to Mike Cooley's book *Architect Or Bee? The Human-Technology Relationship* (Boston: South End Press, 1980). Harley Shaiken makes similar arguments for the auto industry in *Work Transformed: Automation and Labor in the Computer Age* (New York: Holt Rinehart and Winston, 1984).

109. *TEAM-UP: Building Technology Experience to Accelerate Markets in Utility Photovoltaics*, a proposal to the Office of Solar Energy Conversion, Conservation and Renewable Energy, U. S. Department of Energy, submitted by the Utility Photovoltaic Group, Washington, D.C., September 27, 1993. See also Rannels, "Photovoltaics at the Crossroads." Mr. Rannels is Director, National Photovoltaics Program, U.S. Department of Energy.

110. This section consists mainly of a letter from Christopher J. Keavny to Dan Berman, August 8, 1994. Keavny has designed solar cells for Mobil Solar and the Spire Corporation, and is now working with the Jobs and the Environment Campaign and Greenworks in Cambridge. The authors have rewritten it slightly and added some new material.

111. Letter from Keavny, August 8, 1994. Mr. Keavny has over a decade of experience in the United States, Central America, and China.

112. These two points are not new: back in 1978 Denis Hayes wrote, "The Third World may enter the solar era before the industrial world does," for some of the reasons described here. See Denis Hayes, "Third World Options," in *Sun! A Handbook for the Solar Decade* (San Francisco: Friends of the Earth, 1978), pp. 95–124.

113. See, most recently, Bruce Rich's *Mortgaging the Earth: The World Bank, Environmental Impoverishment, and the Crisis of Development* (Boston: Beacon Press, 1994). For another critique of the World Bank see articles by Pratap Chatterjee, Catherine Beasley, and Martin Khor in *terrain* 24, no. 3 (March 1994). *terrain* is the monthly publication of the ecology center, 2540 San Pablo Avenue, Berkeley, CA 94702.

114. See J. G. Vera, "Options for Rural Electrification in Mexico," *IEEE Transations on Energy Conversion* 7, no. 3 (September 1992): 426–433. Vera is an engineer with the Compania de Luz y Fuerza of Mexico City.

115. "ASE GmbH of Germany takes over Mobil Solar Energy Corporation." According to *Germany's Top 300* (Frankfurt am Main: Frankfurter Allgemeine Zeitung, 1991), RWE, Germany's eighth largest company, is heavily invested in atomic and coal-fired power plants and in petroleum and chemicals. Daimler-Benz is Germany's largest company. Best-known for the Mercedes-Benz automobile, Daimler-Benz is heavily involved in "generating, relaying, and exploiting of energy."

116. Paul Maycock, "Photovoltaic Technology: Performance, Costs, and Markets, 1975–2010," *PV Energy Systems 6* (May, 1996).

117. Clóvis Brigagão, *O Mercado da Segurança [The Defense Market]* (Rio de Janeiro: Editora Nova Fronteira, 1984), esp. ch. 3, "Politica Nuclear Brasileira: Opcoes e Dilemas," pp. 119–146. Professor Brigagão is the Director of the Centro de Estudos Norte-Americanos in Rio de Janeiro.

118. See, for example, Deborah Gordon, *Steering a New Course: Transportation, Energy, and the Environment* (Cambridge, Massachusetts: Union of Concerned Scientists, 1991), p. 90. See also Michael Renner, *Rethinking the Role of the Automobile*, Worldwatch Paper No. 84 (June 1988), esp. pp. 18, 19. Renner points out that the value of oil imports as a share of total imports to Brazil rose from 11 percent in 1970 to 43 percent in 1986. Increased use of prime cropland for sugar cane forced Brazil to import more foodstuffs.

119. Personal communication with former Greenpeace staff member Tom Lent, June 1996, who has produced a slide show on a recent visit to Cuba.

120. "Utopia Rises Out of the Colombian Plains," *All Things Considered*, National Public Radio, Washington, D.C., program of August 29, 1994, write NPR, 635 Massachusetts Avenue, Washington, D.C. 20001-3753 for a transcript. See also Weisman's article, "Techno-Topia" in *Los Angeles Times Magazine*, Sept. 25, 1994. In 1997 Chelsea Green will publish a book by Alan Weisman about Gaviotas.

CHAPTER 8 Fighting for a New Solar Society

1. Theodore Geisel, *The Lorax* (New York: Random House, 1971).
2. Cited in Ray Reece, *The Sun Betrayed* (Boston: South End Press, 1979), p. 116.
3. October 1976, pp. 65–96.
4. A typical product of 1970s enthusiasm for solar is architect Edward Mazria's delightful *The Passive Solar Energy Book: A Complete Guide to Passive Solar Home, Greenhouse and Building Design* (Emmaus, Pennsylvania: Rodale Press, 1979). More recent editions exist.
5. Lovins, "Energy Strategy: The Road Not Taken," *Foreign Affairs* (October 1976): 79.
6. Anna Gyorgy & Friends, *No Nukes: Everyone's Guide to Nuclear Power* (Boston: South End Press, 1979).
7. See, for example, Geoffrey Rothwell, "Can Nuclear Power Compete?" *Regulation* (Winter 1992): 66–74, published by the Cato Institute in Washington, D.C. Rothwell argues that federal assumption of liability for nuclear disasters under the Price-Anderson Act is one of the major reasons nuclear power was even remotely competitive, since private insurance companies refused to handle the risk.

8. John R. Emshwiller, "Firm's Profit Is Generated by Energy Efficiency Chaos: E-Source Sells Consumers Advice About the Puzzling Plethora of Products," *Wall Street Journal*, May 11, 1993.

9. The Electricity Consumers Resource Council (ELCON), headquartered in Washington, had twenty-eight member companies in December 1995 that consumed over 5 percent of all U.S. electricity, and was the principal force behind a 150-member organization of heavy electricity users called the Coalition for Consumer Choice. See "1995 Accomplishments Report," ELCON, Washington, D.C., December 1995.

10. On April 24, 1996, FERC finally issued over one thousand pages of regulations dealing with the conditions of wholesale wheeling.

11. For a sense of the national scope of the state-level discussion over electricity, see *LEAP Quarterly Legislative Letter,* Winter 1996, written by Bill Spratley and Jacque Bracken for the Legislative Energy Advisory Program, William A. Spratley & Associates, Inc., Columbus, Ohio.

12. Most important are the Public Utilities Holding Company Act of 1935 (PUHCA), which prohibited interstate utility monopolies and created a system of public financial reporting for utilities; the Public Utility Regulatory Act of 1978, which required utilities to buy a certain proportion of their power from independent producers and required each state to set standards to promote energy efficiency and renewable energy production; and the previously mentioned Energy Policy Act of 1992 (EPACT).

13. Jack McNally and Eric Wolfe, "The Dark Side of Deregulation," *Utility Reporter,* February 1994, published by Local 1245 of the International Brotherhood of Electrical Workers, Walnut Creek, California.

14. If "deregulation" costs ratepayers $28 billion in California, or about $850 annually per inhabitant, is $264 billion nationwide an extravagant number?

15. Philip Covell, et al., *Sacramento Municipal Utility District: Surviving in a Changing Marketplace,* manuscript, March 1996, draft of seminar paper, Graduate School of Management, University of California, Davis.

16. Eric Aldrich, "Cheaper Power vs. Dirtier Air: Electric Market Could Mean More Pollution Borne on Winds," *Keene Sentinel,* Keene, New Hampshire, March 19, 1996.

17. Ilana DeBare, "Environmental Coalition Issues 'Earth Day' Challenge to PUC," *Sacramento Bee,* April 23, 1996, p. C6.

18. Eugene Coyle, April 1996, "Basic Speech" notes.

19. As PG&E has laid off thousands of service workers, customers have to wait days and even weeks for restoration of service after big winter storms. See, for example, Nolte Carl and Jonathan Marshall, "PG&E Power Woes May Worsen, Consumers Want Utility Punished," *San Francisco Chronicle,* December 15, 1995, p. A1.

20. By late summer 1996, utility restructuring had turned into a nuclear bailout for privately owned utilities in California. Assembly Bill 1890, passed in late August 1996 and signed into law by Governor Wilson on September 23, will allow Pacific Gas and Electric and Southern California Edison to bill ratepayers (rather than shareholders) for over $27 billion in bad nuclear-plant investments at Diablo Canyon and San Onofre. The price tag will be $844 per capita or $3,375 per family in California through the year 2001. AB 1890 slipped through with very little press coverage, after a massive utility/industrial-customer lobbying barrage directed at lawmakers and utility shareholders. "This is pork bailout at its worst," said Wendy Wendlandt, coordinator of Californians Against Political Corruption, the organization sponsoring Proposition 212, the campaign-finance reform initiative slated for the November 1996 ballot. While residential ratepayers are supposed to get a 10 percent reduction in rates in January 1998, that reduction will be paid for with $5 billion in state-guaranteed California Infrastructure bonds. In effect, the legislature was forcing taxpayers to lend money to the utilities in order to give those same taxpayers a rate cut. AB 1890, according to economist Eugene Coyle, constitutes a "a 2-to-4-cents-per-kilowatt-hour hidden tax that all Californians will pay directly to private-utility stockholders over the next decade." Coyle claimed that "the entire $27 billion is a liquidation of California assets, which PG&E and SCE would be likely to invest overseas in places like Australia and Indonesia, where subsidiaries of both companies are already active." Albert Vera, founder and chair of the Southern California Cities Consortium, an alliance of local governments banding together to bypass Southern California Edison, considered the new law an "outrage" because it added billions of new state debt in order "to bail out a couple of companies." Even John Anderson, director of ELCON, a Washington DC lobbying group that represents the interests of large industrial power users, declared to the *Wall Street Journal* (August 29, 1996) that it was "not fair to require customers to pay for 100 percent of the mistakes that have been made by the California electric utility industry." Because there were a few concessions to renewables in AB 1890's restructuring scheme, it was already being touted by Ralph Cavanagh of the Natural Resources Defense Council as a model agreement appropriate for emulation in the rest of the U.S.

21. McNally and Wolfe, "The Dark Side of Deregulation."

22. Scott Ridley's "Rolling Thunder Over Retail Competition," *Public Power* (March/April 1995): 15–19, has a very helpful discussion of the issues surrounding utility deregulation. Ridley is the co-author of *Power Struggle*, and is a consultant to the Senate Energy Committee of the Commonwealth of Massachusetts.

23. According to a June 1, 1978, Harris Poll cited in *Blueprint for a Solar America* (Solar Lobby, January 1979), p. 2, 94 percent of Americans favored the rapid development of solar power. Forty-two percent of 1,000 randomly selected voters polled December 7–12, 1994 indicated that renewable energy sources like solar, wind, geothermal, biofuels, and hydroelectric should have the highest funding priority in federal energy R&D spending, and another 22 percent believed renewables should receive the second highest priority. The poll was funded by the Sustainable Energy Budget Coalition, and cited in "Public Prefers Renewables," *Florida Energy Reporter,* March/April 1995, reprinted from the March/April 1995 issue of *Solar Today.* Similarly, market research at Arizona Public Service Company, Arizona's largest private utility, revealed that "63 percent of customers prefer solar power," reported Ed Fox, Vice President for Environmental, Health, Safety, and R&D at the utility. The APSC survey was cited in *The UPVG Record,* March 12, 1996, p. 4, a free newsletter of the Utility Photovoltaic Group, Washington, D.C.

24. The survey conducted for the Sustainable Energy Budget Coalition concluded that 19 percent of consumers expressed a willingness to pay a premium of up to 10 percent for power from renewables, 23 percent would pay a premium of 5 percent, and another 23 percent would pay a premium of 2 percent on their electricity bills. For a longer discussion see Edward A. Holt, "Customer Choice Creates Market Pull: The Case of Green Pricing," *Proceedings of the First International Conference on Solar Electric Buildings,* volume II, pp. 63–68; published by the Northeast Sustainable Energy Coalition, Greenfield, Massachusetts; poll results cited in *Electric Utility Week,* January 22, 1996.

25. It is likely that the participants are paying at least three times for SolarCurrents: first as taxpayers who partially fund the Department of Energy's Utility Photovoltaic Group, second as ratepayers, and third with their contributions of $6.59 a month.

26. Utility Photovoltaic Group, "Detroit Edison Begins Popular PV Program for All Customers," *The UPVG Record* (Washington, D.C.: Utility Photovoltaic Group, March 12, 1996), p. 4.

27. "Minnesota's Net Billing Law," *Wind Energy in the U.S.: A State by State Survey* (Washington, D.C.: American Wind Energy Association, May 1995), pp. 241, 242.

28. Based on relatively expensive Pacific Gas & Electric retail electric rates of $0.12 per kilowatt-hour. Data on energy usage from refrigerators is calculated from "Refrigerators for a Wiser World."

29. Natural Resources Defense Council, "NRDC Applauds Whirlpool and SERP on New Source of Clean Power in Super-Efficient Refrigerator,"

Media Advisory, June 29, 1993; also interview with Jill Stauffer at NRDC, San Francisco, July 1, 1993.

30. The Energy Foundation, *1991 Annual Report*.

31. Telephone interview with Weston D. Birdsall, June 11, 1992. See also text of speech by Birdsall, "Conservation: A Success Story in Osage, Iowa," at the Mid-America Conference, Maryville, Missouri, May 7, 1991.

32. Dennis Fannin, who took over from Wes Birdsall as general manager in 1992; telephone conversation with Daniel Berman, September 25, 1996.

33. Most of the discussion in this section is adapted from Paul Gipe's *Wind Energy Comes of Age* (New York, John A. Wiley, 1995, esp. chs. 2 & 3) and from Robert W. Righter's *Wind Energy in America: A History*, Univeristy of Oklahoma Press, Norman, Oklahoma, 1996, as informative and charming about the the geniuses who have tried to capture "the breath of the sun" as Perlin and Butti's *A Golden Thread: 2500 Years of Solar Architecture and Technology*. For a quick, optimistic summary of windpower, Executive Director Linda White of the Kern Wind Energy Assocation recommends Dawn Stover, "The Forecast for Wind Power," *Popular Science*, July 1995, pp. 66-85.

34. Gipe, *Wind Energy Comes of Age*, pp. 52, 53.

35. Telephone interview with Linda White, Executive Director of the Kern Wind Energy Association, August 15, 1996, which has 650 MW of capacity and sold 1.2 billion kwh to Southern California Edison in 1995, about half of California's total. Sixty-five of the wind turbines in the Ringkobing district are owned by a publicly-owned electric utility, and 35 are owned by a cooperative of 508 local families, according to Gipe, op. cit., p. 65.

36. Compared to 30 percent for the American machines, according to Gipe, *Wind Energy Comes of Age*, op. cit., Table 3.1, p. 73. Though it is not clear from Table 3.1, Gipe told Dan Berman that the percentages refer to cumulative world generating capacity installed through 1993. For 1974-1992 the United States spent about $900 million on capital subsidies above and beyond R&D, compared to about $150 million by Denmark, according to Table 3.1.

37. On the difficulties of transitioning from the military to the mass civilian market, see Ann Markusen and Joel Judken, *Dismantling the Cold War Economy*, Basic Books, New York, 1992, esp. ch. 4: "A Wall of Separation."

38. Gipe, *Wind Energy Comes of Age*, p. 71.

39. Ibid., p. 69.

40. Ibid., pp. 60-63, for more details about buyback rates.

41. Ibid., p. 60.

42. See Amy Linn, "Whirly Birds," *SF Weekly*, San Francisco, California, March 20-April 4, 1995, pp. 8-15.

43. Gigi Coe, "California's Experience in Promoting Renewable Energy Development," *State Energy Policy: Current Issues, Future Directions*, Stephen W. Sawyer and John R. Armstrong, editors, Westview Press, Boulder, Colorado, 1985, p. 194; and Gipe, *Wind Energy Comes of Age*, op. cit., Figure 1.22, p. 35.

44. Note from Paul Gipe, August 16, 1996.

45. The Crude Oil Windfall Profits Tax Act (WPTA) of 1980, which was best-known for taxing domestic oil production, also increased the federal energy tax credit from 10 to 15 percent.

46. The 25 percent state tax credit was in effect from January 1978 through December 1985, was reduced to 15 percent through 1986, and was cancelled at the beginning of 1987; see Thomas A. Starrs, "Legislative Incentives and Energy Technologies: Government's Role in the Development of the California Wind Energy Industry," *Ecology Law Quarterly*, v. 15, No. 1, 1988, pp. 103-158.

47. See Righter, *Wind Energy in America: a History*, op. cit., pp. 209-218 for these examples. See also Gipe, *Wind Energy Comes of Age*, op. cit., Figure 1, 22, p. 35.

48. Figures from Paul Gipe, August 16, 1996.

49. Letter from Professor A. Goetzberger, Fraunhofer-Institut für Solare Energiesysteme, April 30, 1993, to Dan Berman. See also Steven J. Strong, "An Overview of Worldwide Development," op. cit.

50. Dr. Olav Hohmeyer, letter to Dan Berman, april 22, 1993. Dr. Hohmeyer is the director of the Fraunhofer-Institut in Freiburg.

51. The address in August 1996 was Solarenergie Förderverein, Hertzoastrasse 6, 52070 Aachen, Germany; for an English-language version of the German rate-based incentives program, contact Solarenergie's web site: http:\\www.fto.de\sfz\sostarte.htm.

52. Strategies Unlimited specializes in market research in optoelectronic, photovoltaic, and RF/microwave electronic components systems.

53. "Rate-Based Incentives Spreading Rapidly in Germany," *Solar Flare*, 94-5, September-October 1994, pp. 4-7; and "Rate-Based Incentives Expected to Reach 1.8 MWp in 1996," *Solar Flare*, 96-1, January-February 1996, especially pp. 10, 11, which list the cities which have adopted and which are considering the adoption of rate-based incentives.

54. Tom Jensen, telephone conversation with Dan Berman, August 28, 1996.

55. Tom Jensen, ibid. See also "Can Rate-Based Incentives Make It to the States? Variation of Aachen Model to Be Proposed in California," by Tom Jensen, printed in Photon, the journal of the Solarenergie Förderverein in Aachen, May/June 1996, and translated into English.

56. Jensen, "Rate-Based Incentives Expected to Reach 1.8 MWp in 1996," op. cit.

57. Jim Trotter, telephone conversations with Dan Berman, August 27 and 28, 1996.

58. Tim Townsend, Tom Jensen, and Thomas A. Starrs, telephone conversations with Dan Berman, autumn 1995 and August 1996.

59. "Photovoltaics: Clean Energy Now & For The Future," Rocky Mountain Institute, Spring 1991, pp. 5-7. The cost of a 10-foot by 10-foot PV panel is extrapolated from RMI estimates. See *Evaluation of Roof-Integrated PV Module Designs and Systems, Final Report*, Florida Solar Energy Center, Cape Canaveral, Florida, for the National Renewable Energy Laboratory, July 15, 1992; also Mark D. Uehling, "Sunroof for Houses," *Popular Science*, January 1993, p. 37, which discusses Sanyo's designs for photovoltaic roof shingles.

60. Joan Ogden and Robert H. Williams, *Solar Hydrogen: Moving Beyond Fossil Fuels* (Washington, D.C.: World Resources Institute, October, 1989).

61. Based on estimated present annual household consumption of 3,700 kilowatt-hours (San Francisco) to 6,000 kilowatt-hour (Los Angeles). The Rocky Mountain Institute argues that this total could be reduced to 1,500 kilowatt-hours, not including power for an electric car.

62. Ogden and Williams, *Solar Hydrogen*, pp. 38, 39.

63. Ibid., esp. pp. 77-96.

64. Stephen Lyons, ed., with a Foreword by David R. Brower (San Francisco: Friends of the Earth, 1978). *Citizens' Solar Blueprint* was called "The Official Book of the First International Sun Day."

65. See environmental attorney Thomas A. Starrs in "Utility Investment in Resident Photovoltaic Systems," typescript, October 1992. Starrs finds the National Energy Conservation Policy Act and the California Public Utilities Code ambivalent about the proper role for utilities in residential photovoltaic systems. He is now a doctoral student with the Energy and Resources Group at the University of California at Berkeley.

66. *Advice XXXX-G/XXXX-E*, PG&E, before the California Public Utilities Commission, October 1, 1993.

67. Al Gore, *Earth In the Balance: Ecology and the Human Spirit* (Boston: Houghton Mifflin, 1992), p. 327.

68. Bill Clinton and Al Gore, *Putting People First: How We Can All Change America* (New York: Times Books, 1992), pp. 89-93.

69. See any newspaper for November 18, 1993, the day after NAFTA passed the House of Representatives by a vote of 234-200, after heavy lobbying and horsetrading by the administration and corporate and financial interests from the U.S. and Mexico.

70. See twin tables: "Imports of Leading Commodities" and "Exports of Leading Commodities" in *Information Please Almanac, 1992* (Boston:

Houghton Mifflin, 1992), p. 69; ultimate source of the data is Foreign Trade Division, Bureau of the Census, U.S. Department of Commerce; "Oil Users Seen More Vulnerable to Supply Jolts," *Oil & Gas Journal*, November 27, 1989, p. 32.

71. Robert A. Blecker, *Beyond the Twin Deficits: A Trade Strategy for the 1990s* (Armonk, New York: M. E. Sharpe, Inc., 1992), esp. pp. 77, 78 and 127–129.

72. The Schlesinger quotation is from Andrea Ricci, "Analysts Say Imported Oil Has Crucial Role in U.S. Trade Deficit," *San Francisco Chronicle*, January 13, 1992, p. B2.

73. See Ann Markusen and Joel Judken, *Dismantling the Cold War Economy* (New York: Basic Books, 1992), esp. ch. 9, for ideas on how to reconvert the economy. On where Clinton's subsidies and tax breaks are going in energy, see "Bill's Recipe: How He Plans to Help U.S. Business Cream Rivals."

74. Melissa Healy, "Clinton Drafts Plan to Save Atmosphere: Environmentalists Oppose Voluntary Approach to Cut Greenhouse Gases," *San Francisco Chronicle*, September 20, 1993, p. A2 (reprinted from *Los Angeles Times*).

75. See Clinton and Gore, *The Climate Change Action Plan*.

76. See Ricardo Sandoval, "D.C. Post to PG&E Chief," *San Francisco Examiner*, June 14, 1993, p. D1. The appointment was made just as respected PG&E research director Carl Weinberg decided to take early retirement because he knew the giant utility intended to undo many of the photovoltaic and energy efficiency programs he had convinced them to set in place.

77. Huey D. Johnson, "Feinstein Bows to Big Oil's Demands," *San Francisco Examiner*, June 7, 1993, p. A19. Feinstein is the former Mayor of San Francisco, headquarters of Chevron, which signed a major exploration/production contract in Khazakhstan; Ann Devroy and Eric Pianin (*Washington Post*), "Big Concession By President on Energy Tax," lead story in the *San Francisco Chronicle*, June 9, 1993, p. A1. Feinstein and Senator David Boren of Oklahoma both opposed the energy tax proposal.

78. Quoted by Healy, "Clinton Drafts Plan to Save Atmosphere."

79. John Schaeffer, Editorial, *Real Goods News*, June 1993. Ultimately Schaeffer would like to break down the utility monopoly on electric power and financing, and is working with Congressman Dan Hamburg from his district to make it possible for customers to get publicly-backed financing for off-the-grid PV systems in ten years through existing home finance networks, rather than make the customer "dependent on the utility forever." While such a strategy may work for many of Real Goods' comfortable off-the-grid customers, it will come up short for most people, who are necessarily tied to the electric grid. While Schaeffer praised President Clinton for promising, in an Earth Day speech, to conduct an energy audit of the White House and make it a "model for energy efficiency and waste reduc-

tion" it was also true that neither photovoltaics nor solar water heating were mentioned in the President's speech.

80. Hermann Scheer, *Sonnen Strategie: Politik ohne Alternative [Sun Strategy: the Only Possible Policy]* (Munich: Piper, 1993), pp. 90, 91; also chapter 10, "The People's Energy." An English translation by Peter Hoffmann is available as *A Solar Manifesto* (London: James & James Science Publishers Ltd., 1994). Dr Scheer is also a substitute member of the parliamentary assembly of the Council of Europe, chairman of the board of the European Renewable Energy Centers Agency, and editor of the *Yearbook of Reneable Energies*, published annually by James & James in London.

81. *The Economist*, September 11, 1993. See especially C. Fred Bergsten, "The Rationale for a Rosy View," pp. 57–62.

82. See, for example, the *LEAP Quarterly Legislative Letter*, Winter 1996.

83. The authors would like to thank Lorna Enero Berman, Eugene Coyle, Scott Ridley, Benjamin Watson, and Jim Schley for their insights and information about the issues of utility restructuring. For a more comprehensive discussion of some of the issues in this section, we recommend the forthcoming *Reinventing Electric Utilities: Competition, Citizen Action, and Clean Energy,* by Ed Smeloff and Peter Asmus (Washington, D.C.: Island Press), scheduled to appear in December 1996. Ed Smeloff is a pioneering director of the Sacramento Municipal Utility District, and Peter Asmus is an environmental journalist and author based in Sacramento.

84. See "Reuse and Recycling: Waste Prevention and Resource Savings in Utilitization," photocopied notes, June 21, 1992; also "Hidden Innovation: R&D in a Sustainable Society," *Science and Public Policy* 13, no. 4 (August 1986): 196–203; all by Walter R. Stahel, The Product-Life Institute, Geneva, Switzerland.

Index

E

Earth Day, 1, 242
Earth in the Balance: Ecology and the Human Spirit (book), 43, 241–42
East Bay Municipal Utility District (Oakland, CA), 37
Edison, Thomas, 67, 106
Edison Electric Institute, 76
Edison International, 116, 117, 142, 149. *See also* SCEcorp
Edison Light and Power Company, 68
Eisenhower, Dwight D., 88
Eisl, Holger, 203
Elberling, Lance, 121
Electrical Union, 72
electric automobiles, 41, 62, 95, 177–79, 238–39
electric consumption
in California, 114–16
decreasing demand, 16, 65, 222–24, 235
in Florida, 19
in Hawaii, 24–25
hot water heating and, 20
typical household, 208 (table), 209
Electricity Consumers Resource Council (ELCON), 216
electric meters, running backwards, 184, 222, 238. *See also* net excess power
electric power generating plants
construction, resistance to, 114
in Florida, 19–20
in Hawaii, 26
overseas expansion, 117
See also nuclear power plants

electric power grid, 3. *See also* off-the-grid
Electric Power Research Institute (EPRI)
Memorandum of Understanding with ERDA, 8–9
policies, 46, 115, 176
publications, 106, 172
US Advanced Battery Consortium and, 62
electric power utilities
accountability, 124–25
automobile/oil industries, coalition of, 60–63
banks and, 82–87
customer-located, grid-connected systems, 66, 171–75
developing countries, 209
dominate energy policy, 7–11, 16
energy conservation incentives and, 104–105, 108–10
Energy Foundation and NRDC, alliance with, 113–17, 130–32
energy self-sufficiency, attitudes towards, 3–4, 178, 183
Germany, 231–34
history, 67–69
"license" provision, R&D process, 9
NRDC and the "Collaborative Process," 110–13, 114, 115–17, 127–29
overseas ventures, 117
photovoltaics industry and, 205–207
rates/profits, California, 219–20
rates/profits, compared to public, 98 (table), 99

rates/profits, Hawaii, 25–26
rates/profits, power dumping and, 73
rates/profits, regulation of, 47, 70–71, 73, 83–84, 216–19
rates/profits, time-of-use structure, 185 (table)
regulatory commissions, 69–70
See also buyback rates; demand-side management (DSM)
electric power utilities, municipal
Florida, 19
Hawaii, 26
history, 97–103
Osage (Iowa), 225–26, 238, 240
performance, 46
See also public power movement
electric streetcars, 68
electrolytic hydrogen, 60, 87, 200, 231, 235
electromagnetic fields (EMFs), 152
electronics industry, 165–66, 167, 192–94
El Salvador, 207, 208
Emergency Relief Act (1935), 76
employment. *See* Environmentalists for Full Employment (EFFE); jobs
Employment Impact of the Solar Transition (report), 137
Energan (consortium), 145
Energy, Jobs and the Economy (book), 30, 136
energy conservation
ActSquare project and, 120–22, 126

O

Oahu (Hawaii), 22, 24
Oak Creek Energy
 Systems, Inc., 230
Oakland (CA), 37
Occidental Chemical, 86
O'Connor, John, 164, 165
off-the-grid, 40, 42, 177,
 180, 197. *See also*
 photovoltaics (PVs)
Ogden, Joan M., 235, 236
Ohio, 45, 61, 101–3, 216,
 218. *See also*
 Cleveland
Oil, Chemical and Atomic
 Workers Union
 (OCAW), 86, 87,
 138, 157, 158–61, 167
oil drilling, 48, 114
oil industry
 accountability, 124
 automobile/natural gas
  industries and utili-
  ties, coalition of,
  60–63
 banks and, 82–87
 contracting out jobs,
  162
 federal subsidies and,
  53–54
 finances, 49–50, 53–54,
  82–87
 government policy,
  54–56
 history, 48–54
 NAFTA and, 44, 242
 oil depletion allowance,
  49–50, 82
 unions, 138, 157
oil prices, 34, 55–56,
 58–59, 63, 83–84
oil refining, 162, 167
oil reserves, 61–62, 123,
 208, 210, 242
O'Leary, Hazel, 45, 153
Olney, Richard, 70
Ontario Hydro, 75
Operation Desert Storm.
 See Gulf War

Oregon, 45, 85, 183
Organization of Petroleum
 Exporting Countries
 (OPEC), 53, 56, 83
Ormat Energy Systems, 27
Oryx Energy (company),
 115
Osage (Iowa), 225–26, 238,
 240
Osborn, Donald, 122
Our Common Future
 (report), 114
Our Friend the Atom
 (film), 89
ozone depletion. *See*
 greenhouse effect

P

Pacemaker heater, 17
Pacific Energy Center (San
 Francisco), 122
Pacific Gas & Electric
 (PG&E)
 ActSquare project,
  120–22, 126
 awards, 112, 129
 demand-side manage-
  ment and, 106
 energy conservation
  incentives and, 108,
  110, 112, 239
 headquarters, 104
 Hetch Hetchy dam and,
  78–80
 history, 69
 nuclear power plants,
  217–18
 overseas ventures, 47
 Phoenix Project, 94
 Presidio and, 46, 81
 public relations, 125–33
 rates/profits, 30,
  118–19, 127, 129
 renewables, "bridging
  strategy" for, 122–24
 renewables, opposition
  to, 10, 47
 SMUD and, 46, 88, 92

tree-planting program,
  93–94
 See also buyback rates
Pacific Maritime Associa-
 tion (PMA), 163–64
Pacific Northwest, xii,
 84–86, 136
Palo Alto (CA), 174
Paradis, Alan, 201
Parenti, Michael, 56
Parks, Terry, 94
Patapoff, Nick, 177, 178
patents, 9, 10, 69
Paz, Ray, 182
Peabody Coal, 144, 154
Peevy, Mike, 139
Pele Defense Fund, 27–28
Pennsylvania, 48, 218
Peoples Gas (Chicago), 60
Perez, Richard, 181
Perlin, John, 13
Persian Gulf, 57
petrochemical industry.
 See chemical indus-
 try
petroleum. *See* oil
Pew Charitable Trusts,
 111, 113, 114, 115
Phelps Dodge, 114
Phillips 66 chemical plant
 (Pasadena, TX), 162
Phillips (company), 113
*Photovoltaic Cells: Con-
 verting Government
 Purchasing Power
 into Solar Power*
 (book), 203
photovoltaics (PVs)
 applications, 171, 188,
  236
 building integrated,
  171, 201
 costs, 172, 173, 186–87,
  234–35
 customer-located
  systems, utility own-
  ership, issue of,
  171–75, 180–81

About the Authors

DANIEL M. BERMAN, Ph.D., is a journalist, teacher, and energy director of the Jobs and Environment Campaign. His landmark book, *Death on the Job: Occupational Health and Safety Struggles in the United States,* is still in print after eighteen years. He has been a visiting scholar at the School of Public Health, University of California-Berkeley, and lives in Davis, California.

JOHN T. O'CONNOR founded the National Toxics Campaign in 1983, and in 1986 led the coalition that passed the landmark $9-billion Superfund law. He is the president of Greenworks, Inc., an environmental technology company. He has co-authored three other books, including *Fighting Toxics* (Island Press, 1990); among his published essays is "The American Promise," in *Heaven Under Our Feet,* edited by Don Henley and Dave Marsh. He lives in Cambridge, Massachusetts.